THE COPY-TO-PRESS HANDBOOK

THE COPY-TO-PRESS HANDBOOK

PREPARING WORDS AND ART FOR PRINT

JUDY E. PICKENS
Accredited Business Communicator

Illustrated by
Clare Heidrick vanBrandwijk

JOHN WILEY & SONS
New York   Chichester   Brisbane   Toronto   Singapore

Copyright © 1985 by John Wiley & Sons, Inc.

All rights reserved. Published simultaneously in Canada.

Reproduction or translation of any part of this work beyond that permitted by Section 107 or 108 of the 1976 United States Copyright Act without the permission of the copyright owner is unlawful. Requests for permission or further information should be addressed to the Permissions Department, John Wiley & Sons, Inc.

Library of Congress Cataloging in Publication Data:

Pickens, Judy E.
  The copy-to-press handbook.

  Bibliography: p.
  Includes index.
  1. Printing. Practical—Handbooks, manuals, etc.
2. Publishers and publishing—Handbooks, manuals, etc.
I. Title.
Z244.P52 1984   808′.02   84-7411
ISBN 0-471-89862-7

Printed in the United States of America

10 9 8 7 6 5 4 3

Preface

When I started my communication career, I thought my graduate training had prepared me to be a writer, editor, and all-around communicator. I knew how to find a story, write it, and edit it—but I soon discovered that I needed to know much more if the newsletter, brochure, magazine, or other project for which I was responsible was to get printed satisfactorily.

The experience of learning print production by trial and error is shared every day by editors, corporate secretaries, office managers, production coordinators, and others who must occasionally, or as an integral part of their jobs, manage production of print projects. Even seasoned professionals faced with an unusual assignment may easily trip in the labyrinth of graphic design, typesetting, printing, and other production steps. And getting the work produced is only part of the task. It must be carefully planned so that the resulting printed piece meets objectives and is completed on time, within budget.

This handbook presents comprehensive information on all planning and production aspects of print projects, large or small, simple or complex. It is written from the perspective of the print buyer, not from the more technical, narrow view of the designer, printer, or other supplier. It uses *understandable* language and presents *practical* advice to guide you through that labyrinth, including an extensive section on saving money while maintaining quality. It is a valuable textbook for the beginning or future print buyer to gain a working knowledge of the entire production process or a reliable reference when a specific production question must be answered. The extensive glossary makes it a helpful tool anytime for quick clarification of unfamiliar terms, especially since print terminology is extensive and the many synonyms confusing. Terms are also briefly defined in the text where they first appear.

In addition to the contributions of Clare vanBrandwijk, who has so thoroughly and skillfully illustrated this handbook and been a tenacious reviewer, I acknowledge the assistance of the following persons who reviewed chapters or in other ways shared so generously of their expertise:

Lauren Kurowski, production coordinator with LithoCraft, Inc., in Seattle, who was a type supplier to and a reviewer of this project. She is an example of how supplier sensitivity and knowledge can contribute so much to production success.

Ellis Collingham, retired Seattle engraver, whose decades of experience and attention to detail are reflected in the image-assembly chapter.

John Nordahl, print salesman with GraphiColor, Inc., in Seattle, whose well-seasoned knowledge of printing is a resource I wish were available to every print buyer.

John DeNure, salesman with West Coast Paper Company in Seattle, who reviewed the section on paper and reproduced printer's marks from the extensive type museum he so enthusiastically assembled at West Coast.

Patrick Neary and David Cobaine of Sinclair and Valentine and Joyce Schwartz of Pantone, Inc., who assisted with the complexities of printing ink.

George Bayless, owner of Bayless Bindery in Seattle, whose thorough knowledge of finishing and binding contributed to that chapter.

Michael Ziegler, Seattle photographer, and Ron Nix, Omaha designer, who contributed to both text and illustrations.

The staff of the Seattle Public Library whose patient assistance was tapped on many occasions.

Finally, I acknowledge the unique contributions of Lynn Carson King, whose dread of print production as a beginning editor inspired this handbook, and of Phillip Sweetland and Jacob vanBrandwijk, whose encouragement and understanding allowed it to be prepared with the priority it deserved.

JUDY E. PICKENS, ABC

Seattle, Washington
January 1985

Contents

1.	INTRODUCTION TO PRINT PRODUCTION	1
2.	PLANNING PRODUCTION SPECIFICATIONS	9
3.	PLANNING THE PRODUCTION BUDGET	25
4.	PLANNING THE PRODUCTION SCHEDULE	63
5.	WRITING AND EDITING COPY	75
6.	DESIGNING FOR PRINT	95
7.	SELECTING AND USING ART	165
8.	ART SOURCES	213
9.	TYPESETTING METHODS AND TYPOGRAPHERS	239
10.	GETTING TYPE SET	259
11.	PREPARING MECHANICAL ARTWORK	299
12.	IMAGE ASSEMBLY FOR THE PRESS	339
13.	PRINTING METHODS AND PRINTERS	383
14.	PAPER AND INK	413
15.	FINISHING, BINDING, AND DOCUMENTING THE JOB	437
APPENDIX A.	TRADE CUSTOMS	455
APPENDIX B.	MODEL RELEASES	469

viii   CONTENTS

SELECTED BIBLIOGRAPHY   **473**
GLOSSARY   **477**
COLOPHON   **531**
INDEX   **533**

THE COPY-TO-PRESS HANDBOOK

CHAPTER ONE

Introduction to Print Production

COMMON PRINTING NEEDS 2

THE PRODUCTION PROCESS 4

PRODUCTION RESPONSIBILITIES 4
    The Print Buyer, 5
        Planning, 6
        Budgeting, 6
        Scheduling, 6
        Coordinating Production, 6
        Evaluating, 6
    The Supplier, 6
        Pricing the Job, 7
        Providing Production Advice, 7
        Working to Specifications, 7
        Working on Time, 7
        Billing, 7

Long before the invention of movable type, people were communicating words and concepts by various predecessors of printing. Writings and drawings on stone, whalebone, or rough paper helped usher in civilization itself. Now our media range from simple business letters sent through the mail to complex satellite transmission of telephone and television signals.

Despite fascination with the speed and capability of electronic media, we still rely heavily on the printed word and picture. Organizations as large as multinational corporations or as small as community councils count on the versatility, permanence, and completeness of printed messages. Brochures, newsletters, annual reports, and other commercially printed pieces are the mainstays of the neighborhood lettershop and the biggest printing plant in town. They represent a large portion of the tremendous volume of public and private material being printed today throughout North America.

Such pieces don't just happen. They must be produced—guided from collections of words and pictures into cohesive, finished materials that meet objectives, budgets, and deadlines.

This introductory chapter provides an overview of the types of printed pieces being produced and the production process itself. It then outlines responsibilities of the print buyer and the supplier during that process.

COMMON PRINTING NEEDS

While your print-production need today may be for a single-sheet flyer, tomorrow it may be for a 16-page booklet or 100-page catalog. Here are definitions of common commercially printed pieces you may at some point be called on to produce. Planning and production information in this handbook may be applied to each.

Business Stationery and Cards: Sometimes referred to as a "business cabinet," letterhead, envelopes, calling cards, and mailing labels are required for virtually every business. They are usually printed on one side only but, because they are high-image pieces, may have design details that make them complex to produce.

Flyers: These announcements of events or services are printed on one side of a small sheet of paper. They are throwaway pieces for quick communication.

Folders: Also intended for quick communication, a folder is a single printed sheet folded at least once. Information is usually brief and related to a product or service or may be an announcement of an event. A folder may be a throwaway or, if printed on a large, heavy sheet, it becomes a "presentation folder" with one or two pockets to hold smaller pieces (such as sales brochures).

COMMON PRINTING NEEDS

Posters: Used to announce an event or service, a poster is printed on one side of a large, heavy sheet. It is similar to a flyer but bigger and usually of much higher quality.

Brochures: Commonly used to describe a service, program, or facility, a brochure is a small piece, usually printed on both sides and folded into a compact unit. It has only a few pages or panels of information and is smaller in length than a booklet.

Booklets or Pamphlets: Both words apply to a printed piece with fewer than 48 pages, bound together in some way. A booklet has a similar purpose to a brochure but presents more detailed information.

Books, Catalogs, and Directories: These pieces may be much longer, perhaps into hundreds of pages. Their volume requires sophisticated binding and often a hard cover.

Forms: The workhorses of business, forms organize and standardize information with a fill-in-the-blank format. They may be as simple as a two-part receipt to keep track of sales or as complex as a lengthy, computer-generated summary of employee benefits.

Newsletters: Whether for company supervisors or for club members, a newsletter is usually 12 or fewer pages unbound. It is smaller than a daily city newspaper but, like a newspaper, is dominated by brief, timely news items.

Newspapers: Similar to a newsletter in that it presents timely news items, a newspaper may also include more feature articles. It may be as large as a city newspaper or half that size ("tabloid") and may be printed on lower-quality paper than a newsletter.

Magazines: These bound publications have at least eight pages and contain primarily feature articles of interest to their particular audiences (such as company employees or customers).

Magapaper or Newzine: These jargon words are used to describe a publication having format qualities of both a magazine and a newspaper.

Proposals: Written to solicit business or financing by presenting in detail procedures and costs of a project, a proposal may be typeset and printed, typewritten and printed, or simply typewritten, photocopied, and secured into a looseleaf binder.

Reports: Like proposals, reports may be simply produced or elaborately designed and printed. Their purpose is to convey results of studies or, in the case of annual reports, to review the year's activities.

Financial Statements: These pieces may be folders or brochures prepared quarterly for shareholders or extensive booklets to meet legal requirements for more detailed information.

Specialty Items: Based on similar steps used in ink-on-paper printing, specialty items may be such things as vinyl binders imprinted with an association's name, magnetic decals for company vehicles, or T-shirts for sales promotion.

THE PRODUCTION PROCESS

Regardless of what the piece is or how refined the design, the production process for printed materials is similar. Some steps may be abbreviated or deleted for simple jobs, however, or elaborated on for more complex projects. The basic steps (treated in detail in later chapters) are as follows and as shown in Figure 1.1:

1. *Determining Specifications.* Size, quantity, colors, paper, and all other requirements of the job.
2. *Budgeting.* How much is to be spent, including any adjustment of specifications so that the job is in line with funds available for it.
3. *Scheduling.* The start date, delivery date, and timing necessary to accomplish each step in between.
4. *Preparing Copy.* Writing and/or editing the words that make up the editorial portion of a printed piece.
5. *Designing.* All graphic considerations (such as format, paper and ink color, and placement of type) that contribute to how the finished product will look and feel.
6. *Preparing Art.* Execution and selection of illustrations and/or photographs to be part of the printed piece.
7. *Typesetting.* Conversion of copy into type suitable for printing.
8. *Preparing Mechanical Artwork.* Dummy, paste-up, cropping of art, and other preparation for image assembly.
9. *Image Assembly.* Preparation of negatives and/or press plates.
10. *Printing.* The actual transfer of images onto paper (or other medium in the case of specialty items).
11. *Finishing and Binding.* Folding, trimming, binding, and any other additional work necessary to complete the job.
12. *Documenting.* Filing of background information, art, artwork, production details, cost figures, and evaluations for future reference.

PRODUCTION RESPONSIBILITIES

Print production is a cooperative undertaking. You have certain responsibilities, as does each of the suppliers on whom you call. The more informed you are, the better able you will be to execute your responsibilities

PRODUCTION RESPONSIBILITIES

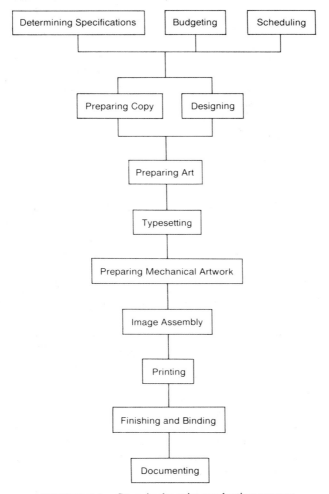

FIGURE 1.1. Steps in the print-production process.

and select and work with competent suppliers who will effectively and efficiently execute their responsibilities.

THE PRINT BUYER

Whether your print project is business cards for sales representatives or an annual report to financial contributors, it places certain responsibilities on you as the buyer of printing and related services. If these responsibilities are waived or inadequately met, your ability to get the job done on time, within budget, and in line with quality expectations and objectives is severely impaired. Current and future relations with suppliers will also be jeopardized.

Planning

As the purchaser of services and the various parts that go together to create your newsletter, brochure, or other printed piece, you must plan thoroughly. Consider each step in the production process just outlined before you launch the project. You may still have unanswered questions (such as color of ink), but at least you will know which questions remain to be answered and when final decisions must be made.

Budgeting

As the buyer, you are responsible for estimating all costs and for modifying specifications so that the total matches the funds on hand. A well-conceived budget should not deplete funds for a future job or expect suppliers to give you more for less.

Scheduling

You are responsible for allotting a reasonable amount of time for each step, based on when the job can start and when it must be finished. Faulty scheduling will push suppliers and threaten quality and supplier relations as well as budget. If the schedule is tight, you are responsible for working out timing and budgeting for any rush charges necessary to meet the deadline.

Coordinating Production

The print buyer is also the project coordinator or manager. Unless you are willing to take your chances and pay a supplier extra to keep your job on track and on time, you are responsible for approving each phase and moving the job along, including informing suppliers of any change in schedule or specifications.

Evaluating

Like any buyer, you are responsible for evaluating how well the job was done and whether or not charges are reasonable. Suppliers should understand and agree to the performance you expect. If the finished piece varies from original requirements making a higher or lower bill reasonable, you must decide if it *is* reasonable and negotiate for any price adjustment. You are also responsible for timely processing of supplier bills.

THE SUPPLIER

Anyone supplying a service or element for your print project also has certain responsibilities. As the buyer, you should expect these responsibilities to be met because when any supplier falls short, you and possibly subsequent suppliers in the process will have to compensate.

Pricing the Job

A supplier should provide complete cost estimates, taking into account all specifications, quality requirements, and the schedule to be followed. Estimates or bids should be ready by the time requested, and suppliers should be willing to put verbal prices in writing. In a noncompetitive situation, you have a right to expect a reasonable price, as if that supplier *were* competing with others for the job.

Providing Production Advice

A competent supplier will not watch you drown if you get in over your head. Be as informed as you can be; but you should be able to rely on a supplier for advice. A responsible printer, for example, will work with you from the start of a job to see if corners and costs might be cut without sacrificing quality. Providing such advice helps assure the supplier of an end product worthy of your repeat business.

Working to Specifications

A supplier is responsible for meeting all your specifications or for negotiating any changes with you (such as using a slightly different size for a brochure to make more efficient use of paper). A competent supplier will not arbitrarily change specifications and wait for you to discover the fact.

Working on Time

You and your supplier share responsibility for completing a project by the deadline. If you have conscientiously involved a supplier in setting a schedule and met all your deadlines, expect a competent supplier to do so, also, or to let you know immediately if something unforeseen jeopardizes that schedule.

Billing

A supplier is responsible for giving you an accurate bill as soon as possible after the work is completed. Unless you have negotiated an adjustment, the bill should be as estimated or bid, with no hidden charges or surprises. In addition, it should be in accordance with any special accounting requirements you have reviewed with the supplier.

CHAPTER TWO

Planning Production Specifications

FACTORS INFLUENCING SPECIFICATIONS   11

   Editorial and Graphic Considerations, 12
      Objectives, 12
      Audience, 13
      Image and Related Materials, 13
      Budget, 14
   Production Considerations, 14
      Press Run, 14
      Paper and Ink, 15
      Finishing and Binding Processes, 15
      Delivery and Schedule Requirements, 16
      Budget, 16
      Supplier Capabilities, 16

PRODUCTION SPECIFICATIONS   16

   Type Specifications, 17
      Body Type Size and Leading, 17
      Column Width, 18
      Estimated Depth, 18
      Headlines and Subheads, 18
      Other Type, 18

Special Requirements, 18
Schedule, 19
Engraving Specifications, 19
Photograph Number and Maximum Reduction, 19
Photograph Treatment, 20
Other Art and Screens, 20
Condition of Artwork, 20
Proof, 20
Schedule, 20
Print Specifications, 21
Finished Size, Flat and Folded, 21
Number of Pages or Panels, 21
One Side or Two, 21
Color, 22
Press Run, 22
Condition of Artwork, 22
Paper Name, Color, and Weight, 22
Special Requirements, 22
Schedule, 23
Finishing and Binding Specifications, 23
Folding and Other Finishing Work, 23
Binding, 24
Packaging and Delivery, 24
Schedule, 24

The foundation of successful production of any printed piece is thorough planning. Comprehensive planning brings words, type, art, and graphic techniques into harmony. The result is a functional, attractive piece at a reasonable cost, by the deadline.

Because of this close relationship of elements, the planning process has at least three steps. They are quickly accomplished on a simple project and take more time with a job that is complex. Those steps are (1) editorial, design, and production considerations leading to tentative, then final, production specifications allowing you to (2) budget for the job and (3) schedule it.

Experienced print buyers realize that this order of planning cannot always be followed. Steps must be taken simultaneously or one factor (such as budget) given precedent, forcing a different sequence of considerations. Keep this fact in mind as you study this and subsequent information about planning a print project. The objective of this chapter is to get you started in one logical sequence, even though the chicken-and-egg analogy is quite appropriate to apply to the process of planning a print job. You may need to refer to later chapters before your list of specifications is adequate for budgeting and/or scheduling purposes.

This chapter guides you through the first step—the drafting of tentative and final production "specifications" or those instructions to a typographer, printer, or other supplier that indicate what is to be done. It opens with an overview of editorial, design, and production considerations that should be explored in the planning phase, followed by an explanation of the many elements in a full set of tentative or final specifications. Terms will be briefly defined here to give you a basic understanding of the production process for planning purposes. Refer to the glossary for complete definitions and to later chapters to understand these terms more fully.

FACTORS INFLUENCING SPECIFICATIONS

As will be presented in detail in later chapters, many factors influence the specifications you plan for a given print project. To assist you in tentative planning, general editorial, graphic, and production considerations are previewed here and summarized in checklists shown in Figures 2.1 and 2.2, respectively.

EDITORIAL AND GRAPHIC CONSIDERATIONS

Printed materials communicate with two categories of ingredients—editorial and graphic. The first refers to the words, copy, or text and the second refers to the way a printed piece looks and feels. One may be more important than the other on a given project; but both are early planning considerations. Careful thinking through of content clarifies what the printed piece is to do and helps set a tone to be carried out by the design.

Objectives

The objective of a printed piece is the purpose served by that piece. It may have only one or more, depending on how many uses it can reasonably be put to without diluting its success. Editorial objectives influence production planning by helping to determine the length of copy and the number of sections into which it must be divided for presentation. If copy is to explain results of a study, for instance, it will be extensive and probably be divided into clearly defined sections. If it is to reinforce information presented during a sales call, however, it will probably be much shorter and more compact.

Graphic objectives go hand in hand. The size, shape, and general appearance of a printed piece (its "format") must be in accordance with what it is to do. Production specifications may have to take into account multiple- or

```
┌─────────────────────────────────────────────────────────────────┐
│         Planning Considerations Checklist: Editorial and Graphic │
│                                                                  │
│   For _____   │
│   Objectives _____   │
│              _____   │
│              _____   │
│              _____   │
│                                                                  │
│   Audience — Primary _____   │
│              Secondary _____   │
│              Characteristics _____   │
│                              _____   │
│                              _____   │
│                                                                  │
│              Distribution Method _____   │
│                                  _____   │
│   Image/Related Materials _____   │
│                           _____   │
│   Budget _____   │
└─────────────────────────────────────────────────────────────────┘
```

FIGURE 2.1. A checklist to assure that editorial and graphic considerations are taken into account when planning a print project.

full-color printing of a point-of-sale brochure so that it attracts attention. On the other hand, a business form that is to organize information would probably be simply printed but may require extensive typesetting or binding into sets.

Audience

The people who are to read the piece you are planning and how they are to receive it influence its specifications. An audience especially interested in the subject will expect more detail than one only marginally interested, thus affecting copy length. If the audience is likely to change over time, general wording will allow the piece to have a longer life than one with specific wording that will soon be dated. A long life means you can plan a piece that is well produced and permanently bound, while a short life may require utilitarian production and a binding that will allow pages to be removed and added.

Who will see the piece also has a great deal to do with how it is designed and the amount of art it should contain. An annual report might look conservative and inexpensive or flashy and expensive, depending on which would be better received by shareholders. A brochure directed toward prospective tourists would require extensive photography to attract readers and perhaps different photographs for young people versus those who are retired.

FACTORS INFLUENCING SPECIFICATIONS

A piece must be produced to get to its audience. A brochure to be enclosed with a business letter must be planned to fit into a business envelope. A newsletter to be mailed without an envelope or wrapper must be large enough to accommodate all the copy and sufficient space for a mailing label. Before you plan to use a particular paper, consider how it will hold up in the mail or in a display rack and how it should be folded.

Image and Related Materials

The piece you are planning should be consistent with the image you wish to project and with related materials. If your company wants to be known as a solid, no-nonsense organization, plan specifications accordingly using subdued colors and textures for a traditional, clean look. Leave bright colors, unusual papers, and novelty folds to more free-spirited enterprises.

The printed piece should relate in both image and content with other materials already being used or being developed. Perhaps editorial content must tell the whole story at length because no related materials are available or maybe it must tell only part of it, backed by other pieces that will be part of the package. Similarly, the piece should look like part of the package. Layout, paper, and colors should match or harmonize.

Collect all related materials as you start planning a project so you can take into account their editorial and artistic content and overall design. If you are to produce several new pieces, try to plan them simultaneously so they will be effective together.

Budget

Because of the many design choices available to you, budget is also an important consideration when planning what techniques you will include in production specifications. Expense doesn't necessarily increase in direct proportion to design complexity; but the two are related. Because of production capability, you may be able to incorporate several design elements for the same cost on the press.

For instance, you may get two colors on both sides of a small newsletter for the same press cost as two colors on only one side, given the way the paper feeds through the press. Or you could use six black-and-white line illustrations for less than using two photographs because the illustrations cost less to prepare for reproduction. If you have a low budget, draft a conservative list of design specifications so you can proceed with planning, then expand it later if you find your budget will allow.

Illustrations and photographs must be considered in production specifications in relation to what is needed to support editorial content and what art might be available. Art on hand should be evaluated for appropriateness and its ability to reproduce satisfactorily (as covered in Chapter 7).

The cost of securing new art must be taken into account as you draw up production specifications as well as the cost of reproducing it on the press.

PRODUCTION CONSIDERATIONS

Once you have considered editorial and graphic factors that influence planning specifications, turn to how actual production of your printed piece will further impact those specifications. How you would like it to read, look, and feel may be affected by several aspects of the production sequence.

Press Run

The number of copies you want ("press run") helps determine how complex your specifications might be. In general, the higher the run, the more detail becomes financially feasible on a per-unit basis. Multiple colors, special techniques with art, and intricate folding cost far less per unit on a run of 10,000 than they do on a run of 500. If your press run is 1000 or less, plan a relatively simple design with one or two colors and limited special techniques—unless cost is no object.

To determine press run, take into account both immediate and long-term use of the printed piece. Every letterhead, form, or report has a "life" or length of time during which it is current and usable. For example, letterhead is current and usable only as long as all information on it is correct. If your company is to move in six months, the letterhead you need delivered next

---

Planning Considerations Checklist: Production

For _____

Press Run — Immediate _____
          Long-term _____
          Waste and File _____
          Total _____

Paper _____

No. Colors _____

Finishing/Binding _____

Delivery _____

Schedule _____

Budget _____

Supplier Capabilities _____
_____

FIGURE 2.2. A checklist to assure that production considerations are taken into account when planning a print project.

week will have a short life. In this instance, press run should be sufficient for immediate use only.

A form to keep track of responses to customer phone calls, however, may have a long life. In figuring its press run, calculate how many copies will be needed to provide an initial supply, and then how many you will need on the shelf throughout the year to supplement your initial supply. That figure, then, will be your annual press run.

Also take into account waste and file copies. If you need 100 copies of a report to distribute, don't plan a press run of just 100. Some copies will be spoiled in printing and others in finishing and binding. Also, you will need a few extra for your files. On a short run, allow at least an extra 4% for each color to compensate for waste in printing, plus at least another 4% for waste in finishing and binding. On long runs, the percentages are less, especially with experienced suppliers.

If you need 100 reports to distribute and 10 reports to file, plan to specify a press run of *not fewer than* 110. Your printer will work backward from that figure to determine how much paper to have on hand to allow for waste, depending on size of run, number of colors, type of press to be used, type and weight of paper, and complexity of finishing and/or binding requirements.

Paper and Ink

Previous considerations about objectives, audience, and image (among others) will influence what paper you plan to use. Some papers will produce what you want; others will not be appropriate because of quality, characteristics, or cost. You will have a broader choice of papers if a job is to be run on a sheet-fed press than if roll paper for a web press is required.

A piece is printed with one or more colors of ink. Black is the most common choice for one-color jobs. From there, the combinations are almost unlimited. The number of colors you plan influences production specifications because the more colors you use, the longer production may take, the higher the cost may be, and the more engraving and printing skill may be required. Especially if your job is to be produced on a small, one-color press by a quick printer or by an in-house printer, plan just one color for the best results. Plan two or more colors only if the printer can assure you that time and capability are available to do a satisfactory job.

Finishing and Binding Processes

Intricate, time-consuming finishing details will influence both schedule and cost of a job. If you are short on either or both, keep finishing requirements to a minimum. Also, some processes require certain paper characteristics to be effective, which is another cost consideration.

In addition to cost, the binding process you must use to match related materials or might choose to use can impact how many pages the piece will have. A process requiring a wide center margin will have less room for text and art per page than a method that allows pages to lay flat and be fully visible.

Delivery and Schedule Requirements

How a job must be delivered can affect your planning. Pieces to be shipped by air, for instance, may need to be packaged up to a maximum weight per box, or you may want the bindery to help you with distribution by packaging a certain number of copies per bundle.

The more you plan to have done to produce your project, the longer production will take. If you have a firm and not-too-distant deadline, simplify specifications now so the job can be finished on time or be prepared for rush charges to meet your schedule.

Budget

As with editorial and design details, production considerations strongly influence budget planning. Specifications that are complex and time consuming in production are expensive to execute. Length of press run, choice of paper, number of colors, intricacy of finishing and/or binding, and special delivery requirements all add up to higher production costs than more conservative specifications. (See Chapter 3 for advice on ways to economize by moderating production requirements.)

Supplier Capabilities

Specifications may need to be changed because of supplier capabilities, even if your job is to be printed by the biggest shop in town. The task may be too complex for your printer (or any printer) to accomplish by your deadline, within the budget set, or up to the quality you require. Every printer has at least one story to tell about a "designer's delight" that called for an odd size with a novelty fold in six closely registered colors with an unusual die-cut, embossing, and perforation, all by next Wednesday!

Involve your suppliers in planning before investing time and money toward a piece with difficult or impossible specifications. Discussing and altering complexities before production starts will smooth planning, budgeting, and scheduling as well as promote a positive working relationship with your suppliers once production is under way.

PRODUCTION SPECIFICATIONS

Editorial, design, and production considerations lead to the listing of tentative and, eventually, final production specifications. The information you

PRODUCTION SPECIFICATIONS

have already compiled may be adequate to take to a copywriter, designer, illustrator, or photographer to start development of those elements. All production work from typesetting on, however, requires detailed specifications for realistic cost and schedule estimates before any work is begun.

If you want firm production bids on your job, you must have firm specifications. Don't expect a supplier to bid on tentative specifications to help you plan a project, then bill you that same amount even when requirements are later increased. A common practice on a job for which extensive planning is involved is to ask for "budgetary" estimates based on tentative specifications to get an idea of cost. Requirements are then adjusted and suppliers approached again for final estimates or bids.

TYPE SPECIFICATIONS

Develop tentative, then final, type specifications to the point at which you can give a typographer at least the following information, as illustrated in Figure 2.3. (For comprehensive coverage of type specifications, see Chapter 10.)

Body Type Size and Leading

"Body type" is type usually no larger than 12 points used for the text or main part of a printed piece (such as the articles in a newsletter). "Leading" is the space between lines of type. Type may be "set solid" with no extra space or air between lines, making it compact, or set spaced apart. Thus both size and leading affect the length of a column of type. In Figure 2.3, 9/10 is specified, meaning that body type is 9 points of type on 10 points of space, allowing 1 point of leading between lines.

Specifications for Type Estimate

For _Orientation brochure_
1. Body Size/Leading _9/10 Souvenir_
2. Column Width _13 picas_
3. Estimated Depth _10 inches_
4. Headlines/Subheads _2 heads - 18 point type_
   no subheads
5. Other Type _4 cutlines - 8 point type_
6. Special Requirements _kerning of heads_
7. Schedule _all copy in 4/11 - need on 4/13_

FIGURE 2.3. An example of type specifications necessary to receive an estimate.

Column Width

In Figure 2.3, column width is specified as 13 picas. "Pica" is a common unit of measure in the printing industry and is closely equivalent to $\frac{1}{6}$ of an inch (0.4216 cm). "Column" is a vertical block of type made up of lines all set to the same maximum width. In the figure, then, type is to be set 13 picas or just over 2 in. wide (or 5.48 cm).

Estimated Depth

"Depth" is approximately how long a column of type will be. (See Chapter 5 for detailed information on estimating copy length.) In Figure 2.3, depth has been estimated at 10 in. (25.4 cm). Add to your estimate of depth extra space between paragraphs or additional lines to compensate for anticipated changes after type is set. Do not include headlines or subheads; they are accounted for separately.

Headlines and Subheads

A "headline" (or "head") identifies the subject of an accompanying article or block of copy. It is usually larger than the body type and often of a different weight or type style. A "subhead" is smaller than a headline and appears with it as further introduction to the text or within the text to break it into sections.

To estimate cost and time, a typographer needs to know how many heads and subheads you expect to need and of approximately what size. Subheads within the text confined to a body size may be set along with that text and require very little extra effort. Heads, however, are usually set separately because of their larger size and possibly different style and weight. In Figure 2.3, they are to be set in 18-point type, and no subheads are expected to be required.

Other Type

Your typographer must also be told about any other type you will need (such as page numbers or "cutlines" to describe photographs to appear in your printed piece). Tabular matter is time consuming to set and is especially important not to overlook when listing requirements for other type. In Figure 2.3, four cutlines in 8-point type are the only other type planned for the job.

Special Requirements

Let your typographer know of any special requests you anticipate. In Figure 2.3, "kerning" of headlines is expected. It is the adjusting of space between

PRODUCTION SPECIFICATIONS

type characters so that they will fit into a given space or appear evenly positioned. Such extra work takes time that the typographer must add to the cost of your job. A rough idea of special requirements is sufficient to give you a feel of cost; but be specific and comprehensive when you go back to your typographer for a final estimate or bid.

Schedule

In Figure 2.3, all copy will come in at once. Piecemeal typesetting takes more time to do because the typographer has to switch from one job to the next and back again. Also state how many working days you think the typographer will have to do the job and when the project might be ready. If you expect a tight schedule, you may need to plan for rush charges.

ENGRAVING SPECIFICATIONS

The process of image assembly (or "engraving") may be done by a specialist in this phase of production or by your printer prior to making press plates. (For details on this process, see Chapter 12 and refer to Figure 2.4 for an example of the specifications covered.)

Photograph Number and Maximum Reduction

For a realistic idea of cost and timing, you need to indicate how many photographs ("halftones") you anticipate (four are indicated in Figure 2.4). If color photographs are to be reproduced, state that you are asking for separations for full- or four-color process printing. Indicate the maximum amount you expect photographs to be reduced; a shop with limited camera equipment may not be able to reduce to less than 33% of the original size.

Specifications for Engraving Estimate

For _Recruiting Brochure_

1. Number and Maximum Reduction of Photographs _4; 50%_
2. Treatment of Photographs _special treatment of cover photo_
3. Other Art/Screens _one screen – art supplied_
4. Condition of Artwork _camera ready_
5. Proof _yes, one_
6. Schedule _need by May 15_

FIGURE 2.4. An example of engraving specifications necessary to receive an estimate.

Your choice, then, is either to go to a different shop, provide photographs that don't require as much reduction, or alter your specifications.

Photograph Treatment

Let your engraver know of any special treatment you might require on photographs (such as eliminating the background). You have many choices, as presented in Chapter 7. For a rough idea of cost, you don't have to know exactly what might be needed (as in Figure 2.4). However, exact decisions on photograph treatment must be made before going back to an engraver for a price on final specifications.

Other Art and Screens

Indicate if you plan to have "continuous-tone" illustrations (black, white, and shades of gray) that must be treated the same as photographs and/or "line illustrations" that are black and white only. Also tell the engraver if you expect to require "screens," any of several patterns used to create a gray or lightly colored area or a special effect with halftones or line art.

Condition of Artwork

"Camera-ready art" is the term now commonly used to identify type and black-and-white illustrations prepared to size on mechanical artwork and ready for line conversion for the press, in addition to photographs ready for halftone conversion. (Strictly defined, the term refers to all elements ready for conversion into letterpress plates.) Any work that the engraver must do to get artwork to a camera-ready state for the printing process to be used will cost extra, as will special requirements such as close registration of multiple colors.

Proof

A proof of some sort is your check of placement, color, and other details of the project before press plates are made and the job is run. If the project is simple (say a one-color flyer with no halftones), you won't need the extra expense and time of having a proof made. However, a proof is an essential production step for every job with halftones, multiple colors, or a high price tag. Specify how many copies of a proof you will need for simultaneous approvals because each must be made separately.

Schedule

Indicate if art will come in all at once or if some can be done early. Advance work may help the engraver meet a tight schedule without overtime and lower your cost.

PRODUCTION SPECIFICATIONS

PRINT SPECIFICATIONS

Plan print specifications to the point at which you can tell a printer enough to give you a budgetary estimate, then make final decisions on those requirements for a more exact price, should specifications change. (For detailed information, see Chapter 13.) Figure 2.5 illustrates a set of print specifications.

Finished Size, Flat and Folded

Specify size of the piece as printed and as folded and trimmed. In Figure 2.5, the brochure will be $8\frac{1}{2}'' \times 11''$ (216 × 279 mm) flat, folded and trimmed to $3\frac{1}{2}'' \times 8\frac{1}{2}''$ (89 × 216 mm). With this information, the printer knows the format of the piece and size of sheet or roll going through the press.

Number of Pages or Panels

Indicate how extensive the piece will be, using number of pages or "panels" (folder or brochure sections). Here is where basic editorial and graphic considerations are especially pertinent. In Figure 2.5, the brochure is expected to be six panels of printed material.

One Side or Two

Specify how many sides of the paper are to be printed. Even if you want only one line of type to appear on the second side, you must plan for printing on both sides.

Specifications for Print Estimate

For _Benefit Brochure_

1. Finished Size Flat/Folded _8½" x 11" folds to 3½" x 8½"_
2. Number of Pages/Panels _6 panels_
3. One Side or Two _2 sides_
4. Color(s) — Standard/Mixed _2 - black, mixed green_
5. Press Run _1,500_
6. Condition of Artwork _fully assembled for platemaking_
7. Paper Name/Color/Weight _Kilmory Pewter - 70# text - self cover or comparable sheet_
8. Special Requirements _none_
9. Schedule _need by August 16_

FIGURE 2.5. An example of print specifications necessary to receive an estimate.

Color

Tell the printer how many ink colors you will need and on how many sides. Figure 2.5 calls for two colors on two sides. Also indicate what those colors might be, especially if they will require mixing (remembering that black counts as one color). Also state, if possible, whether ink coverage will need to be light, medium, or heavy to reproduce the amount of image you expect to have. One or two lines of type will require light coverage, while a full page of body type or a large photograph will require heavy inking.

Press Run

State the number of copies you will need. See earlier information in this chapter for advice on reaching a figure that takes into account current and future uses. If you must have delivery of "not less than" a certain number, include that stipulation.

Condition of Artwork

If the printer is to do image assembly, the job will come "camera-ready" (as covered earlier). It may, however, come to a printer fully assembled by an engraver and ready for platemaking. If you specify the latter, you must be prepared to guarantee that the engraver's work will correspond with the printer's requirements.

Paper Name, Color, and Weight

These points will probably not be definite until the design is complete and art is in hand. However, a printer can't give you an estimate without at least a realistic guess about paper. In Figure 2.5, name, color, and weight of paper for this self-cover job are specified with the qualifier "or comparable sheet." If you have no particular paper in mind now but know what characteristics you want, let the printer suggest one on which to estimate. For example, "I want something white, heavy enough to be mailed by itself, opaque enough so photos won't show through, and with a little texture" is an adequate statement of characteristics for this purpose if you can't identify a sheet by name.

Special Requirements

Indicate any special press work that your job might require, such as tight "registration" (close alignment of elements on the press). Special work may call for extra preparation time, larger paper, or close press monitoring to achieve what you want, all adding to the job's cost.

PRODUCTION SPECIFICATIONS

Schedule

Tell the printer when to expect the job and whether it will be run all at once or in a split run. In Figure 2.5, a single run is specified. Indicate the deadline for delivery to you (or to the bindery) so that, if time is short, overtime hours or rush charges can be planned into the estimate, should they be considered necessary.

FINISHING AND BINDING SPECIFICATIONS

While your printer may do elementary finishing and binding, complex work will probably be sent to a bindery. If supervised by your printer, the work will be included with print specifications. If done by a bindery directly, a separate set of specifications must be provided. (See Chapter 15 for details on this phase of production and Figure 2.6 for an example of specifications.)

Folding and Other Finishing Work

Specify how the job is to fold, if it is. In Figure 2.6, the brochure is folded only, using a simple six-panel barrel fold. Be as specific as you can about method of folding because cost and time vary, depending on which you select.

If you want your piece to have holes along one edge to fit into a three-ring notebook or to be cut and trimmed, you must plan for such finishing work. Perforating, embossing, and several other finishing techniques may be done by a bindery, all contributing to production costs and schedule requirements.

Specifications for Finishing and Binding Estimate

For _Orientation Brochure_

1. Folding _six panel, barrel fold_
2. Other Finishing _no_
3. Binding _no_
4. Packaging/Delivery _boxed anyway / local - to basement, Shipping and Receiving_
5. Schedule _in June - need by June 22_

FIGURE 2.6. An example of finishing and binding specifications necessary to receive an estimate.

Binding

A printed piece may undergo one or more finishing processes, and then be bound using one of several techniques covered in Chapter 15. Cost and time vary for each; so you need to be specific.

Packaging and Delivery

Tell the bindery when, where, and how you need the job delivered. In Figure 2.6, delivery is local with no special packaging required. Boxing a job a particular way or delivering it out of town may cost more and take extra time.

Schedule

The time required to complete finishing and binding procedures depends on length of press run and on whether the work can be done by machine or must be done by hand. Another factor influencing when your job can be completed is other work in the bindery at the same time. A regular job with standing specifications will be processed ahead of one that is not regular or that requires special work. Indicate when you need delivery; the bindery will tell you if that date can be met.

CHAPTER THREE

Planning the Production Budget

APPROACHES TO BUDGETING   27

   When Budget Is Fixed, 27
      Need Versus Want, 28
      Comparative Planning, 28
      Revising Specifications, 29
   When Budget Is Open, 29
      Need Versus Want, 29
      Appropriate Quality, 30
   The Annual Production Budget, 30
      Regular Projects, 30
      Special Projects, 31

PRICING PRODUCTION   31

   Methods of Pricing, 32
      Estimates, 32
      Bids, 32
   Getting Estimates and Bids, 32
      When To Ask for an Estimate, 33
      When To Ask for a Bid, 33
      How Many To Get and From Whom, 34
      "Politics" of Pricing, 34

Evaluating Estimates and Bids, 35
  Specifications, 35
  Working Days, 35
  Price, 36
  Trade Customs, 37
  Additional Considerations, 37

THE ECONOMICS OF CHANGING SPECIFICATIONS   38

When To Revise, 38
  Planning, 38
  Writing and Design, 39
  Illustration and Photography, 39
  Typesetting, 39
  Mechanical Artwork, 40
  Image Assembly, 40
  Printing, 41
  Finishing and Binding, 41
Avoiding a Surprise Bill, 41

UNDERSTANDING PRODUCTION BILLS   42

Content of a Bill, 42
Verifying Charges, 43

ECONOMICAL PRODUCTION   45

Working With Suppliers, 45
  Economical Scheduling, 45
  Matching Capability to Project, 46
  Contracting Directly for Services, 46
  Placing Related Jobs, 47
  Repeat Business, 47
  Doing More Yourself, 47
Revising and Adapting, 48
  Copy, 48
  Design, 48
  Art, 48
Design, Illustration, and Photography, 49
  Saving on Design, 49
  Saving on Illustration, 50
  Saving on Photography, 50
Typesetting, 51
  In-House Typesetting, 52
  Commercial Typesetting, 52
Mechanical Artwork and Image Assembly, 53
  Saving on Mechanical Artwork, 53
  Saving on Image Assembly, 53
Press Work, 55
  Matching Press to Project, 55
  One Run, 56

APPROACHES TO BUDGETING

Ink Colors, 56
Masters, 56
Press Proofs, 57
Purchase and Use of Paper, 57
Direct Purchase, 57
Buying Ahead, 58
Matching Paper to Project, 58
Comparative Shopping, 59
Using Leftover Paper, 59
Modifying Shape and Size, 60
Finishing and Binding, 60

Determining how much will be spent and for what is the basis for planning the production budget for a specific project or for all projects expected during a budget year. This chapter looks at approaches to the budgeting process, and then at ways of pricing production. Advice is presented on how to protect yourself against undue revision charges and how to understand your bill. The chapter closes with an extensive section on ways to cut corners to save money on print production.

Planning a production budget cannot be isolated from planning specifications or the production schedule. Once editorial, design, and production factors have been taken into account to compile tentative and final production specifications, the budget is planned and then the job is scheduled—the last step in the planning process.

APPROACHES TO BUDGETING

Your approach to budget planning will be either forward to a dollar figure or backward from it, depending on whether you have a definite spending limit or a definite set of specifications. A fixed budget is common; but even when budget is open, careful planning can help you make efficient use of funds.

WHEN BUDGET IS FIXED

A fixed budget has a definite limit. A figure is selected as adequate for the job or as all that can be allocated in relation to other spending priorities. When a project lands on your desk with a maximum cost attached, you must work within it at that time, regardless of when and by whom funds were set aside. You will need to work backward to draft tentative specifications you think will match the fixed budget, secure prices, and then make adjustments so that production cost is within the limit.

Need Versus Want

"Need" and "want" are two distinct factors in budget planning. When the budget is fixed (and strained), that distinction is your first concern—after you are assured that the printed piece is really necessary. What is needed may be less than what is wanted or vice versa. "Need" can usually be stated simply: "We need something to explain our conference facility to potential users." "Want," however, is probably a list of things: "We'd like it to be colorful, very visual, and provide some means for response."

If after pricing basic specifications (needs) you have money left over, start incorporating the extras (wants) until your budget is exhausted. Rank them in order of importance for image, convenience, and/or internal considerations. Two colors would be nice but not as important as quality paper, for example. A reply card could be printed separately instead of as a perforated panel of the brochure, allowing the convenience of more simple production of the brochure. To accommodate a drawing of the floor plan, use of less important photographs could be limited to only a few, one being of the director (as expected by the director).

In some situations, what is wanted in a printed piece may be less than what is needed. For instance, a marketing manager may have budgeted for a simple brochure for use by sales representatives in the field, only to have them report that the competition is using a very attractive, high-image piece. Your job as the producer of the brochure is to present enough production options to get as much as possible from an existing or cautiously expanded budget when such a mismatch of need and want becomes apparent.

Comparative Planning

To help you see how your needs and top-priority wants add up in relation to your budget, compare this project with similar ones before you ask suppliers to do detailed pricing. Your objective is a list of tentative specifications worth a supplier's time to price for you, not a pipe dream that is too unreasonable to justify the effort.

Information about previous projects you have done will help you narrow down specifications to match budget. Check your production files (as covered in Chapter 15) for a project as close as possible to what you have in mind concerning press run, size, number of photographs, paper, finishing, and/or binding. The cost of reproducing similar specifications can be comparable, even when end products are different. For example, the cost of typesetting a four-page brochure will be much like the cost of typesetting a four-page newsletter of the same size if both have a comparable amount of

APPROACHES TO BUDGETING

type. Take into account how long ago a similar project was done because production costs vary year to year or even month to month.

Another way to compare costs is to ask a supplier about any similar project that is currently (or was recently) in production in that shop, then the supplier doesn't have to do original calculations. Enough information can be extracted from another job to tell you if you are close to your budget limit. Such planning shorthand is reliable only if projects are comparable in specifications and in proximity of production.

Revising Specifications

Distinguishing between needs and wants, and then comparing your current project with others of known cost will tell you if specifications must be altered to correspond more closely with a fixed budget and, if so, in which direction. Using as an example the facility brochure mentioned earlier, you may find you can afford to meet basic needs and have a floor plan drawn if you drop the idea of a second color and a perforated reply card. Or needs alone may be over budget, leaving you the choice of paring specifications even further and not fully meeting objectives, looking for ways to increase the budget, or reappraising the merits of doing the project at all.

The more experienced you are in print production, the more accurately you can weigh the cost of one specification against another before seeking help from suppliers. When business for them is lean, however, they may be eager to work with you to make adjustments on a project that otherwise might not be done.

WHEN BUDGET IS OPEN

An open budget has no specific dollar limit. Certain specifications are required, regardless of cost. Even in times of austere budgets, such spending latitudes are not unheard of. Some jobs are too small and/or essential to justify extensive comparative pricing and adjusting of requirements (e.g., a short run of company stationery). Others have such potential for generating income that price is secondary (e.g., spending several dollars per copy for materials to impress and inspire sales representatives to market aggressively) is not unusual, given the possible payoff.

The fact that budget is open does not, however, preclude the need for responsible planning. It is simply a forward approach when requirements are more firm than expenditure limits.

Need Versus Want

As with a set budget, look closely at what is needed versus what is wanted. An annual report to employees that is expensive in every detail may be interesting to produce and win awards but may be poorly received by an audience more concerned about job security and pay than about high image. Plan specifications to meet all needs, then add only those wants that will bear up under scrutiny. Also, don't forget the basic question about how necessary the piece really is. A project done "because we've always done it" or as simply a showpiece of production talent is a waste of budget.

Appropriate Quality

To help you determine if tentative specifications will lead to a piece appropriate to your audience and quality requirements, review its objectives. If it's for quick information (a throwaway), re-evaluate using odd-sized paper and a specially mixed ink; a standard size and color make more sense, given the limited purpose of the piece. Consider if an insert in an existing publication could carry your message, rather than designing and producing a separate piece for the same audience.

On the other hand, your tentative specifications might not be up to the quality expected. If planning is thorough, you can readily make changes to gain quality when objectives are re-examined.

THE ANNUAL PRODUCTION BUDGET

Planning the annual production budget for a regular printed piece, an assortment of special pieces, or a combination of both can be a detailed and time-consuming process. Each project has several elements to be priced, any or all of which may change by the time you're ready to do the work, as may production costs. A regular project is more predictable than a special job, and budgeting for the two types of printing is somewhat different.

Regular Projects

Budgeting for a regular publication (such as a monthly magazine) is comparatively detailed and exact because it is a repeat project; any budgeting error will be amplified each time the publication is done. Start planning the annual budget for regular jobs by pricing every specification already familiar to you, based on past experience. Frequency of publication, size, press run, method of distribution, and other major requirements will most likely not change greatly on a regular job over the course of the year.

Next, incorporate any new requirements (such as additional copies or an increase in postage). When you ask suppliers for prices, confirm for them when and how often the piece will be produced. Their cost to do the first issue will probably be less than it will be in six or nine months, and their per-issue price for the year will have to take into account the difference.

If specifications for any aspect of production will be the same as in the past, simply ask the supplier to predict a percentage change in price. This shortcut can be used when budgeting for such projects as a monthly newsletter or an annual membership directory. For the same amount of type per issue, for example, your typographer might say, "Just add 10% for next year." That figure should be adequate for budgeting purposes. When requirements are to change, however, suppliers will need to do more precise pricing, as covered later in this chapter.

If you are considering changing suppliers on a regular project, now is the time to shop around. The figure on which you base your annual budget should be sufficient to cover any switch in suppliers because of unsatisfactory service or cost, either at the start of the new budget or midyear.

Special Projects

A realistic idea of costs to be met by an annual production budget is not as easily compiled when the plan includes special projects. Their specifications may be vague and, because each job is an unknown, costly production pitfalls or money-saving shortcuts may not be easily forecast. The more experience you have with budgeting and print production, the better your chances for a realistic picture of costs. The budget for specific projects will have to vary throughout the year, however, as requirements vary from what they were at the time you projected costs.

Base your planning on as much information as you can gather. Talk with the originators of special projects to define as precisely as you can what they have in mind. Review your production files for similar projects to round out a list of specifications. As pointed out in Chapter 2, suppliers will be able to give you only budgetary estimates when specifications are tentative. When a particular project comes up later in the year, they will have to price it again, based on final requirements.

Build in a contingency to the estimated cost of each special, unfamiliar project to attempt to cover a later change in specifications. How much that contingency is will depend on how vague requirements are now. Basing estimates on your best information about maximum needs, then adding at least 10%, is wise. If that 24-page booklet turns out to be only 16—or 32— your budget will probably still be adequate.

PRICING PRODUCTION

Whether your budget is fixed or open, it must be based on a realistic idea of the cost of producing a particular printed piece. You can get a rough idea with a list of tentative specifications; but a supplier can give a formal price only on final specifications. You must know what you want, with very little or no margin for change, before a typographer, printer, or other supplier can make a realistic calculation of cost.

METHODS OF PRICING

With final specifications in hand, you are ready to ask for estimates or bids. Deciding which type of price to request on a given project starts with knowing the difference between the two.

Estimates

An "estimate" is a proposed price for your job, based on the information you have provided. While it may be the figure that appears on your bill when the job is finished, the supplier is under no obligation to bill exactly as estimated. If the job comes in when and as specified, a reputable supplier will charge the estimated amount. However, every change in specifications or schedule jeopardizes an estimate, even though suppliers may protect themselves with some allowance for change so that they can bill as estimated in good faith. Consider an estimate as a minimum charge; unless production costs appreciably less than estimated, you will not be billed for less.

Bids

A "bid" (or "quotation") is a firm price for your job, based on the information you have provided. It is the amount that will be billed, even if the supplier makes mistakes or if costs go up between the time of the bid and actual production. As with an estimate, a safety margin is built into a bid *for the supplier*. The supplier is entitled to add to the bill the cost of mistakes or changes for which you are responsible.

To bid a job, a supplier will want very detailed specifications. "I'll have 20 pages of copy" may be close enough for a typographer to give a ballpark estimate but not a firm bid; you'll need to be more specific about point size, column width, and other requirements covered in Chapter 2. A bid leaves a supplier very little room to negotiate charges later. In exchange, you will have very little room to revise specifications if you want a supplier to work on a bid basis.

GETTING ESTIMATES AND BIDS

Estimates and bids are used in print-production trades for different purposes. Knowing when to ask for one instead of the other will help you avoid uncertain production bills and maintain amicable relationships with suppliers. The pricing process may take suppliers several days, depending on complexity of a job.

PRICING PRODUCTION

When To Ask for an Estimate

Estimates are the more common method of pricing. Reputable suppliers give them as close attention as they do bids and make a practice of keeping to them unless customer changes are excessive.

If the project is small, ask for an estimate, even an off-the-top estimate. Your cost may be a bit lower—and supplier relations higher—if you don't make a habit of requiring extensive and expensive calculating on every little job. "Can you give me an idea of how much 500 sheets of our current letterhead would cost?" is adequate for short runs of basic work. However, if even a small job is to require intricate production (such as close registration or finishing detail), get a formal estimate to protect your budget from unexpected charges and spell out to your supplier full specifications for the work.

Ask for only an estimate if the job is recurring and you are confident that your supplier is reputable. Even if specifications don't change over time, production costs probably will. With an estimate, your supplier can make adjustments when costs change appreciably. While you might save money by holding a supplier to a dated price, you will face a substantial increase when the supplier prices the work again or when you must find a new supplier unimpaired by a history of low pricing. Make a point of pricing any job in a familiar shop at least annually, however. Experienced print buyers can all tell about doing comparative shopping and finding that prices with a long-time supplier had crept up faster than elsewhere while the buyer wasn't looking.

You may not need to ask for an estimate if the supplier is one with whom you work frequently and well. This practice is usually a safe one but can result in a surprise when the bill comes. Get at least an informal estimate if you care at all about cost.

When To Ask for a Bid

Bids serve a useful purpose in print production as they do in other forms of commerce. Request them more selectively, however, than you do estimates.

If you have firm, unchanging specifications, ask for a bid so that price is guaranteed. As protection against unreasonable customer demands, an experienced supplier will not bid *unless* such is the case. If a supplier allows little (if any) margin for change in the bid, the price can be lower than that for a project susceptible to costly revisions.

Consider asking for a bid when a supplier is new and untested and when project requirements are firm. While an estimate leaves some room for supplier error or questionable business practices, a bid does not. If performance and quality turn out to be acceptable, you are probably safe in switching to an estimate on future work with that supplier.

Projects with high press runs or expensive complexities are candidates for bids, even from familiar suppliers. Many businesses and agencies require bids if cost is expected to exceed a certain amount. In the case of a large job, any unexpected per-unit charge can multiply into quite an increase in cost, which a firm bid on the job would prevent.

How Many To Get and From Whom

The number of prices you solicit will vary from job to job. The combination of small project and familiar supplier probably requires only one price because cost is not great and the supplier can be trusted to charge you a fair price. A large job, one that is complex, and/or the need or desire to get acquainted with a new supplier should prompt multiple estimates or bids. If time is short, accept a price by phone and ask that it be backed up in writing.

Three prices are usually sufficient for comparative shopping on even the most expensive and complex project. Pricing a job is time consuming, even when an estimator does it with computer assistance. Many elements go into the total amount, often including prices from subcontractors. Do not make a habit of requesting a half-dozen estimates on every job. Word travels fast among suppliers, and they will soon be reluctant to put out the effort to price your work if they know the chances for its being more than an exercise are slim.

Suppliers will also go the other way if you mix quality and reliability when asking for estimates and bids. Talk with suppliers of *comparable* performance. For instance, a first-class print shop doesn't stand a chance of pricing lower than a "cheap-and-dirty" shop that produces marginal work. The quality supplier may ask you who else is pricing your job and, unless performance is similar, may decline to look at the project to save throwing away time and effort on a sure loser.

Prices from suppliers of similar quality in your area will probably be similar and the cost comparison among them about the same, project to project. Printer A may always be a bit higher than Printer B, but both lower than Printer C. Consider going out of town for a different mix of suppliers on a large job (such as 100,000 copies of a catalog). You may find comparable service at a lower price—but be sure to take into account the cost of long-distance calls, traveling to supervise production, and shipping of finished copies.

If using a union supplier isn't a requirement of your job, solicit prices from both union and non-union suppliers of similar quality and reliability. The difference now isn't as great as it once was, however, as wages have tended to even out for skilled workers, regardless of affiliation.

"Politics" of Pricing

Print production for government agencies must follow a detailed procedure from pricing through delivery, as it must for many private-sector organizations. Because budgeting is the planning step that invites some suppliers to

PRICING PRODUCTION

participate and excludes others, it is the time to take into account internal formal requirements as well as unofficial practice or "politics." Find out about any such expectations before planning your budget. A newsletter for unionized employees may need to be printed in a union shop to avoid labor-relations ramifications. Your employer or client may have personal or historic reasons for wanting to do business with a certain supplier. A cousin of your supervisor who is a designer is not uncommon, nor is a good customer of your company who is a typographer "we've used for years." If you expect to have to use a supplier for reasons other than competence or price, plan your budget and/or schedule accordingly to try to compensate and expect to manage the project closely.

EVALUATING ESTIMATES AND BIDS

When you contact suppliers for prices, make sure each is figuring on the same set of specifications. Put these specifications in writing, along with a due date, and stipulate if a price by phone is sufficient or if you need a written, signed estimate or bid. When prices come in, you will have a standard for evaluating each and comparing one against another.

Figure 3.1 is an example of an estimate for a print job. Evaluation and comparison focus on three types of information it presents: specifications, working days, and price. Although these categories will vary with the kind of production being priced (design, photography, typesetting, etc.), the process of evaluation and comparison is the same.

Specifications

On the sample estimate (Figure 3.1), specifications for printing and binding are repeated (Section A). Check to see that they are the same specifications you gave the printer. Any variation in quantity, paper, number of colors, finishing, binding, or other details means that the price may be different when corrected to correspond with your requirements. An estimate based on inaccurate or incomplete specifications won't compare exactly with those from competitors. A supplier who misses the mark here either misunderstood what you wanted or is lax in pricing (and possibly in performance as well).

Working Days

Section B of the estimate (Figure 3.1) indicates how many working days will be required to do the job. In this example, days *from receipt of camera-ready artwork* and art until delivery were counted. An estimate that quotes working days *from receipt of proof* doesn't tell you the same thing; it doesn't include the number of days the printer will need to convert your camera-ready artwork into press negatives or plates for proofing. If you will require fast turnaround, look critically at this section. Pin down how many working

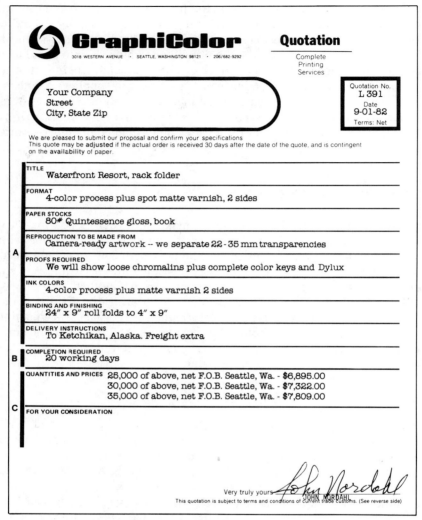

FIGURE 3.1. An example of a print quotation showing specifications (A), working days (B), and price (C). (Courtesy of GraphiColor, Seattle.)

days will be needed for the *entire* job. A supplier whose price is based on eight leisurely days to do the work can do it for less than one whose price is based on meeting your stated turnaround time of five days.

Price

The total price is Section C on the example (Figure 3.1). It does not include tax unless a figure for tax is specifically itemized. On a recurring job (such as

PRICING PRODUCTION

a quarterly newspaper), the estimate will stipulate how long the stated price will hold. Because paper costs are especially susceptible to increase, a printer may want to refigure an estimate midyear to take into account such increases. On a full-color project or one that will have several halftones, the quotation may include a "subject to seeing copy" clause which gives a supplier the latitude to alter the price, should the work be more extensive than planned. The estimate or bid may also indicate how long you have to accept the price; 30 days is standard unless otherwise stated.

Prices from suppliers of comparable quality and reliability should be within close range of one another if figured on the same specifications. One that is substantially high or low may be the result of a misunderstanding. That supplier will appreciate a call from you to review the specifications and a second chance to price your job if something was overlooked. Such consideration is especially important if you are dealing with familiar suppliers. Realizing when your job is in production that a serious pricing error was made creates a difficult situation for both buyer and supplier.

Trade Customs

Pricing and other procedures of printing and related trades are guided by long-established "trade customs" or terms and conditions. Reprinted in full in Appendix A, these customs are used voluntarily by artists, typographers, printers, and binders. Most commonly seen are printing-trade customs, often appearing on the back of a printer's pricing sheet. It may be a duplicate of the version in the appendix or a variation developed by a given print shop.

Review this "fine print" because it is part of your price quotation and stipulates such things as who owns what and terms of payment. Ask for a copy if one doesn't accompany your price from an unfamiliar supplier or when cost of your project is very high.

Additional Considerations

Especially if yours is a large project, ask each supplier if price is based on subcontracting any of the work. This practice is not uncommon and is one way a supplier can schedule more business and/or give you a better price. On a series of pamphlets, for example, a typographer may "farm out" some of the sections to another shop so all the work is finished by your deadline. A printer may send a simple part of a big job out to a small shop and retain more complex parts to cut costs and provide adequate quality. The supplier who subcontracts work is responsible for it in both schedule and accuracy. If you know before you award a job that parts of it will be subcontracted, you may want to divide it up yourself so you have full control over who performs the work at what price.

Check references to verify quality and reliability before accepting a price from a new supplier. The cost may be less because of poor performance or low this time to get your business and much higher once you are a regular customer. A check of references can save you many headaches later—or confirm your enthusiasm for a new supplier. (See later chapters for details on selecting suppliers.)

Make a practice of informing winning suppliers as soon as possible so work can be scheduled and any supplies ordered. As a courtesy, let everyone pricing a project know if a decision will be delayed, especially if it is a big job of particular interest. Common practice is for major suppliers to assume that a project went elsewhere unless they hear otherwise. You may wish, however, to inform losers (especially small or freelance suppliers) that their estimates or bids weren't accepted.

THE ECONOMICS OF CHANGING SPECIFICATIONS

The specifications of your print project influence cost twice—when they are determined and if they are changed. Economies can be built into your initial specifications (as summarized later in this chapter). Those specifications can also be changed with an eye toward keeping the financial impact to a minimum.

WHEN TO REVISE

Any change during production potentially influences both schedule and cost. Unless the change *reduces* requirements, your schedule will get tighter and the job will cost more with every change that comes as a surprise to suppliers. The cost of changing specifications depends on what outside services you are buying, how you are paying for them (hourly or project), and the stage at which a revision is made. A change in work done in-house will cost you time but probably not cash out of pocket. A minor change may be absorbed by an outside supplier working on a project basis but be added to the bill if work is by the hour. In general, the further along your job is in production, the more expensive each change in specifications becomes.

Planning

Changes on paper cost no more than the time involved to rethink and refigure if no supplier is yet being paid and if you are doing the planning yourself. A photographer quoting a price for working a half day can readily give you a new price for a full day if your requirements change during planning, for example. A printer estimating your job can refigure doing it on a different paper at no cost to you, as long as the paper you originally selected hasn't been delivered to the print shop.

THE ECONOMICS OF CHANGING SPECIFICATIONS

Writing and Design

Changing your mind about objectives, audience, length, or other factors related to copywriting will increase the cost of this element if you are buying outside writing services on an hourly basis. A copywriter who is charging by the project will quote a price to complete the job to your satisfaction and build in an allowance for revisions. If changes are extensive (especially if basic objectives are altered so that the writer has to start over), expect to be billed for an additional fee, even on a bid project.

Similar backtracking when working with an outside graphic designer will increase your bill for this element. A flat fee is common with designers but will not cover changes that could reasonably be considered excessive.

Illustration and Photography

Because illustrations are commonly charged by the project, making minor revisions in this type of art will be figured into the project fee. Expect to pay more, however, if you want major revisions in finished art. For example, if you are shown rough sketches and want minor changes, they should be covered by the flat fee. However, if the artist makes final drawings based on the rough sketches you approved and then you change direction, expect to pay more.

Photography, film processing, and prints are commonly charged by the hour and piece. Therefore, expect a corresponding increase in total cost if you change your mind about the amount and type of photographs you need after initial work has been done.

Typesetting

Like other suppliers, a typographer will allow a margin for change when a job is estimated or bid. Errors made by the typographer will be corrected within that price, no matter how extensive they are. "Author's alterations" (changes made at your discretion) may be reset and absorbed within the original price if they are minimal. A common typesetting practice is to start charging extra when type changes to accommodate author's alterations exceed a certain percentage of the total originally set. If you anticipate numerous changes on galleys, say so when the job is priced and ask about alteration charges.

Any major revision after type is set (such as size, typeface, or column width) may be costly, both in time and money. How costly depends on the typesetting method and equipment used. If you or another person who must approve the job has trouble visualizing how the copy will look when typeset, ask for a sample of type to avoid having the entire job set before you realize that the typeface is hard to read or the column too wide. Include a charge for sample type in your budget; the additional cost will be far less than starting over after being surprised at how copy converts into type.

Understand the limitations and capabilities of the equipment on which your type will be set (as explained in Chapter 9). On certain systems, changing typeface is simple and inexpensive; on others, costly resetting is involved. If you know the alternatives for changes in type and their corresponding costs, you can make better judgments, should you not like what you see on galleys.

Mechanical Artwork

Once type is set, paste-up or page make-up begins to execute mechanical artwork according to the layout you have specified. Most paste-up is still manual, usually tedious because of the precision required, and quite time consuming. These qualities add up to an expensive step in production. Any change in the position of text, headlines, art, rules, or other elements means moving and repositioning.

Paste-up through a typographer or printer will probably be billed by the hour so that any changes beyond correction of supplier errors will result in added cost. Freelance mechanical artists may bid a project; but that price won't cover major elective changes (such as switching a story from one page to another and rearranging both pages).

If you must make changes at the paste-up stage, try to make them all at once. Every time a section of mechanical artwork is handled, it must be squared on a drafting table or board before the change can be laid in. Consolidating the work by section will cost less than repositioning artwork for sporadic changes.

Image Assembly

Recropping or substituting photographs, changing special effects on art, and altering type or use of color add to the cost of your job if done at the time of image assembly and proofing. Comparatively, you have more opportunity to change or move elements inexpensively during paste-up or page make-up than you do in negative or other assembled form.

To keep down alteration charges on negatives, try to revise in blocks of type or art so that the film can be cut apart and the change taped into place. Or confine essential changes to as few negatives as possible to avoid remaking the entire job; anything that can be salvaged intact represents money you don't have to spend again.

If you must get approval from someone before a job goes on the press, don't wait until you have a proof. Clear your work before it reaches that stage—as comprehensive design, copy, galleys, art, and/or mechanical artwork—then clear the proof to show the fully assembled piece. The more changes you can anticipate and make before you proceed to a proof, the less your job will cost. (Chapter 12 presents specific information about changes on the negative.)

Printing

By the time a project gets on the press, you have almost no latitude for making inexpensive changes. As just stated, do all you can to assure that a job is accurate and approved before plates are made, paper is prepared, ink is mixed, and the press is started. If you have any doubts, see the paper you ordered to confirm that it is what you want and clarify what the color(s) will look like on that particular paper. On an especially complex or big job, check press proofs so you can make whatever adjustments might be necessary and possible before the entire job is printed.

Finishing and Binding

Theoretically, you can change specifications up until a project is delivered to your door. The bindery is its last stop, and the finishing/binding stage is your last chance to change your mind.

A piece designed and printed to be finished or bound one way can be finished or bound another at a similar cost if the methods are comparable and the design allows a switch at this point in production. For example, a brochure intended to be barrel folded can be accordion folded. Forms specified to be single sheets can, instead, be glued into tablets ("padded"). They can be drilled or punched for insertion into one of several types of looseleaf binders if the binding edge has a wide enough margin.

If layout permits, these and other requests of a bindery can be added to specifications or one procedure substituted for another, possibly at additional cost and time. This stage is not the one, however, at which to realize that the special fold you wanted cannot be accomplished quickly and inexpensively by machine or that the document for which you purchased special-order ring binders is too thick for them.

AVOIDING A SURPRISE BILL

As you can tell, the chances are great that a change in specifications after your job is in production will result directly in a change in your bill. Simple precautions, however, will prevent your being surprised when the bill does come.

Before you change specifications, go over your reasoning with affected suppliers. Perhaps the effect you now want can be achieved via a less expensive route. Because your suppliers are specialists, they may see shortcuts not apparent to you. Don't assume, however, that your graphic designer or typographer fully appreciates the impact of one change on other phases of production. Check it through all production phases to make certain that a cost-saving solution now doesn't become an expensive problem later.

When you and a supplier agree to a change in specifications, ask for an estimate of how much it will cost. This elementary question allows you to gauge the merits of an additional charge before it is incurred. Also, you will then know how much of the eventual bill can be attributed to the change so that the true cost of doing the job again, without revisions in midstream, can be recorded in your production files.

UNDERSTANDING PRODUCTION BILLS

One result of thorough planning of a print project is a satisfactory product that meets objectives, budget, and deadline. The other is a bill from each supplier involved. Understanding those bills is part of successful budgeting for the job just completed and for any similar one that comes along later.

CONTENT OF A BILL

Information on a bill differs from what is included in the estimate or bid on which it is based. It is a summary only, not a detailed listing, of specifications that went into pricing and producing the job.

Whether handwritten on plain paper by a freelance illustrator or computer-typed on a special form by a printer's accounting department, a bill should clearly indicate its source by name, address, and telephone number and be dated. It should state a price, including applicable sales tax and shipping charges. The description of the work performed should be sufficient for you to identify the job. A bill for composition of a four-page newsletter may be as simple as that illustrated in Figure 3.2. A print bill may be slightly more detailed, as shown in Figure 3.3.

A bill may include one or more customer or job numbers—the buyer's and/or the supplier's. Print buyers who work for agencies or large companies with varied and numerous projects assign a number to each job to identify it from inception through completion; that number may be repeated on supplier bills for easy reference. Similarly, a supplier may assign a "job ticket" number to your project when it comes in and repeat it on the bill. This number identifies it throughout production in that shop and is used by that supplier for filing information about your job for future reference.

A bill from an established company (as opposed to a freelance supplier) will include an invoice number which may also serve as a job number. An invoice may be your only bill or may be sent when the project is finished, followed at the end of the month by a statement showing the total of all charges owed.

The statement (or invoice if no statement is used) may include a reference to payment (such as "1% per month after 30 days"). As spelled out in trade customs for print-related industries, payment within 30 days is the norm, although one supplier may request faster payment and another be lax about the 30-day limit. Stating terms on the bill reinforces payment expectations

UNDERSTANDING PRODUCTION BILLS

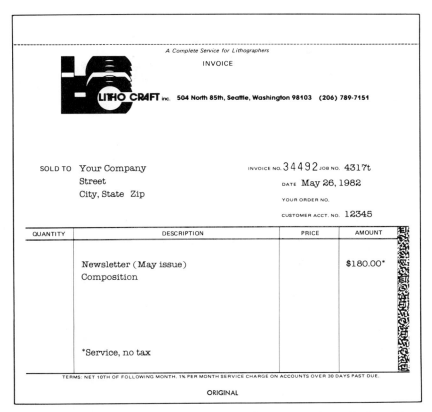

FIGURE 3.2. An example of a typesetting bill. Note that no state sales tax is charged because composition doesn't result in a finished product. (Courtesy of LithoCraft, Seattle.)

and lets you know in writing of any interest that will be added if payment is late.

Your bill from a copywriter, designer, illustrator, or photographer may also include a reference to terms of the sale of artistic material. A photographer may stipulate, for instance, "one-time use only" or "negatives on file." Such a statement reiterates the terms of your agreement about purchase of rights to artistic work. (See Chapters 5, 6, and 8 for details on rights.)

You may request that your bill include a breakout of charges incurred on straight time versus any overtime or the cost of extra work you might have requested. Noting these special charges on the bill makes them a matter of record for future planning.

VERIFYING CHARGES

What you are charged for a print job will be no surprise if price was part of your planning process. Bills for all elements of a project should be exactly as

GraphiColor

3018 WESTERN AVENUE • SEATTLE, WASHINGTON 98121 • 206/682-9292
TOTAL CAPABILITY • PRINTING AND LITHOGRAPHY

INVOICE DATE		INVOICE NO.
10-01-82		22987

SOLD TO:
Your Company
Street
City, State Zip

SHIPPED TO:
Waterfront Resort
Ketchikan, Alaska

CUST. NO. 3267   SLSM 05/Nordahl   TERMS: NET THIRTY DAYS
1½% Service Charge on amounts past due.

CUSTOMER ORDER	SHIPPING INFORMATION
9-07-82	9-28-82 via Alaska Air

ITEM NO.	QUANTITY SHIPPED	JOB DESCRIPTION	AMOUNT
	30,145	Waterfront Resort	$7322.00
		Air freight	525.00
		Freight tax	26.25
		INVOICE AMOUNT	$7873.25

Payment of invoice due in full 30 days after invoice date.
A service charge will be added for each month the amount is delinquent.

☐ GOV'T   ☐ RESALE #   ☒ OUT OF STATE

FIGURE 3.3. An example of a print bill. Note that no state sales tax is charged because shipment is out of state. (Courtesy of GraphiColor, Seattle.)

bid and close to or exactly as estimated, unless you agreed to an increase or decrease prior to billing.

Some charges are taxable and others are not, depending on sales-tax provisions in your locale. In general, something that is supplied as a finished product is subject to sales tax, whereas something that only contributes to a finished product is not. An illustration or photograph is considered finished, even though you may crop it, enlarge it, or otherwise change it for use in your printed piece; the fact that it has a retail value as is makes it taxable.

Copy or galleys of type are not considered finished (and taxable) because they are not end products; they must undergo further work before they can have a retail value. A job printed "not for resale" (such as a newsletter or annual report) has sales tax added. If it is a consumer magazine or other piece intended to be sold, sales tax is not added until the time of sale.

For additional information on taxes, consult your state or provincial revenue office. Long-established suppliers in your area or municipal authorities can confirm local practices.

ECONOMICAL PRODUCTION

A print project need not be expensive in every detail to meet its objectives. Altering specifications in some way (such as cutting back on the number of ink colors, simplifying design and production complexities, or printing fewer copies) are obvious methods of saving money. Several others, however, can be considered during planning to take advantage of as many ways as possible to make the piece you are planning less costly. To understand each fully, refer to later chapters that cover in detail each phase of production referred to in these suggestions.

WORKING WITH SUPPLIERS

Efficient dealings with suppliers, whatever their roles in your print project, can result in savings. Part of that efficiency is common courtesy, paired with enough knowledge of production on your part to single out yours as a considerate, reasonable account that doesn't require exceptional time to service. Look for additional savings by applying one or more of the following procedures.

Economical Scheduling

Planning ahead for a leisurely schedule will save production dollars. A job that is ready to go into production far in advance of its deadline can be used by suppliers as "filler" when immediate work of the day is finished. Another economical procedure is to place work in an off-season when business for suppliers is slow. Such a project lets the supplier keep employees busy and cuts the cost to you.

Schedule your job with another project immediately ahead of it on the press, using the same color(s) to avoid a "washup" charge (cleaning of the press) and/or the same paper to save on makeready charges. Running your project on a web press (more economical than a sheet-fed press for a long run) may become feasible for a short to medium run if you can use the same paper and ink already set up for a job to run immediately ahead of yours.

Work that pushes suppliers is expensive, especially if it requires overtime. Even when overtime will be needed to meet your deadline, however, you can avoid spending more than necessary. Ask about overtime charges for each major step in production, then invest in overtime at the most economical point to do so. For example, overtime printing of your job may cost far more than overtime typesetting. Therefore, to pick up a day in a tight schedule with the least impact to your budget, choose the lower charge of overtime typesetting.

Copywriters, designers, and some other suppliers may figure time charges portal to portal. To save money here, go to them, rather than asking them to come to you.

Matching Capability to Project

The cost in time and money of using a supplier not fully geared to your type of project or one that is overqualified can be higher than the cost of a better match of supplier and project. Chances for error will also be higher because of lack of supplier skill or attention to a small or odd job.

Suppliers challenged beyond their standard level of performance—in quality as well as turnaround time—charge more for the extra effort needed to meet your expectations and deadline. For example, asking a printer with a two-color press to run your four-color job is asking for additional skills and time not required in the two-color market to which that printer may normally sell. You may get what you want, but also a higher bill than would have resulted from going to a printer who concentrates on a four-color market.

On the other end of the scale, services can cost more than they need to if your job is too small for a given supplier or outside the regular flow of that shop's business. For instance, an order for business cards is better placed with a printer set up for and interested in doing such work. A large print shop not geared to production of business cards will have to charge more for the extra trouble caused by such an odd job and may give it low priority. Certain projects are ideal for a quick printer or even more efficiently done on office duplicating equipment; production will cost less when jobs are placed accordingly.

Contracting Directly for Services

To avoid markups, make your own arrangements for services whenever possible and have bills sent directly to you. For example, a photographer who finds the models needed for your catalog project will probably add a "finder's fee"— a percentage markup of the models' hourly charges as compensation for finding them for you.

Similarly, a printer asked to produce a brochure from rough copy will charge on a fee-plus basis to subcontract for editing, design, typesetting, and any other services needed to get your brochure refined and ready for the press (especially if the profit margin for printing alone is low). Finding and managing each supplier yourself will save you money as well as increase your control over the project.

Placing Related Jobs

You may also save if you award related jobs of comparable quality to the same supplier, rather than distributing your jobs among numerous suppliers. Have one designer do a series of brochures, for instance, so that they look related and so that they have elements (such as photographs) in common, thus reducing production costs. Place the brochures with one engraver or printer so that only one set of negatives of the art will be required. If quantities are the same, group them on the press to give your budget the benefit of having the job run on larger and more efficient equipment, possibly also at a savings on paper.

When quality of related pieces is not to be similar, however, choose your suppliers accordingly. Go to a quality printer for a high-image poster to publicize a tour, for example, and to a lower-quality printer for a simple brochure that provides additional tour information.

Repeat Business

Regular use of suppliers improves the service you get from each. Working together long enough to get to know one another's expectations and capabilities means each job goes more smoothly and takes less time to coordinate than one that brings buyer and supplier together for the first time. Being able to count on repeat business saves a supplier the overhead expense of continual marketing to a series of one-time customers and allows that supplier to keep down charges.

Doing More Yourself

Asking less of suppliers at various stages of production will save you out-of-pocket money. As you plan a job, look critically at your skills and/or those available to you. Maybe on a simple job you can do paste-up or your company editor can write copy. The more you can do yourself or have done in-house, the less you will have to purchase outside.

Suppliers may charge extra for services that in the past were considered overhead (such as picking up and delivering jobs). Though individually small, these charges add up and may be unnecessary drains on your budget. Ask before accepting a supplier's price about any such extra costs, then avoid them if you can or keep them to a minimum by, for instance, consolidating pick-ups and deliveries into one per day or using a company vehicle to make them yourself.

REVISING AND ADAPTING

Instead of producing your printed piece "from scratch," consider how you might revise or adapt existing copy, design, or art. This procedure will save money if you would otherwise pay outside suppliers to do the work and may also save you time.

Copy

A complete rewrite of copy for a printed piece may not be necessary. A close look at how much of the copy is still current will reveal the extent of change needed. Revising only the affected copy will save you time and money if very little is to be altered.

Being able to revise existing copy also means a savings in typesetting if some of what is already set can be used again. Reset only the blocks of type that have changes if the old type is not faded or damaged (ask your typographer or engraver if you're not sure). Save on paste-up, too, by replacing only the affected type on mechanical artwork. Save even more if printing is to involve negatives by having the new blocks of type set, converted into negatives (bypassing artwork entirely), and substituted on the old negative. Such revising of copy will save money when changes are minor and when layout remains the same.

Design

Once copy for a new piece is written, save designer time and expense by adapting the design of any related piece to the new project. Follow the same format used on similar materials, for example, a new brochure to be added to a series of brochures on employee benefits. (Choose an adaptable format in the first place so it can be used again and again.)

If related pieces aren't available, find another format that will be appropriate and adapt it. (Some samples are presented in Chapter 6.) Any design produced in-house or done for hire by an outside designer for you is fair game for adapting to a new project. Any other *basic* format you see and like in the public domain may also be adapted if the material to be presented is not in direct competition with that presented in the original.

Art

Making use again of illustrations, photographs, and even the original mechanical artwork can incorporate significant savings into your production plan. If a table in last year's annual report, for example, presents a list of products that is still current, simply update the corresponding sales figures and column heading and leave rules and all other headings in place.

Illustrations and photographs you have used before can be called back

ECONOMICAL PRODUCTION

into service. Change their size or shape for a different look for less than the cost of more illustration or photography. Save even more by using art the same size and shape to avoid re-doing negatives; create a fresh appearance by adding or changing color when the piece using such art is printed.

Transfer designs, ornate rules, and other purchased camera-ready art can also be adapted to save the cost of original art. As covered in Chapter 8, such art is available by the roll, sheet, or in themed packages, ready for adding directly to mechanical artwork or for reducing or enlarging for single- or multicolor production.

Original artwork can be reused if it is not damaged (see Chapter 11). Compare old type with new type of the same size, face, weight, and style, however. With some phototypesetting processes, the type diffuses over time into the paper, causing characters to enlarge and blur. New type is crisp; if the difference is noticeable, old type must be reset to match new.

DESIGN, ILLUSTRATION, AND PHOTOGRAPHY

Overall design, illustrations, and photographs contribute substantially to the success of a printed piece and may contribute a major portion to its cost. Planning ahead can keep that portion to a minimum.

Saving on Design

Discuss objectives, audience, budget, and other aspects of your printing project with a designer at the onset. Budget is especially critical since design has a strong influence on production costs. Ask to see "thumbnails" (small sketches) before the graphic designer fully develops format and layout. A "comprehensive dummy" (or "comp") is done actual size to show all elements of a printed piece the way they will look in the finished product. Doing a comp is time consuming and should be reserved for only the most likely design(s). Keep down the design budget by being very selective at the thumbnail stage, preferably picking only one idea to be executed in a comp. When working on a simple job with a graphic designer in whom you have confidence, skip the comp phase if you don't need a comp to visualize how a thumbnail will translate into a finished product.

To avoid inadvertently choosing a design that will be unnecessarily costly to execute, take the most promising design idea to the suppliers who will be producing it. The more intricate the design, the more you will likely pay in typesetting, paste-up, engraving, printing, finishing, and binding. Check it out ahead of time so that any revisions can be made before production starts to keep your project on schedule and within budget.

Thorough planning goes beyond production to the user of the piece if it is letterhead or business forms. Ask a secretary, for example, to test a proposed letterhead design to see if it is easy to use and if it results in a nice-looking letter. Also, vertical spacing of lines that corresponds exactly to standard typewriter or word-processor spacing saves time for every person using a form. Such considerations may not be apparent to the graphic designer or to you but will be to the user. Asking in advance of production will save the expense of revisions and assure that piles of forms don't go to waste because they aren't fully functional.

Saving on Illustration

Make sure your illustrator understands the tone and layout of your job. Fully explain what you have in mind and provide all items to be drawn, photographs from which illustrations are to be made or which will be printed with the illustrations, and access to places and people to be illustrated. Review budget limitations so that the illustrator will know if you can afford "line art" (black and white only or simple color) or more expensive "continuous-tone art" (black, white, and shades of gray or full color).

Ask to see rough sketches of your illustrations before detail work is started to avoid paying for finished art that isn't what you had in mind. Confirm at the time you see sketches the printed size of each illustration. Drawing twice as large is customary; an illustration any larger may require the expense of two reductions to get it down to size (see later information about negatives).

Protect all illustrations with a sheet of tissue paper over the front and a heavy paper backing adhered to it with an adhesive formulated for this purpose. These simple precautions can save you the cost of replacing art damaged in shipping or handling during production. Using a special adhesive will prevent discoloration over time that might otherwise occur with rubber cement or another glue.

Saving on Photography

Getting the most from your photo budget starts with being selective about what you have and what you need. Survey what is already on hand for possible use. Perhaps you can use a photo again, or maybe other shots taken on the same roll of film could be used for the first time. Consider "stock" photos; the cost will be less than the expense of quality original photography (see Chapter 8). Especially if your need is for general photography (such as scenics), you may save a significant amount by buying one-time use of stock photos.

When original work must be done to meet all or part of your need, pay only for the quality and quantity of photography you require. Informal portraits of employees for a company newsletter may not demand the same caliber of photography as similar portraits of the board of directors for an annual report. If the layout calls for five photos, select the best and most appropriate five. Don't order more than you can use just because they are well done. Resist the temptation; you will spare your budget now and can

ECONOMICAL PRODUCTION 51

always buy the extra shots later if the cost can be justified for some other purpose.

Order prints from a "contact sheet" (which shows each shot in miniature) to avoid buying useless prints. Out of 20 shots, you may want only two—and may want them printed a certain size or with a particular finish. Be selective and save yourself the expense of unwanted prints or prints not made to your specifications. As covered later, confirm with the engraver or printer who will be making the negatives for your print job what size and finish are preferred.

Organize your needs by time and location so that you use a photographer efficiently. Especially if you are charged by the day or half day, keep the photographer fully occupied; any unused time is wasted budget. Have all photo work done at one location or studio whenever possible. Getting set up to start shooting can require considerable time, particularly if the photos are to be high-quality color. You will save money if you can move people and objects to the photographer, lights, and camera equipment, instead of the other way around. Also, using a standard setting will save the expense of doing shots over ("retakes"), should use of some locations result in poor photos.

Make sure your photographer understands the nature of the job. If you have a layout of the piece for which the work is intended, go over it with your photographer so that position, size, tone, ink and paper color, and any other pertinent details are understood. In addition to shooting to fit your need, the photographer will also select a corresponding film so that, for example, print film is used in the first place to save you the cost of converting slides into the prints you need.

Pay only for the use you expect to give the photography you buy. Many photographers charge one price if you use a print or slide once and a much higher price if you plan multiple or advertising uses. Buy one-time use of shots of buildings for an engineering brochure, for instance, and multiple use of shots of the firm's principals that you expect to need for other purposes later.

Protect photographs from damage during production by mounting them with special adhesive on cardboard and covering the face with tissue paper. Avoid writing on the back or otherwise marring them so they can't be used again.

TYPESETTING

At the typesetting stage of production, you can save the entire out-of-pocket cost by setting type in-house or save a portion of cost by efficient use of outside services. The choice depends on the magnitude and quality of the job. A small, simple project may allow in-house production, while a large, complex one is better done by a commercial typographer. Also a possibility in the balance of budget and quality is a combination in which headlines are done outside and the text produced in-house.

In-House Typesetting

Consider saving money by setting copy on a typewriter, word processor, or other in-house equipment. Such methods are adequate for small jobs not requiring the quality and appearance of commercial typesetting (such as flyers, routine mailers, or informal newsletters).

Commercial Typesetting

Going outside for type will be your choice when size, quality, and appearance of the job require it (such as on magazines, formal announcements, or annual reports). Your approach to commercial typesetting has a lot to do with the cost difference between it and in-house production.

Method of Typesetting. Select a typesetting process that matches the project to be set. As covered in Chapter 9, you have various methods from which to choose. Phototypesetting is now the workhorse of the industry, suitable for many kinds of jobs. A very large project, however, may be better done by electronic composition, and hot type still has its uses. Before sending a job to one typographer over another, consider volume and nature of the copy, quality desired, and future need for efficient updating of the type. A typographer geared to your kind of job will work more efficiently than one poorly matched in capability of equipment and/or personnel.

Condition of Copy. The way your copy comes into a typographer significantly impacts the time and cost of doing the job. As detailed in Chapter 5, provide clean copy, neatly typed with few (if any) insertions or handwritten changes that can slow setting, especially if the job was priced on an hourly basis. Error-free copy (whether on paper or a computer disk) is imperative for efficient electronic composition.

Provide full specifications for all copy so that the typographer doesn't have to guess how a headline or footnote is to be treated. Try to send in everything at once, grouped according to type size, face, weight, and style so that the typographer can proceed without interruption to set large blocks of copy, thus saving time and money.

When a project permits, set type in position to simplify paste-up. For instance, set subheads used to divide text into sections as the text is set so they don't have to be pasted into position separately. Some typesetting equipment can produce type galleys in sequential columns. If you want text to appear in three even columns across a magazine page, for example, you can so specify at the time type is set and not have to arrange the columns by hand during paste-up.

Keep special work to a minimum, however. Intricate placement of type to wrap around an illustration takes time. You may pay more to have a typographer try to do such work than you would pay to have it done by hand during paste-up. Explore the tradeoffs when you have special work; much depends on the skill and equipment of your typographer.

Plan Ahead. Read copy carefully so that changes in galleys are minimal. Each author's alteration on galleys beyond what the typographer considers allowable adds to the price. If you must make changes, remember that exact substitutions at the end of paragraphs cost less than extensive alterations in the beginning or middle. For example, changing "sink" to "lavatory" in a paragraph may mean that several lines or the whole paragraph must be reset to accommodate the longer word, especially if it appears early in the paragraph or partway through. Also, the least amount of resetting is required if type is unjustified (such as "flush" or even on the left and "ragged" or uneven on the right). Because each line doesn't necessarily fill the entire column, a longer word can often be worked into existing space. If you anticipate numerous changes at the galley stage, consider using an unjustified column for the least disturbance of type.

Have portions of copy set at the same time layout is planned to check appropriateness of typeface, style, size, weight, leading, and column width. Such sample type will add to the cost but be much less expensive in the long run than having to reset a job because of dissatisfaction or miscalculation.

Avoid having your budget exhausted by minimum charges. A typographer will charge a basic fee to set one word, a headline, or a paragraph. Plan ahead to group small jobs so that you get the most for your typesetting dollar.

Make a photocopy or carbon of all copy before you send it to the typographer. Following this simple practice can avoid the cost of reconstructing that copy, should your material be misplaced, lost, or damaged at the typographer's or enroute.

Ask your typographer to save the tape, forms, or disk file of your job if you expect to re-do it later with minor changes. (Different systems have different methods of retrieval and update.) Whatever can be retained for future use will save the expense of starting over.

MECHANICAL ARTWORK AND IMAGE ASSEMBLY

The most tedious phases of print production are when mechanical artwork is pasted up and when all elements are assembled in negative form prior to making press plates. Though the work is detailed and time consuming, it need not be overly costly if well thought out.

Saving on Mechanical Artwork

Providing clear instructions will save effort and cost when the layout of your job is executed in camera-ready artwork. Asking a mechanical artist who charges by the hour to stop to decipher what you mean is asking for a higher bill. Indicate exact measurements and placement, mark where corrected type is to go, and write in the first three words of headlines and cutlines so the mechanical artist won't have to read the text to determine what goes where.

Another time and budget saver is doing paste-up according to how the job will run on the press. If your 16-page brochure will run "four up" (four pages at a time) through the press, arrange pages on mechanical artwork in the correct sequence, by fours. Ask your printer how many up is preferred. Artwork that grows too large is awkward to handle and invites misalignment from one side to the other.

If your project is to run in two or more colors, also ask your engraver or printer if areas to appear in the additional color(s) are to be pasted on an "overlay" (a separate plastic sheet over the base artwork). Often the preference is to separate colors on the negative; knowing so during paste-up will save the extra time and expense of doing unnecessary overlays.

Saving on Image Assembly

Clear and complete instructions will also save money on image assembly, whether for offset, letterpress, or any other printing method. Resolving confusion takes time, as does re-doing engraving work because of inadequate or uncertain instructions. Indicate photo cropping and percentage of enlargement or reduction, call out what is to be printed in which color, and specify the line screen you want used to convert photographs into halftone negatives.

Organize engraving work to take advantage of savings from doing the whole job at once, instead of piecemeal. Photos to be reduced or enlarged the same percentage and at the same exposure may be "gang shot" (grouped) for considerable savings over individual attention to each. On color work, you might save by having quality prints made to size, then positioning them as they are to appear on a page ("pre-assembly") and making one large set of separations, as opposed to small, individual ones. This procedure is especially efficient if you are using several color shots on a page or spread because it saves stripping time.

Avoid unnecessary charges by having illustrations drawn or photos printed for one-step resizing. Ask your engraver or printer about reduction and enlargement capability because it varies according to what camera equipment is available. Commonly, line art can be reduced down to half or enlarged up to double the size of the original in one step. If your illustration needs to be reduced to one-fourth, you may have to pay for two reductions to get it down to size. Halftone reduction also varies with equipment. A small print shop may not be able to reduce a photo more than 33% of the original, whereas a large shop or an engraving specialist will have a wider range.

ECONOMICAL PRODUCTION

When you have control over the quality of illustrations and photographs, make sure they require no special labor or camera manipulation to reproduce. An engraver will charge to go over fine lines in a drawing, for instance, so that they will not be faint or nonexistent when reduced. Likewise, getting the most out of a dark or grainy photograph takes extra time. Avoid such expensive salvage work with attention to quality in the original art. Also recognize the limitations of what can be done. Even in skilled hands, a dark or grainy photograph can't be made to appear light or smooth, and color correction can change the balance of colors only so far. Asking an engraver to achieve the impossible is frustrating, time consuming, and costly.

PRESS WORK

You may save in several other ways by planning press production with budget in mind. Because of the relatively high cost of press time versus other steps in production, savings here may be appreciable.

Matching Press to Project

Economical pairing of your job and the press on which it will be printed involves consideration of size and speed. Press time is charged by the hour; so your objective is to keep running time and hourly rate as low as possible.

Unless yours is a simple one-color project with a short run, you will save money by considering press size. In general, the more pages that go through a press at once, the more economical the job. For example, if you are printing 5000 eight-page newsletters, select a printer with a press big enough to run four or eight pages at a time, rather than one with a small press that will accommodate only one or two. The difference in hourly rates between the two presses will probably justify using the more expensive press for a shorter period of time. Get comparative figures, however, because a newer, smaller press requiring only one operator may be more efficient for a particular job than an older, larger press requiring two or more operators.

Two presses of comparable size billed at the same hourly rate may not run at the same speed. Ask for an estimate of press time, which should reflect both press speed and the ability of its operator. Make sure that time estimates include the total hours that your project ties up the press (makeready, printing, and washup), not just printing time alone because your bill will reflect all these activities. Even if the hourly rate for a faster press is higher, you may save because your job will be on it for less time.

A printer who can keep a press in operation over two or three shifts a day can afford to charge less per hour than a printer with only one shift. Take into account the fact that an idle press adds to overhead. When all other factors (quality, skill, press size and speed, turnaround time) are equal, you may find an additional source of savings at a print shop with more than one shift.

One Run

Once your job is on the press, you will save money if the press can keep running until the job is completed. Printing 100 copies of an annual report this week for a shareholders' meeting, and then printing another 2500 next week for mailing is poor and expensive planning. You will pay twice for all the makeready work needed to get the job on the press and twice for washups. If overtime is a factor in the equation, however, consider a split run. Meeting only emergency needs on overtime and delaying the bulk of the run until it can be done on straight time may be a more efficient approach.

Ink Colors

Using more than one color of ink on your job can be done with economy in mind by conferring with your printer in advance about how the project will run through the press. A simple example is a four-page newsletter.

If imposition is to be "work-and-turn" or "work-and-tumble," it will be laid out so that an outside and an inside spread are on one side of the sheet the first time through, and the corresponding inside and outside spreads on the other side of the sheet the next time through. Therefore a second color can be used on both spreads for the same cost. If imposition is to be "sheetwise," printing will be one page at a time, or two outside spreads will be run the first time through, and then two inside spreads. With sheetwise imposition, you will save by restricting a second color to either the inside or outside page or spreads. (See Chapter 12 for more information on imposition.)

Regardless of printing process, avoid the expense of a washup that is necessary if you switch colors. Use blue and black on the front, for example, as well as on the back of a piece. Switching to any other combination on the back would add the cost of cleaning the press in between passes.

In selecting color for a one-color job, keep in mind that black is the most commonly used ink color. A job printed in black will cost less because a printer may already have that color on the press. Also, any other standard color of ink that doesn't have to be mixed will cost less than one that does.

Masters

Preprinted sheets ("masters" or "shells") can save you money if a recurring job has limited use of a second color or if the same piece is to be personalized for different locations or time periods. Assume you print 100 copies on a quarterly basis of a small rate sheet listing products and prices in black below your company's green logo. A year's supply of heading can be printed in green (a one-color job) in a quality shop and each quarter's edition printed in black (also one color) in a lower-quality, lower-priced shop, probably for less than a short, two-color job would cost each quarter.

ECONOMICAL PRODUCTION

Versatile masters are also economical when information must be personalized by location or time. For instance, you may have 10 offices that want varying quantities of a sales brochure, each with the local address and phone number in a second color. To avoid a costly short run for each office, print masters in one color, followed by individual runs with local information in the second color. A supply of masters on hand also allows a quick response to orders for additional copies or revised versions when an address or phone number changes. Another example is a rate brochure with standard information on the outside and prices on the inside. Print a quantity on the outside only, come back as prices change to print a supply on the inside, and then have those folded. For a recurring job that would look well on colored paper, consider having masters printed in several colors (such as a different color for each new edition of a price sheet).

Press Proofs

Check press proofs at your printer's shop rather than having them brought to you. As long as your job is tying up the press, you will be charged, whether or not the press is running. On any job you consider complex (especially if it involves an exact matching of colors) or expensive, do a press check as it starts to come off the press so any adjustments can be made early. A job that is not checked at this stage and later found to be unacceptable means an expensive rerun and possibly refinishing and rebinding if the error is not caught until delivery.

PURCHASE AND USE OF PAPER

Paper is another ingredient that can cost less if planning is thorough. Whether the job is large or small, consider these possibilities for paring your bill.

Direct Purchase

Buying through a printer is a common method of acquiring paper; but it can also be purchased directly from a paper house to save the printer's mark-up (possibly 10%). Take advantage of this savings if ordering and delivery are convenient for you to do and if the job is large enough to justify the extra effort on your part.

A paper house will deliver an order to your printer according to your specifications. You are responsible for guaranteeing that the paper meets the printer's requirements (such as size, quantity, and condition). You own any paper that is left over and must make arrangements for its removal from the print shop. You are also responsible for storage if you buy paper direct for a recurring or future job. To secure a large project, your printer may provide free or low-cost storage of your paper for later use. Otherwise, you must see that it is properly stored and delivered in a timely manner to the printer.

Direct purchase may also be a cost-saving option for a small job if you can find suitable paper at an office-supply outlet. Such a business catering to in-house reprographic departments will stock a variety of precut sheets from which to choose. You can simply buy it off the floor or from the warehouse and take it to your printer.

Buying Ahead

Purchase of paper in bulk by either you or your printer protects your budget against price increases and is efficient for such projects as a monthly newspaper. A predicted price increase, however, must be greater than the interest your money could be earning if it were otherwise invested. If not, you are better off buying paper only as needed. Proper storage of bulk paper is important because the portion you use in November must be as satisfactory for printing as the portion you used in May.

If you have funds left in your budget that won't carry over to a new budget, consider purchasing paper ahead of time so that the residue from this year's budget cuts the cost of production next year. With predictable projects (such as the monthly newspaper), you can buy in bulk this year for next year through your printer or directly from a paper house and keep the paper in storage until needed.

Matching Paper to Project

When selecting paper for a project, save on press time and cost by suiting the paper to the press on which the job will be run. Any paper that has a high moisture content, is too lightweight, or has a finish that is difficult to feed through the press slows production. Since you are charged by the hour for press time, the more the paper slows the press, the more expensive the job. Always ask your printer about paper quality before making a final selection to save money and avoid production delays.

A "coated" paper (with an enamel finish) generally costs more than an uncoated paper. Unless your job is to run on a letterpress that requires coated stock, select an uncoated sheet to meet your needs and budget. Even if your objective is quality photo reproduction, you can now find an uncoated paper that may be as effective as the coated papers that used to be required to achieve that objective.

Consider "skid paper"—stock that is so frequently requested that your printer buys it in bulk to have readily on hand. The printer saves money buying such paper in large quantity (by the skid) and can pass that savings along to you. If such paper is adequate for your need, you reap the cost advantage of a bulk purchase, even on a small job.

ECONOMICAL PRODUCTION 59

Comparative Shopping

Shopping around can save you money on your paper purchase, whether you're buying direct or through a printer. Know the characteristics you want (such as color, texture, or opacity), look for a sheet to match, and then look for any similar sheet at a lower price. A given paper house may have more than one sheet suitable to your purpose, especially if yours is a common use (such as letterhead). The selection gets even larger among paper houses. Comparative shopping may reveal that the paper you like has a close equivalent at a lower price or that Supplier A will sell it to you for less than Supplier B would.

Ask about ease of delivery of the paper you have in mind. A paper readily available locally will cost less than a paper that has to be specially ordered and shipped across country—and it will be a more practical choice in the long run if you expect to have to match it later to print the job again or for a related project.

Using Leftover Paper

For a short press run when choice of stock is open, look into using leftover or discontinued paper. Leftover paper may be odds and ends your printer or a paper house has on the shelf or a portion of a sheet that would otherwise be discarded from another print job. Discontinued paper is offered in odd lots directly by a paper house or a supplier specializing in marketing odd lots for several paper houses.

Paper off the shelf will probably be priced according to when it was purchased, not according to the current price. For example, if you need 500 flyers and require only that the paper be light colored and of a minimum weight, ask your printer to show you samples of leftover paper, either all of one color or of an assortment the same weight and texture adding up to 500.

Also consider shopping for odds and ends or discontinued paper from a paper merchant or clearinghouse. Again, specifications must be fairly open, and if you buy a discontinued sheet, the project must be one time only because you probably won't be able to match the paper later. Even if your job is large, you may find a suitable sheet at a bargain price by checking for odd lots. Remember, however, that paper is a bargain only if it is in printable condition.

Paper to be trimmed from another print job may be just what you need for a smaller job. For instance, assume that you are about to print a large brochure that takes up two-thirds of a sheet. Rather than waste the other third, use it to print a smaller piece at the same time, in the same color. Even if you are running 1000 large brochures and need only 500 of the small piece, you can afford to throw away the excess because you're not paying for the paper or separate makeready and washup on the press. When you buy paper, you own the whole sheet or roll. If the surplus is large enough to go back on the press at a later date, ask that it be turned over to you or properly stored until an appropriate job comes along.

Modifying Shape and Size

Sometimes a slight change in the shape or size of a printed piece can mean big savings on paper. For example, a presentation folder designed to read horizontally may print only one per sheet. Changing the design so that it reads vertically or making the pocket slightly shorter might allow two per sheet, thus saving on paper and press time. If you have some latitude in shape or size, ask your printer or paper merchant early in planning about choices in size of paper, and then design for the most economical cut from the sheet or roll.

Avoid costly special orders of paper or envelopes unless your project or press run warrants considerable extra expense. An odd size of paper must be specially cut for your job by the paper mill, and an odd size or style of envelope must be specially made.

FINISHING AND BINDING

Thorough planning can result in savings at the bindery, particularly on big jobs. As a general rule, the larger the printed sheet, the more efficient are finishing and binding. A sheet with eight pages on a side can be finished and bound faster and cheaper than two sheets of four pages. As you plan a job, ask your printer or bindery about the most economical layout from a finishing and binding point of view.

Also check capability before deciding on a fold that is not standard ("novelty"), matching a rule across the center gutter of a publication, or other complexities. Intricate work in the bindery adds to the cost and is also more likely to be achieved on a small job than a large one (unless you pay extra for close monitoring). Registration in binding will drift on a long run and can be kept exact only by stopping and adjusting the equipment. Select a simple design for such a job so that a little variation in registration isn't noticeable.

Finish and bind all your job at once. Piecemeal work involves considerable expense to set up, dismantle, and reset the project. Split a run only when doing so is cheaper than processing the entire job on overtime.

Make sure your printer delivers your project to the bindery in satisfactory condition, according to the bindery's requirements. Sheets not properly protected from the rain, for example, will be rejected by a bindery as too high in moisture content to be processed. A job delivered in anything but prime condition may still be completed by the bindery, but at a much slower and costlier pace.

ECONOMICAL PRODUCTION

On a large project, look into providing your own boxes for the bindery to use. You may spend less buying them yourself than having the bindery buy them for you and add a markup.

Plan distribution as you plan production to avoid costly delays. A small expenditure in the bindery may save time and money when your job is distributed, whether by an outside mailing service or by employees. A few dollars for machine folding a newsletter for mailing may cost less (and get it out faster) than hand folding, even by employees working at minimum wage. The bindery can help you count by "slipsheeting" (inserting paper of a different color) at predetermined intervals (such as every 100 copies) to eliminate much of the manual counting needed to fill orders. Your job can also be boxed by number or weight to save the time and cost of recounting, reweighing, and repacking materials for shipment to other locations.

CHAPTER FOUR

Planning the Production Schedule

COMPILING A SCHEDULE   64

    Scheduling Considerations, 64
       Start-Up and Deadline, 64
       Size and Complexity, 65
       Quality, 65
       Previous Experience, 66
       Workload, 66
    Drafting and Revising the Schedule, 66

STAYING ON SCHEDULE   69

    Before Production Begins, 69
    During Production, 70
    When Production Is Finished, 72

SCHEDULING SHORTCUTS   72

    Simplifying and Dovetailing, 73
    Cooperating With Suppliers, 73
    Subcontracting, 74
    Using Overtime, 74

A print project is fully planned when it is on the calendar—when each production step is scheduled so that the job can proceed efficiently from start to finish to meet the deadline. You are ready to plan a schedule after you have decided on specifications, solicited prices, and selected suppliers. A draft schedule can be made before selecting suppliers so that those contending for your work will have an idea of timing; but it may need to be revised later according to the requirements and capabilities of those who are eventually chosen to do the work.

This chapter covers what to consider as you prepare a schedule, how to stay on it, and ways to save time. This information may be applied to any print project, although schedule complexity will vary with the job. Many times, budget and schedule are closely tied: Cutting cost may require slower production and delayed completion, or pushing suppliers to meet a deadline may increase your costs. As you plan a schedule, take into account when the job must be finished in relation to the production economies presented in Chapter 3.

COMPILING A SCHEDULE

Your schedule may be firm or loose, depending on how predictable a project is. At a minimum, it must be sufficient to give you and affected suppliers a timing guide for each phase of production. A schedule that meets this objective reflects information about the job that influences timing and the turnaround capability of both in-house and outside suppliers.

SCHEDULING CONSIDERATIONS

Before you start attaching dates to production activities, consider at least the following factors that impact timing. Prior planning experience and a few phone calls should be adequate to gather such information, the base on which your schedule will be built.

Start-Up and Deadline

A job's time span is defined to be when a job can get started and when it must be finished. Therefore they are your first considerations. The easiest jobs to schedule are those that can be started at any time and have an open or liberal deadline (such as a re-order of business forms needed sometime next month). Careful scheduling is required, however, when the deadline is absolute and when start-up and delivery dates are close together (such as a weekly newspaper).

Before putting activities on the calendar, find out if all elements of a project will be ready to go at the same time or if the schedule must accommodate a staggered start-up. Copy may be ready by the 15th, for example, but photos not until the 20th. Your schedule for final design and later production steps must then allow for this disparity.

Also, delivery of a printed piece may not be the last activity you must schedule. "Deadline" may, instead, be the day an annual report must be in the hands of shareholders, not the day it is delivered to the mailing service. Clarify what deadline means to avoid basing the schedule on a faulty assumption.

Size and Complexity

The magnitude of a job is another consideration since size and complexity have an influence on how long production will take. They impact total time required as well as how much time is allotted to specific production steps.

Size is a factor in both number of pages and press run. An eight-page newsletter will take longer to produce than a four-page version, even if they are the same in page size, press run, and complexity. Time for writing, editing, design, art, typesetting, dummy, paste-up, and engraving increases as amount of material increases. Once on the press, production time depends on how many copies are printed and on size and speed of the press. In the bindery, the quantity to be processed and what must be done influence the delivery date.

Complexity is a factor that affects the production schedule at every phase. If you select a design with precise details, you must allow for precision from paste-up through binding. Close registration, many photographs, and embossing will add hours or days to an otherwise quick job. Copy that must be compiled from several sources, heavily edited, and approved by a committee requires more time to prepare than copy that is rewritten from last year's brochure. Formal portraits of company executives in a fully controlled setting require more photography time than informal shots of them at their desks. These and other factors that add to production complexity may be justified for a given project; but the time to execute them must be allocated in the schedule.

Quality

The quality you expect in the finished product also influences the production schedule. High-quality work may take longer than medium- to low-quality work. A form could be typewritten and mimeographed this afternoon, for example, but would require days to produce if typeset and commercially printed.

The fact that one piece takes longer than another to produce doesn't necessarily mean that it is of higher quality. You may wait two weeks for a low-quality, low-priced print shop to finish the same kind of job that a quality, higher-priced printer would produce in half the time.

Previous Experience

Keep a record of the schedule that worked on a regular job (such as a monthly publication) to use as you schedule it again. On a piece that you do frequently with the same suppliers, you can predictably transfer to a current calendar the number of working days required for each production step. Confirm the timing with suppliers, however, if the repeat job doesn't happen regularly enough for them to have time already set aside for it. Their workloads or staffing may be different now. Production records will tell you roughly what timing to expect; but check with suppliers as you plan to avoid schedule conflicts.

Workload

Scheduling a print job must take into account how much work you and your suppliers can handle. Before you select a supplier based on price and quality alone, ask specifically about turnaround time. "Working day" is longer for a typographer or printer operating extended hours, for example, or for a freelance illustrator with the flexibility to put in extra hours on your job. When you ask suppliers for prices, be as specific as possible about when the job will come in and when you will need it. If business is slow, a supplier can devote full and immediate attention to your project; if business is brisk, your work will have to wait its turn. With careful planning, your project can stay on schedule with the busiest supplier; if you can ensure its predictable arrival, you allow the supplier to set aside time to do it and coordinate all other work around it.

Workload and timing will influence how fast your project gets done, even on overtime. Typesetting that would normally require five working days probably can't be cut to one day, no matter how much surcharge you are willing to pay. The option of overtime may not always be available either; few suppliers will work overtime on Christmas Eve or before or on other holidays. Neither will they be willing to push your job through if employees have already been working considerable overtime for other customers.

Also consider your workload and that of anyone else in-house who must approve proofs or other aspects of the job. During an especially busy time, the project may have to wait its turn on your desk or be left sitting while a key vice president is away from the office. Look ahead on your own calendar and inquire about the calendars of other in-house people involved to avoid throwing the production schedule and outside suppliers off with unforeseen internal delays.

DRAFTING AND REVISING THE SCHEDULE

Like a budget, a schedule can be built forward (from start-up to deadline) or backward (from deadline to start-up). The direction is determined by which

COMPILING A SCHEDULE

date is more certain or even mandatory. Given what you know about timing, size, complexity, and quality of your project, experience with similar projects, and workloads of all concerned, draft a list of how many working days will probably be required for each production step.

A report project illustrates how this drafting process works. The piece to be scheduled is a quality, 24-page document typeset according to an existing format with a black-and-white file photograph on the cover. Text and cover are to be offset printed in one color on different papers. The 500 copies must be saddle-stitched, trimmed, and delivered locally. Production is to occur during June. Following is a draft of the schedule by activity and estimated working days to complete each:

Copy	
Editing	2 days
Retyping and proofing	2 days
Approval of edited copy	1 day
Final revisions and specing for type	$\frac{1}{2}$ day
Photo selection	$\frac{1}{2}$ day
Typesetting	$2\frac{1}{2}$ days
Galleys	
Proofing	$\frac{1}{2}$ day
Approval	1 day
Correcting	2 days
Paste-up	
Pages and cover	4 days
Proofing	$\frac{1}{2}$ day
Approval	$\frac{1}{2}$ day
Revisions	1 day
Engraving	3 days
Proof approval	1 day
Printing	
Cover	1 day
Pages	2 days
Binding	1 day

According to this list, production will require 27 working days if taken one step at a time. Now you are ready to put these steps on a calendar, based on a stated deadline, start-up date, or both.

If a delivery deadline of June 30 is absolute, start building your schedule backward from that date, corresponding with the number of working days required for each step (as shown in Figure 4.1). This approach leads to starting production on May 30.

JUNE	Copy editing		Retyping and proofing copy	Approval of edited copy	
	30	31	1	2	3
Final revisions to copy and specing for typesetting; selecting cover photo	Typesetting		Proofing of galleys	Approval of galleys	
4	5	6	7	8	9	10
	Galleys in for correction; paste-up of pages, cover		Final type corrections ready		Proofing and approval of mechanical artwork	
11	12	13	14	15	16	17
	Revisions to mechanical artwork	Preparing negatives			Approval of blueline	
18	19	20	21	22	23	24
	Printing of cover	Printing of pages		Trimming and binding	Delivery	
25	26	27	28	29	30	

FIGURE 4.1. An example of a calendar showing production steps by day.

If a start-up date of June 1 is firm and your objective is to produce the report as soon as possible thereafter, build the schedule in the forward direction, incorporating the ideal timing for each step. Figure 4.2 shows a linear version of such a calendar.

You will not often have the luxury of an ideal schedule, whether approached forward or backward. In each of the examples, production requires more than five calendar weeks. If delivery must be by June 30 and work can't start before June 1, you will have to revise your schedule to decrease production time (as illustrated in Figure 4.3). You may be able to take advantage of timesaving shortcuts (such as overlapping activities) and/or may find that your budget will have to absorb the additional cost of overtime. Determine what production steps are to be taken on overtime by considering both cost and when overtime will best improve a schedule. In general, pushing yourself or employees in preparation of the raw materials for a printed piece will save more time at less cost than pushing outside suppliers once production is well under way.

Whether working forward or backward, you may build your schedule using either the calendar (Figure 4.1) or the linear (Figure 4.2) method, depending on which is better for you. Or you may start with a linear schedule to clarify where overlaps can occur, and then convert it to a calendar form that can be a useful reference for all suppliers.

STAYING ON SCHEDULE 69

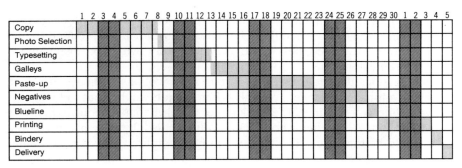

FIGURE 4.2. An example of a linear calendar showing where production steps overlap.

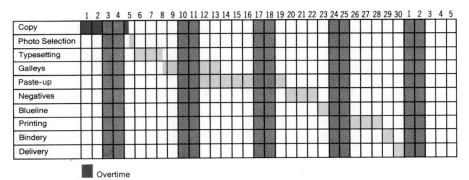

FIGURE 4.3. Hatched areas on this linear calendar indicate when overtime must be used to stay on schedule.

STAYING ON SCHEDULE

Compiling a schedule on paper is one thing; staying on it is quite another, especially when a print project is large and complex, when it involves multiple approvals and suppliers, and/or when it is subject to change at any stage. Even in difficult situations, the job can get out on time if you carefully plan a schedule that gives you the best chance for success. If you exert control over what you can foresee, production efficiency will not be as seriously impaired by anything unforeseen.

BEFORE PRODUCTION BEGINS

Start with a realistic schedule. A schedule that doesn't take into account what should be obvious is destined for failure. Allowing suppliers only half the time they need to do your job won't make them work any faster. Thorough planning that calls for realistic turnaround and builds in extra hours when changes can be anticipated starts a project off on the right foot. Trying to meet an unrealistic schedule invariably leads to a poor product and/or poor relations with suppliers (who may not work with you again under similar circumstances).

Include schedule requirements in your specifications. When you solicit prices on a job, tell suppliers roughly when the job will be sent to them and by what date you will need the work done. Confirm that schedule when you award the job. A supplier who can use your job as filler may price it lower than would be the case if fast production were required. Having full timing information at the onset will get your project on each supplier's calendar to help avoid misunderstandings and resultant delays.

Try to avoid especially busy times for suppliers. Help assure that production will stay on schedule by avoiding days when a supplier is particularly busy. For example, a typographer will have to leave your booklet on the shelf if it comes in on a day always reserved for another customer's monthly magazine. Schedule your work at a different time or look for another supplier who isn't as busy when you need service. Regular projects create predictable workloads and income for suppliers. They will, therefore, be very willing to help you plan a schedule for your regular project that makes production of it easier to manage.

Allow time for the unexpected. Equipment failures, illness, or labor disputes are always impossible to predict and can easily throw off your production schedule, as can overlooking holidays when counting working days. Deadline permitting, plan an extra day for the unexpected. On a job that must move along with precision, ask suppliers to notify you immediately of any possible delay. A reputable supplier who understands your urgency and wants your future business will help you locate someone else to finish the work, should a problem not be resolved in time to keep a project on schedule.

Work with reliable suppliers. Especially when a production schedule is tight, you can't afford unreliable suppliers. Ask for and check references of prospective new suppliers and periodically review the performance of those you have used before. If a key person at one shop is replaced (such as the production manager), your next project with that supplier may not proceed as smoothly as in the past. With a new supplier or new personnel, reiterate your requirements and monitor performance until you are certain that standards will be maintained.

DURING PRODUCTION

Give clear, complete instructions. Failure to communicate leads to time-consuming confusion and mistakes. If you want photographs converted into halftone negatives using a particular line screen, say so before engraving work begins, not after a different screen has been used. If a typographer has agreed to give you complete galleys by Thursday, underscore that agreement at the time copy goes in; don't wait until Thursday afternoon when little can be done to meet your deadline. When timing is especially tight, put instructions in writing for even a small or regular job. They reinforce the urgency of the work and leave no room for guessing or lapse of memory.

Confirm what you expect. Do not assume that a regular job will proceed through production like clockwork. Suppliers all have other things going on and may forget details that will set back your schedule. Also, when a long-standing specification changes, a phone call from you might not be sufficient to break an old habit. For example, if a bindery has always folded 200 copies of your newsletter for mailing, a verbal change to 300 relayed to the production manager via the bindery receptionist may not register. If you receive only the usual number, you will have to remedy the oversight at the last minute and risk missing your mailing deadline. As a simple but effective reminder, attach a list of specifications for the job pertinent to the supplier executing the next phase of production and circle in red any requirement that is new or different. For an eleventh-hour change, talk directly to the person in charge to make sure the message gets across.

Change specifications only when absolutely necessary. A change in requirements may or may not affect the production schedule, depending on how extensive it is and on when the change is made. Careful planning of specifications won't always preclude the need to make changes during production but should keep changes to a minimum for the least impact on your schedule. Deleting or simplifying requirements will probably put you ahead; but a major change (such as wanting a different typeface after the entire job has been set) will put you far behind. The sooner you decide on an additional requirement and inform suppliers, the less it will influence production timing and, depending on its magnitude, may even be absorbed into your original schedule.

Do your part on time. Taking an extra day to read galleys or to react to a design is a day lost and one you can't reasonably expect suppliers to make up for you. As project manager, the onus to perform on time is on you as much as on your suppliers. If you slacken on the job, the schedule begins to fall apart.

Proof thoroughly. Starting with copy, check all work thoroughly before moving to the next phase of production. Catching a typographical error on the press and wanting it corrected is a sure way to miss your delivery date. Check copy for facts, figures, spellings, titles, completeness, and tone. Check galleys and corrections carefully against copy. Proof all tables and charts and check drawing details. Confirm how photographs will be cropped and check them for details also, in case something in the photo might distort your message or be embarrassing once in print. (For more information on what to look for when proofing, see later chapters on individual production steps.)

Make sure everything is approved. Unless you're having a piece printed for personal use, you will probably need copy, design, art, galleys, dummy, mechanical artwork, proof, and/or press pages approved by your supervisor, the head of the department that originated the project, the club president, or someone else who shares responsibility for the finished product. To stay on schedule, ask during planning what approvals will be needed and get them at the appropriate times. Don't proceed to a proof before you ask the vice president of marketing to read the content of a sales brochure. Many times a person who gives approvals will need help visualizing how copy will convert to type or how photographs will look when reduced. You may, then, need to have sample type set or art reduced on a photocopier to assure an educated approval and decrease the chance for complications during production. If you anticipate delays in approvals because several people are involved or because someone will be out of town when needed, plan your schedule accordingly. Failure to get timely approvals results in costly revisions, production delays, and a slip in your credibility with suppliers and your employer.

Monitor each supplier's performance. Every step in print production affects the steps that follow, much as the falling of the first domino in a row. To stay on schedule, make sure each supplier performs on time and to the full extent expected. A paragraph missing from copy will take time to insert later on the galley, for instance, and a photographer who doesn't deliver contact sheets on time holds up final layout decisions, jeopardizing every subsequent phase of production.

WHEN PRODUCTION IS FINISHED

Make sure delivery is properly set up. A printed piece that's off the press and out of the bindery as planned can still miss the deadline if it's delivered to the wrong place. In a large company, boxes of brochures may sit in the shipping and receiving department for days for want of a name or phone number for timely notification of their arrival. Specify delivery requirements when a job is priced, when it is awarded, and when it goes on the press or into the bindery. Verify that packaging will meet distribution requirements to avoid the delay of re-boxing and re-labeling.

Pay suppliers on time. Sitting on production bills won't influence the print project just completed but may have a bearing the next time you need service. For example, a copywriter who puts out extra effort on this year's annual report, only to wait three months for payment, may not give the job top priority next year or be willing to do it at all. Uphold your end of the bargain through to payment for services, and suppliers will be interested in meeting your schedule next time.

SCHEDULING SHORTCUTS

When timing is tight, you may be able to take advantage of one or more procedures to cut hours or days off your production schedule, without cut-

SCHEDULING SHORTCUTS

ting quality. Look for additional ways as you gain production experience, especially with regular projects.

SIMPLIFYING AND DOVETAILING

Simplify the design. It influences the time required to do most (and often all) subsequent production steps. The more complex the design, the more time-consuming detail work will be needed. If your objective can be met by decreasing the number of ink colors or photographs, eliminating a die-cut, or in some other way reducing production requirements, you will probably save time.

Take production steps simultaneously. You will save precious days when a schedule is close if you can dovetail steps. Don't wait until the design is final to start writing copy. Have final illustrations drawn while type is being set. Print a separate cover while negatives are being made for inside pages. Some work, however, has to wait until a preceding step is taken to avoid wasted effort or mistakes. You probably would be foolish, for example, to do a detailed dummy before all galleys are in hand unless you can precisely predict length of text. But your schedule will get shorter the more organized you are about when stages of production can be safely started and completed.

Seek approvals in stages. On a job involving lengthy copy, ask for approvals in stages so that typesetting can start, rather than waiting until all the copy is ready. The person approving it won't need days to plow through to the end before typesetting can proceed. Send the copy for approval section by section if what is said on page 1 isn't influenced by what is said on page 20. Use the same approach on galleys and proofs if sequential approvals can be safely given.

COOPERATING WITH SUPPLIERS

Keep suppliers informed about changes. If photographs are ready a day early, the dummy can be started a day early. If they are late, the dummy will have to wait. Let suppliers know quickly about such schedule changes or about modifications that will simplify or complicate production. They can then plan to accommodate your job early or possibly put out extra effort to get a flagging project back on schedule. Informing suppliers is doubly important when your deadline is moved up or you have additional requirements. On a very tight project, delivery in the morning instead of the afternoon can make a difference, as can folding a poster instead of leaving it flat. The sooner a supplier knows about a change, the more likely that the production schedule can be altered in your favor.

Work with known suppliers. The better you know a supplier, the more you can count on that supplier's meeting or beating your schedule. When timing is very tight, work only with those on whom you can rely to give the quality you need by your deadline.

SUBCONTRACTING

Disperse a large job to several suppliers. A big print project may be completed faster by using more than one supplier for production steps. Your management skills may be put to the test doing so; but you can get work done this way when an individual supplier might have to meet your schedule on overtime, if at all.

Using multiple suppliers can be more expensive if duplicate sets of art or other elements are needed—and very costly if one shop's product doesn't match that of another. Have all negatives made in one location, then pay for duplicate press plates if the job is to be run in more than one shop. Ask for and compare sample galleys from two typographers before awarding each a portion of a big job.

An alternative to juggling multiple suppliers yourself is to award the job to one supplier and ask that portions be subcontracted, as required, to meet your schedule. You will, however, pay a markup as compensation for such supplier coordination.

USING OVERTIME

Consider "rush" service. You will pay more for service performed on a rush basis; but it may be your best choice when timing is absolutely critical and cost no object. Rush service will pare hours or days off a schedule, depending on how much of a surcharge you are willing to pay. A 50% rush may save a day and a 100% rush two days; a 200% rush may pull out all the stops on your job, including night, weekend, and holiday work to get it completed.

Ask each supplier what "rush" means in time savings because definitions vary. It may mean moving your job ahead of others on the priority list and doing it on straight time, or it may mean overtime. Also ask cost so that, if you must push to meet a schedule, you will know what production step taken on overtime will gain you the most time for your money.

CHAPTER FIVE

Writing and Editing Copy

COPY CONSIDERATIONS   76
    Objectives, 76
    Audience, 77
    Design, 79
    Existing or Related Material, 79
    Schedule and Cost, 80

IN-HOUSE COPY SOURCES   80
    Staff or Members, 80
    Rewritten and Reprinted Copy, 81
    News-Service Copy, 81

COPYWRITING TALENT   82
    Finding a Copywriter, 82
    How Copywriters Work and Charge, 83
        Fees and Expenses, 83
        Use Rights, 84
    Evaluating the Work of a Copywriter, 84
        Range of Skills, 85
        Style, 85
        Flexibility, 86

Effectiveness, 86
Cost, 86
Schedule, 86
Checking Copywriter References, 87
Working With a Copywriter, 87

EDITING TO CHECK AND CLARIFY 88
Editorial Style, 89
Editing as Checking, 89
Accuracy, 89
Completeness, 90
Editing as Clarifying, 90
Comprehension, 90
Appropriateness, 91

GETTING CLEARANCES 92

A printed piece carries its message in words and graphics. These two elements communicate together, though one may be more important than the other in a specific piece. This chapter concerns preparing the copy portion of the message of a print project, whether column after column of text or just a few words. It covers points to consider when writing; in-house copy sources; finding, evaluating, and working with a copywriter; editing copy; and how to get copy cleared before publication.

COPY CONSIDERATIONS

When the words that are written, revised, and edited add to the effectiveness of a printed piece, the whole message gets across. When they are poorly written, revised, and edited, they detract beyond the ability of an enticing design, eye-catching photographs, or costly paper to camouflage. The road to printing words that are effective starts before you take pencil and paper, typewriter, or computer keyboard in hand or engage a copywriter. The following factors shape the copywriting and editing process.

OBJECTIVES

To be effective, copy must directly support the overall objectives of a printed piece. If it is to sell or propose, words must be persuasive ("You

COPY CONSIDERATIONS

will . . ." instead of "You could . . ." or "You might . . ."). If it is to inform, words must provide details in a logical order, not a smattering of information and unsubstantiated statements. The same subject approached from different directions will have a different look on paper. For example, the subject of a new office building will be treated one way if the objective is to sell its merits to prospective tenants and quite another way if the objective is to inform structural engineers about how it was constructed.

While the broad objectives of selling and informing may be applied to most writing projects for print, they may be broken down into more specific purposes to guide copywriting. Following are some examples:

To *explain* a program, service, or facility.
To *educate* (such as a resource booklet).
To *persuade* the recipient to take action.
To *reinforce* a sales call or other medium.
To *legitimize* a program, service, or product by presenting it in print.
To *support a desired image* of an organization or product.
To *organize* information (such as a business form).
To *represent* in print a person or organization.
To *inspire* to high principles or achievement.
To *update* previous information.
To *call attention* to a source for more information.

AUDIENCE

The person who will read a printed piece is another factor in determining how copy should be written and edited. One audience may be interested in a lot of details, while another would be better addressed with an overview only. In the previous example, the engineers would expect details about the building's stability, while prospective tenants would accept a general statement but expect details about floor plans and amenities.

The audience for a printed piece may be on two levels: primary and secondary. The former is the more important; the latter is less important but should also be taken into account when copy is written. For example, prospective customers may be the primary audience of a sales brochure and employees who need to know about the new product line the secondary audience. A related consideration is whether or not the audience is likely to change somewhat over the life of the piece. If so, copy will need to be more general to bridge the time span.

Another factor is location; the broader the geographic distribution of the audience (local, regional, national, international), the more universal words should be (and those words may need to be printed in more than one language). The challenge is to position a piece editorially so it does the job intended, without excluding, offending, or conveying outdated or inappropriate information to its readers.

An informed, interested audience will understand technical or complex language, while simpler words should be used for an uninformed audience not keenly interested in the subject. Education level is a related factor. In general, the more educated the audience, the more complex the writing can be and still be understood.

One gauge of word choice as it corresponds with education level is the Gunning Fog Index. Though it doesn't take into account training in a specific subject that would permit more complex language, it can reasonably be applied to a broad audience, for instance, a company newspaper for all employees.

To use this index as a test of writing versus education level, select a passage of at least 100 words. Count the average number of words per sentence in that passage. Next, determine the percentage of words in the passage with three or more syllables (excluding proper nouns, numerals, and hyphenated combinations of small words). Add together the average words per sentence and the percentage with three or more syllables and divide by four. The resulting figure is roughly equivalent to the years of education needed to comprehend what was written. If the index were applied to this paragraph, the calculation would be as follows:

A. Number of words in passage: 105
B. Number of sentences: 6
C. Average words per sentence: 17.5
D. Percentage three or more syllables: 16

$$\frac{C + D}{4} = \frac{17.5 + 16}{4} = \frac{33.5}{4} = 8.38$$

Equivalent years of education: 8.38

Choice of words and punctuation have created a passage to be readily understood, leading the reader point by point through a process that is probably not familiar to most people in the audience. If it had been written for an audience of writers or linguists, however, wording could have been more complex, regardless of their overall education level.

Audience also influences style of writing. A year-end report to shareholders would probably use more formal language than a comparable report to employees. A formal tone ("the Company") maintains the investor/staff distinction, whereas an informal tone ("you," "we") reinforces a we're-all-in-this-together management philosophy. Because active voice requires a clearly involved subject ("We reached an all-time sales record"), it is not as formal as the passive voice ("An all-time sales record was reached"). You need to know both objectives and audience to select a style, however. If a stated (or understood) objective is to say as little as possible on a delicate subject, complex language and sentence structure, formal language, and passive voice will effectively deliver a minimum of information.

Also know your audience so that copy includes everyone and doesn't offend. When writing to a mixture of races, sexes, ages, and abilities, select words that are inclusive and sensitive. Phrases such as "black forecast" are potentially offensive to affected racial groups. "He" and "man" exclude women in the audience, and using "girl" or first name only for women and not men is discriminatory. "Older person" is a safe, neutral choice when referring to someone up in years, as is "disabled" or "impaired" when writing about physical or mental limitations.

DESIGN

Ideally, objectives and the subject are primary determinants of copy length. A copywriter writes as much (but only as much) as is necessary to do the job, and then it is edited and the design adjusted so that copy and design meet objectives in concert.

Copy for a piece that is already fully or roughly designed, however, must take that design into account from the onset. Length is obviously a factor; it must be written to fill the space allotted. It may need to fill that space in a particular way, assuming that a design has been worked out with full knowledge of the subject. For example, the design may call for three clearly defined sections to the copy or its narrow columns may dictate short sentences.

EXISTING OR RELATED MATERIAL

Copy for a new printed piece may not be entirely new itself. Any existing material that might be revised or incorporated needs to be considered before writing begins. If sufficient information on a subject has been written already, the project could be an editing job only, rather than one requiring both writing and editing. For example, the annual report to employees may be simply an edited version of the report to shareholders. If copy exists that could be incorporated, the writing task is a selective one to provide missing information, with skillful editing to weave together old and new material.

The relationship of the current printed piece to related materials already being used or being developed must also be considered. If your newsletter is one of a series of quarterly publications, it should be written in a similar style to that of its predecessors, for example. Pieces in an information packet to be distributed to convention attendees should all carry out the convention theme.

SCHEDULE AND COST

Timing and its relationship to cost are the final copy considerations. On a tight schedule, revising copy already in hand may be more feasible than starting over to produce a fresh manuscript. Also to be considered is when information will be available, especially if you plan to pay a copywriter to do the work. When a project is stop and go waiting for printed information and/or resource people to be available, it is more costly in time and effectiveness than one organized into a unit. When the writer can't concentrate on the job through to completion, sections written at different times won't necessarily fit together without considerable editing.

IN-HOUSE COPY SOURCES

Copy for a newsletter, flyer, report, or other print project may originate inside or outside your company or association. In-house sources include yourself for both writing and editing or another employee or member for at least the initial writing. Other sources that may be brought in and treated as in-house are files of material that may be rewritten or reprinted, or news-service copy. This section explores each of these in-house copy sources, keeping in mind that, though copy may originate with someone else, you as manager of the project should plan to fulfill the role of final editor, double-checking facts and the other points covered later in the section on editing.

STAFF OR MEMBERS

The writer and editor of copy may be the person who will also manage its conversion into type and a finished printed piece. Such a do-it-yourself arrangement allows control of a job throughout production. If you know objectives and the budget, have an idea of the format, and set the schedule, you can write, revise, and edit to correspond.

If you are not particularly skilled in this area or able to do all aspects of each print project efficiently, you will have to look to other in-house sources for knowledgeable, cost-conscious copywriting. A department head might write an article for the company magazine, a secretary compile a membership roster, or a member write a report of the annual meeting for the association newsletter. By searching out an in-house source other than yourself, you can then concentrate on editing the material, laying it out, and following through on production.

Treat staff or member writers much as is suggested later in guidelines for working with a copywriter. Specify requirements of the design and set a deadline. Review editorial considerations and style to help assure that contributed copy is in line with what you need, when you need it, and retain the right as editor to accept, reject, or alter the copy you receive.

REWRITTEN AND REPRINTED COPY

The manuscript for a new printed piece need not necessarily be new. For example, information used in last year's brochure on association services may be appropriate this year with only minor changes. The president's speech to shareholders may be rewritten as an article for the employee newspaper.

When considering rewriting or reprinting copy, look into any restrictions of copyright. Unless copyright was expressly established in writing at the time a writer produced the original copy, it is considered to be "work-for-hire," with copyright belonging to the employer or purchaser as publisher. Thus it can be reprinted by the employer or purchaser as is. It may also be rewritten when the writer makes no claim to derivatives of the original copy.

A writer can retain copyright to future use of the material if only first rights were sold. For instance, a writer may have produced an article for your magazine on construction of a new corporate headquarters. If the writer relinquishes or sells you only the right to publish the article first, you cannot rewrite or reprint it for another publication without express permission. The writer *can* do so, however, because the right to use it again was not sold.

As a courtesy, credit should be given when the copy you use is reprinted from another source. If the amount used is more than a few lines of direct or paraphrased quotation, permission must be granted in writing to avoid copyright infringement. Most corporate and organizational writers will readily grant permission to reprint at no cost, in exchange for a credit line (e.g., "Reprinted with permission from *Forestry News*"). Expect to pay a permission fee if your publication is a consumer magazine, marketing brochure, or similar piece for commerical purposes. (For additional information on copyright considerations, consult one of the several books available that cover the rights to written material.)

NEWS-SERVICE COPY

Though news services have for years been a major source of copy for city and national newspapers, they are relative newcomers to corporate and association media. A news service or syndicate provides ready-to-use copy of a general nature (as with the Associated Press or Reuters for news) or in specific subject areas (as with Dear Abby or other columns). This same concept has been adapted by enterprising writers pooling their efforts for major professional and trade audiences.

If as the editor of a newsletter for electrical customers, for instance, you subscribe to a news service specializing in energy-related articles for a lay audience, you will regularly receive a package of articles to reprint as is, revise to suit your needs, or not use at all. The price is the same, with the understanding that only subscribers have access to the articles. If you consider the price reasonable in relation to writing copy yourself or going to another source, a news service can offer a ready source of timely, well-written material. Before subscribing, however, closely evaluate the caliber of writing and know the number and mix of articles you can expect to receive. If only one out of six articles supplied each month is appropriate, the service probably won't be worth the price.

Large companies with branch-location or agent publications may also provide a type of news service. If you're charged with producing an employee newspaper for your plant, articles supplied by the home office can be a valuable source of pertinent copy.

COPYWRITING TALENT

If writing copy in-house or editing file or news-service copy isn't feasible, you will need to turn to an outside copywriter. The benefits of doing so are several; but the cost must be weighed against in-house alternatives. An experienced copywriter brings specialized skills and objectivity to an assignment. Someone who writes frequently for print will have an understanding of production and be able to blend together copy with graphic elements of a project especially well. The principal drawback is unfamiliarity. An outside writer will most likely need more orientation to the subject or situation than would an employee or member doing the work.

FINDING A COPYWRITER

A freelancer who specializes in writing or provides writing as one of several services offered is an attractive outside source when you need copy tailored to your subject and project. By checking with colleagues, suppliers, professional or trade associations, artists' agents, and even the Yellow Pages or work-wanted classified ads, you can locate freelance copywriters. Another way is to call the editor of a publication using a freelancer with whose writing you have been impressed.

A freelancer with a volume of business sufficient to support a writing specialty may focus on annual reports, benefit materials, or advertisements, for example. The higher level of expertise may justify the slightly higher fee such a specialist may command over general copywriters.

A variation on strictly freelance writing is a consortium of freelancers who market themselves as a group. Look for a listing in the Yellow Pages. Be prepared when you call a consortium office to outline your project, including schedule and cost limitations. A freelance writer will then be assigned to pursue the project with you, though you retain the right to evaluate that person and ask for someone else if you're not satisfied. By contacting a consortium, you tap a pool of supposedly prescreened talent, without having to search out individuals yourself, though you will also help pay the overhead costs that accompany such convenience.

An advertising or public-relations agency has employees and/or freelance writers on call to provide copy for pieces to be produced by the agency. They generally won't provide copy only, and if you find one that will through colleagues, suppliers, or the Yellow Pages, an agency's higher overhead costs will certainly be reflected in the fee. An agency is your most logical source for copy if it is also to design and produce the piece. Such comprehensive service provides full control over project quality and schedule. As the client, you are entitled to know the copywriter assigned to your project and to expect that person to be competent, whether the copywriter is an agency employee or a freelancer brought in by the agency.

HOW COPYWRITERS WORK AND CHARGE

Working from a tentative outline, a copywriter uses personal interviews; reports, old brochures, or other documents; or a combination of these sources to collect information. Based on this research, a writer polishes the outline and writes an initial draft, seeking additional information if gaps become apparent in the process. Copy is then revised until the writer is satisfied that the draft meets your requirements or until feedback is needed to be able to move ahead. Your review of the draft with factual corrections and/or recommendations on organization, style, and content is the basis for final copy. Others who must approve copy may see the same draft you do if you consider it close enough to what is needed, or they may wait to see the revised draft as you have edited it. In the latter situation, you may ask the copywriter to do another revision after reviews are in to produce final copy.

Fees and Expenses

A copywriter may charge by the hour or by the project. (An alternative in a buyer's market is for a publication to pay a certain amount per printed word or page, take it or leave it.) If the scope of work is closely defined and the writer experienced with the subject and with estimating, a project fee may be a beneficial arrangement for both parties. It guarantees that the work will cost you no more than the stated charge (assuming you don't change your requirements), and it grants the writer the freedom to work creatively, without the restraints of a ceiling on hours, and to profit from working efficiently. A copywriter using word processing may charge by the project to secure compensation for the hours saved by having such capability, or simply use access to such equipment as a selling point with prospective clients.

A copywriter may want to charge by the hour with a new client. Because the amount of work needed directly relates to how a client manages the review process, a writer is wise to test client efficiency and be paid accordingly. An hourly fee also provides you some protection against paying for more work than was done, should the project not be as time consuming as estimated. Based on initial experience, you and the writer may later agree to project fees.

A writer's hourly fee probably includes travel time to and from information sources and possibly also to and from your office. Since a project fee is based on an estimate of hours, travel time is also included in this pricing option. Out-of-town travel may be charged at half or some other portion of full fee.

Expenses for travel, interviews over lunch, research materials, photocopying, long-distance phone calls, and postage are legitimate charges to a copywriter's clients. Typewriter ribbon and other supplies that the writer needs to be in business are not, unless your project requires major purchases of such supplies.

Use Rights

A writer may not bring up use rights but may simply do the work, turn over the copy, and express no claim to its subsequent use. When no claim to copyright or to derivative or reprint rights is made, those rights belong to you as purchaser of a "work-for-hire." Should the writer wish to retain any or all of those rights, expect to be asked to sign a statement to that effect. Also expect to negotiate a reduced fee in line with any agreement allowing the writer the potential of profiting again from the same copy produced for you.

EVALUATING THE WORK OF A COPYWRITER

As with design, illustration, photographic, and mechanical-art talent covered in later chapters, copywriting is evaluated by example and reference. A copywriter will provide a portfolio of samples and a resume or qualification summary for your review, meet with you to ask and answer questions, and give you names of past clients as references. Even with an in-house source for copy, you should consider the following points to help determine if the person is suitable because an incompetent copywriter, even if free, will exact a high toll in terms of quality and ability to meet your deadline.

Evaluate a writer's skills by portfolio, interview, and references. Don't insult a talented, experienced writer by requiring some kind of writing test. If you can't evaluate competence from these sources, a writing test won't add to your information—but will send talented writers to more enlightened clients. When in doubt about a writer's ability to take on an unfamiliar subject or a major project, start with a small assignment to see how the writer handles the material and how you work together.

You may evaluate a copywriter's work ahead of time so you will know whom to call when you have a need or wait until you have a project in hand. In the latter situation, you should ask that the portfolio include similar projects, though the copywriter may elect to add others to acquaint you with a full range of skills.

Ask to see a portfolio ahead of your meeting or to keep it afterwards. The advantage in the former timing is being able to talk more specifically about how the writer approached a given piece and how it corresponds with your requirements. Plan to hold a writer's portfolio for only a day or two unless the writer doesn't need it back that soon. Someone with a limited supply of samples can't talk with other prospective clients as long as you have those samples. Return a portfolio as the writer requests, either by calling when it's ready for pick-up or mailing it. Don't assume the writer wants to risk loss or damage through the mail or that a writer who doesn't provide postage will not want materials returned at all. Common courtesy from the start will pay off later in an amicable, businesslike working relationship.

Range of Skills

A competent copywriter should exhibit technical writing skills, including grammar, spelling, punctuation, and paragraphing. Material should be organized in a logical, easy-to-follow sequence with headings and subheadings that contribute to understanding what is being communicated. Word choice should be succinct, appropriate to audience level, and inclusive so that it doesn't alienate or offend. Copy should illustrate a thorough grasp of a specific subject if the writer specializes or the ability to put together the who, what, when, where, why, and how of a more general subject. The writer should have interview and research skills to glean the most from information sources.

If the writer can provide original copy and the finished piece, you can gauge how much editing was required by a previous client to achieve desired results. A piece that bears little resemblance to original copy tells you that either the writer missed the mark or that the client failed to provide enough information. Also discuss this point when you check references and realize that project frustrations are as often caused by poor direction as by poor writing.

A copywriter experienced in writing for print production knows how to produce copy appropriate for a given design, including writing to fit the space available and matching what is said in copy with what is shown in the art that accompanies it. You will find that this skill will save rewriting time and costs and make the project easier to manage.

Style

Like other creative people, copywriters tend to develop a style: formal, informal, humorous, and so on. The writing style you see in portfolio samples should be appropriate to your subject and audience. For example, a factual approach may be best when writing about marine biology to marine biologists, whereas the same information might be presented in a more conversational style to tourists at waterfront resorts. If you need a humorous writing style, realize that it requires particular talent not every copywriter has. Ask specifically for samples of it and evaluate them to determine if both humor and the message behind it come across.

Flexibility

A competent copywriter should be able to work from interviews or research documents as well as rewrite existing material for a new audience and purpose. Interview skills are especially important if the writer must rely heavily on other people to provide information. A writer who doesn't know how to plan, conduct, and record an interview will end up with scant material on which to base the manuscript.

Effectiveness

Talk with the copywriter as you go through the portfolio about how well pieces met their objectives. A brochure written to sell should have had an impact on sales; similarly, an employee handbook should have had an impact on knowledge of policies or use of medical benefits. The copywriter should have some idea about effectiveness, though a check of his or her references will provide you much more.

Cost

Regardless of how the fee is set, a writer should have the ability to estimate realistically. If the portfolio includes pieces similar to what you will need, ask for representative copy costs, keeping in mind that the quality of client direction has a lot to do with the time required to complete copy.

When evaluating the cost of one person's services over those of another, also clarify how much of the work was done by the copywriter. If the writer provided a first draft only, the billing will be substantially less than if revisions were required as full information slowly came to light. Again, include cost questions when you talk with the copywriter's references.

Schedule

An experienced copywriter should be able to estimate time and stay on schedule. Because copy is an early ingredient of print production, it must be ready when specified or jeopardize delivery of the finished piece. Extenuating factors can always arise (such as a sudden trip out of town by the primary source of information). Discuss with the copywriter and with the copywriter's references the scheduling circumstances of each portfolio sample that is of particular interest to you.

A copywriter using a word-processing system has schedule advantages over one who isn't (although technology can't replace ability to write). Word processing allows fast turnaround of copy so that revisions can be quickly incorporated. When changes are simple to make, the writer tends to polish a piece more than would be the case if faced with a major retyping of it for small revisions. The copy that results from word processing is easier to make error free, as opposed to typewritten copy that must be retyped and subjected to new errors in the process. Word-processing copy may also save time (and money) in typesetting, as detailed in Chapter 9.

CHECKING COPYWRITER REFERENCES

Ask the copywriter to provide at least two references, either current or past clients you will be able to call to confirm the information and impressions gained during your portfolio discussion. Talk with them about range of skills, writing style, flexibility of sources, charges, and ability to meet deadlines. Ask especially about effectiveness of the pieces written, realizing that timing, design, and distribution are also influences.

Confirm the extent to which the copywriter was involved on a given project. A writer may claim more responsibility than was actually the case (as may a client who doesn't want to admit to having delegated all the work). Another point to check is how well the copywriter got along with information sources, employees, and other suppliers involved in the project. You have a right to expect a cooperative, sensitive attitude, along with writing, estimating, and scheduling competence.

WORKING WITH A COPYWRITER

At the onset of a project, itemize objectives, format, audience, art, and other copy considerations, preferably in writing via a letter of agreement or purchase order. Provide a style guide if your publication has one or, if not, provide several back issues or other examples so the writer can select an appropriate editorial style. Recognize that copy won't be written exactly as you would have done it. A writer who has written "red, yellow, and blue" and watches you change the phrase to "yellow, blue, and red" for no factual reason won't be back to suffer through such nitpicking again.

A writer needs full access to people, places, and documents and orientation to your organization—its products or services, territory, business philosophy, and so on. Put the new piece into context by reviewing with the writer what similar materials have been used in the past, what is now being used and may accompany the new piece, and what the competition is using.

In consultation with the writer, set a realistic schedule. Allow as much time as possible for revisions, especially if the process involves three or four steps. An experienced copywriter will tell you if the desired result can be achieved in the face of a close deadline. Here again, word-processing capability should be taken into account since it can eliminate hours or days in the schedule.

Appreciate that the first draft probably won't be exactly what you want. A skilled, attentive writer will come close but also plan to make revisions to your full satisfaction. You are responsible for giving detailed instructions, approving an outline (if the project is large or complex enough to require a formal one), and giving specific feedback so that revised copy is exactly as wanted or very nearly so.

A competent copywriter serves you best if allowed to serve you through the design phase of production. Because the writer is close to the project (possibly closer than you), you will be wise to confer on last-minute copy changes to avoid inadvertent errors of fact or implication. Involvement in proofing type galleys and in design decisions is additional assurance that the manuscript is correctly set and arranged. In consultation with the graphic designer of a piece, ask the copywriter to write headlines, subheads, cutlines, and any other secondary copy to assure compatibility with content and tone of the text.

Organized feedback for efficient revising of copy will save time and frustration for all concerned. A copywriter needs informed guidance given in a timely manner so that revised copy benefits to the fullest. The more back-and-forth motion needed to revise a writing project, the more costly it becomes and the greater the chance for losing continuity of information and consistency of editorial style.

Finally, provide samples of the finished piece for the writer's portfolio and payment within 30 days of billing. If the experience has been positive, offer to be a reference to the copywriter's prospective clients.

EDITING TO CHECK AND CLARIFY

Writing fresh copy or rewriting something previously used is only part of preparing words for a print project. The other part is editing—the process of checking and clarifying until copy is polished, ready for final approval and typesetting.

To serve its purpose, editing may need to be heavy-handed so that copy is fine-tuned—not just accurate and complete but also understandable. Light editing may achieve this end with copy prepared by a skilled copywriter. However, if copy is written by someone very close to the subject or not experienced in writing, expect to make liberal use of your editing pencil. A writer close to the subject tends to use jargon, technical words, and abbreviated explanations, all stumbling blocks to effective communication (unless readers are equally acquainted with the subject). A novice writer tends to omit facts, use unnecessary verbiage to make up for scant research, and sprinkle dull language with clichés and other non-specific words that obscure the message and tax the reader's patience.

EDITORIAL STYLE

While information earlier in this chapter touched on points of writing style, the purpose of this book is not to teach you how to write. Those details are best left to the many guides available covering how to research, interview, and write. If you are editing copy, however, you do need to be concerned with the editorial style of your publication or organization.

An editorial style guide will help you do a thorough job if you are editing a regular newsletter, a series of marketing brochures, a major directory, or an annual report, for example. Points to include in your internal style guide may be based on a commerical guide (such as *The New York Times Manual of Style and Usage*) but should also include points tailored to your particular situation.

For example, decide if editorial style for names will be formal or informal. The president may be Charlie around the office but prefer Charles or Mr. Jones in print. If you label him Mr., will you also use Ms. for all female employees? Is it "*C*ompany" or "*c*ompany"? Will you always spell out abbreviations the first time they're used or assume everyone knows what they mean? Also consider punctuation, such as comma usage in a series ("Tom, Dick, and Harry" or "Tom, Dick and Harry"). Make a list of these style decisions for your reference and for giving to in-house or outside writers to cut down on the amount of editing you must do of their manuscripts.

EDITING AS CHECKING

As a process of checking copy, editing covers facts, dates, spellings, grammar, punctuation, and overall completeness that may have been overlooked in writing. To check copy, review it with accuracy and completeness in mind.

Accuracy

Even if you have written the copy yourself and consider it accurate, verify names of people and things, figures, and titles about which you have any uncertainty. Unlike verbal forms of communication, the print media offer no hiding places; mistakes made in print are indelible and long remembered. Follow the lead of professional editors of newspapers, magazines, and book publishers who are attuned to the importance of every detail and know that to accept written information without question is to invite a mistake in print.

Look up words that are commonly misspelled and verify that abbreviations and words unique to your subject are correctly spelled and used. An error that might not be noticed by a casual reader may discredit the entire piece in the eyes of a reader very familiar with the topic.

Check that verbs agree with subjects, that sentences are properly punctuated, and that phrasing makes sense. Even seasoned writers construct sentences in a hurry; the editor's job is to make sure that those sentences communicate the intended message. For instance, the head of a university journalism department wrote in an alumni publication: "We on the faculty will shake the hands of one-fourth of our students who are graduating May 15." Loose writing and (presumably) no editing leave the reader to wonder just what faculty members will do with the hands of the other three-fourths of the graduates.

Confirm that proper nouns are capitalized and that use of capital and lowercase letters is consistent. If all items under one heading are to be capitalized, check that they really are. Look for other inconsistencies (such as using "twelve" on one page and "12" on the next).

Completeness

The story may be accurate without being the *whole* story. Facts may be omitted, and no answers given to questions readers will logically ask. For example, brochure copy to sell a new line of windows will be almost useless unless it states standard sizes or indicates whether or not tinted glass or other common options are available. All the steps in a process should be covered in copy so that the reader fully understands.

A simple way of checking for completeness is to see if copy answers who, what, when, where, why, and how. All these questions may not be answered; but each should have been considered as copy was written. Perhaps "why" isn't known—but at least you, as editor, have checked that it wasn't overlooked.

EDITING AS CLARIFYING

A more judgmental aspect of editing is the process of making sure copy can be understood and is suitable for the purpose and format of the printed piece. To clarify copy, review it with comprehension and appropriateness in mind.

Comprehension

Well-edited copy is lean while still being complete. Unnecessary verbiage is eliminated to say what needs to be said, simply and directly. "A decline in customer response necessitated diminished utilization of part-time personnel" may be what the vice president wrote. "Fewer sales meant fewer jobs for part-time workers" may be what your readers will understand. Likewise, an article for a trade magazine about a new product will probably be written differently from an article for a consumer magazine on the same subject because specialists understand technical terms that ordinary consumers don't. Substitute common words for grand but uncommon ones (such as "help" for "facilitate") and choose words that are familiar to your audience.

When in doubt, err on the side of simplicity. If you can't follow the meaning of a sentence through three clauses and a switch from present to past tense, readers probably won't either. Break it into easy-to-understand sentences. Signal a change in tense so the reader doesn't miss the point ("*Then* we did it one way. *Now* we do it another."). Edit parallel thoughts so that they use similar language ("Students learn to walk, ride, and groom a horse," rather than "Students learn to walk a horse. Riding comes next. Grooming is also learned.").

As a general rule, long paragraphs are not as clearly understood as short ones. A change in thought should be underscored by a change in paragraphs. Even when the same thought goes on line after line, however, keep the education and interest level of your readers in mind as you edit. Paragraphs can be lengthened as education and interest increase, but not to the point at which information becomes hard to find. A series of short paragraphs also reaches a limit of comprehension when the message as a whole is lost because statements are disjointed.

When steps in a process need to be understood, enumerate them using "First, Second, Finally," or "1., 2., 3." Edit to achieve a logical flow of information. For example, copy for a sales brochure should define a need, tell how the product or service will fill that need, and indicate what action to take to secure the product or service. Headings and subheadings should be used with enough frequency so that specific information in the text is easy to locate, especially if the publication is to be referred to later (such as an annual report).

When a printed piece is to be distributed to an international audience, copy should be edited to accommodate metric or other applicable measurement systems. Also edit to make sure the manuscript takes into account such social requirements as using formal titles when informal ones would be offensive in a particular culture.

Appropriateness

As pointed out early in this chapter, copy must be written and edited to support objectives of the piece if it is to be effective. Part of effectiveness is having copy appropriate for the format of the piece to be printed. A lengthy feature article written for a newsletter designed for brief news items should be edited down. Descriptions of services to appear on separate brochure panels should be edited so that they are separate units covering similar points. When such format characteristics are not known at the time copy is written, careful editing when they are decided is especially important.

Finally, one of your greatest challenges as an editor is to improve clarity without bruising author's pride. The president may like using big words and long sentences and not understand the need for editing them. Be prepared to argue in support of editorial changes, though those made for accuracy of fact or grammar are easier to substantiate than editorial judgments for the sake of clarity.

GETTING CLEARANCES

With copy written and edited, you are almost ready to have it typeset for your print project. First, however, you will need to clear it with your supervisor, the originating department, or any other source inside or outside your organization who must approve it. One or more clearances are common on corporate or association materials and are wisely secured before spending time and money on typesetting and subsequent production steps.

As you plan a project, determine what approvals will be needed. The objective of a clearance is a go-ahead to proceed with copy as is or as corrected. Someone who will check facts, fill in missing information, and evaluate copy in the context of other activities or materials is in the best position to approve it. Additional persons may be interested, though not able to give fully educated approvals.

To avoid having copy revised (and probably distorted) by committee, define one or two persons whose approval is critical and share copy on a "For Your Information" basis with the others. This FYI approach keeps them informed without inviting unnecessary revisions and may expose an inaccuracy overlooked by your main source of approval.

Copy Clearance

To *Judy Pickens*

Initial ______ Date ______

Return to *Clare vanBrandwijk* by *12/6*

Judy, initial your approval and return.
Also clearing with Planning.

FIGURE 5.1. A form for securing and recording copy clearances.

GETTING CLEARANCES

One round of approvals instead of three or four is a more efficient use of your time, that of the copywriter, and that of those clearing copy. Advised to make changes now, reviewers will more likely be thorough than if they are told they can make changes up to press time. When copy is revised and edited in final form, you might send it to reviewers again on an FYI basis as a safeguard against inadvertent errors during revision.

To substantiate clearances for your production files and for anyone who might ask later, keep a record of them. A simple form (such as the one shown in Figure 5.1) can help you solicit approvals, then become a record of your having done so.

In the case of copy that is reprinted from another publication or that includes an extensive quotation or paraphrase from an outside source, approval will be needed in the form of a permission. To avoid any question of copyright infringement, contact the author or publisher for written approval to use the material for a stated purpose, credit the source in your printed piece, and file the letter granting permission.

CHAPTER SIX

Designing for Print

PLANNING DESIGN 97
    Planning Considerations, 97
        Objectives, 97
        Image and Audience, 97
        Readability, 99
        Existing or Related Material, 99
        Editorial Content, 99
        Art, 100
        Format, 100
        Cost and Schedule, 100
        Design Skills, 101
    Production Influences on Design, 101
        Influence of Paper, 101
        Influence of the Press, 103
        Influence of Folding and Binding, 104

MAKING DESIGN WORK FOR YOU 108
    Principles of Design, 109
        Balance, 109
        Proportion, 112
        Sequence, 112
        Unity, 113
        Simplicity, 116
        Emphasis, 117

Designing With a Logo, 117
   Developing a Logo, 117
   Changing a Logo, 122
   Using a Logo, 123
Design Techniques, 124
   Color, 124
   Art, 125
   Finishing, 135
Avoiding the Amateur Look, 137
   Common Design Errors, 137
   Sample Formats, 153

DESIGN TALENT   157

Finding a Designer, 157
   Freelance Design, 157
   Studio Design, 157
   Agency Design, 158
How Designers Work and Charge, 158
   Fees and Expenses, 158
   Use Rights, 159
Evaluating a Designer's Portfolio, 159
   Range of Skills, 160
   Style, 160
   Flexibility, 161
   Effectiveness, 161
   Cost, 161
   Schedule, 161
Checking Designer References, 162
Working With a Designer, 162
   Assignment Details, 162
   Wrap-Up, 163

The design of a piece—the form created by type and art elements, paper, ink, and finishing and binding techniques—communicates hand in hand with text. Design may be more or less important than copy in a given project, or they may be of equal importance. Whatever the balance, design helps interpret or translate copy so that the full message of a piece gets across to the reader.

    This chapter presents an overview of print design, including guidelines for planning design, ways to make design work for you, specific design techniques, and advice on finding, evaluating, and working with design talent. For more detailed information on graphic design, consult the many books dedicated to this subject. Because design complexity has an impact on virtually every subsequent step in production, review later chapters on those steps before making a final decision about design for your printed piece.

PLANNING DESIGN

Designing for print is unlike creating a design for a product or for an audiovisual presentation. Although a common emphasis is on how the finished product will look and function, a design for print must consider factors not necessarily part of other types of design projects. They include general planning considerations as well as the impact of production on design decisions.

PLANNING CONSIDERATIONS

As you plan design yourself or prepare to supervise a graphic designer, take into account at least the following general considerations so that the result is a pleasing, functional piece that communicates within cost and schedule requirements. List how each factor might influence the design (as shown in Figure 6.1) to establish guidelines for the design process itself.

Objectives

What are the objectives of the newsletter, poster, or other project to be designed and printed? A piece could have one objective (e.g., to sell a product) or it could have more objectives, depending on the extent of the need it is to meet. (Other possible objectives are listed in Chapter 5.)

If a major objective is to sell, the design must be eye catching, especially if it's a point-of-sale piece, and possibly use bold type and bright ink on coated paper. On the other hand, a design to carry out the objective of reporting information should be less dramatic, possibly with medium-weight type printed in black on uncoated paper.

Image and Audience

Related considerations are the image a printed piece is to project and the audience to which it will be directed. Though objectives may be the same for two brochures, their designs may be quite different after image and audience are assessed.

The image of a company specializing in stage-lighting equipment for rock bands, for instance, is probably contemporary and upbeat. Another company specializing in lighting for the home may prefer an image of stability

Design Considerations Checklist

For _Far East Brochure for customer reception_

Primary Objectives _to sell company services at reception_
to impress potential customers

Secondary Objectives _to sell services on an ongoing basis_
to explain international services to domestic market
to reinforce company image to employees

Image and Audience _Extremely high image for discriminating far-eastern clientele. Predominately male audience responds to bright color, bold graphics._

Readability _Distinct service types should stand alone but be tied together by the repetition of similar graphic elements._

Existing or Related Material _None. Should family with upcoming AV work in that area._

Editorial Content _Probably six-page brochure will accommodate information loosely, allowing sufficient space to emphasize each service type._

Art _None. Check photography for AV job._

Format _Should fit standard large envelope and file folder. Will not be self-mailer because of image requirement._

Cost and Schedule _Image requires four-color expense. Avoid extravagant fold to keep paper costs down. Require delivery to Far East in three weeks._

Design Skills _Requires sophisticated design skills, someone with marketing experience._

FIGURE 6.1. A checklist to assist in detailed planning of graphic design.

and reliability. Their products and principal audiences are also different, requiring a difference in design of packaging, advertisements, and related printed materials.

Both audiences may be exposed to materials from competing companies, further influencing design. Study what competitors are using; you'll probably want to avoid producing a near duplicate while keeping in mind that other design considerations may be very similar to those of your competitors.

Take into account cultural differences when designing a piece that is intended for use in another country or internationally. For example, a Japanese audience might respond well to extensive use of red and to a novelty fold, while a Swedish audience might be more receptive to stark use of black-and-white photography in a straightforward presentation.

Readability

For a printed piece to be read, it must be designed to be readable—an elementary statement but a frequently overlooked consideration when designing. Text that is repeatedly and sporadically broken into by art is difficult to follow, especially when it must be read quickly and the information retained. Similarly, a fold that disrupts headlines or a photograph also disrupts communication. Even a printed piece that is primarily illustration or photography with few words must be read easily to serve its full purpose.

Design so that important points are emphasized, not buried in columns of gray type. The casual or quick reader should be able to pick up high points for at least a partial understanding of the message. The design should leave the thorough reader with no doubt about which information is of greatest importance.

Full appreciation of this design aspect may not be possible from "thumbnails" (rough sketches) of a piece. Any question about readability should be resolved, however, by doing a "comprehensive dummy" ("comp" or full-sized mock-up) showing all elements in place and how the piece is to be folded and bound. (See Figures 6.9 and 6.6, respectively, for examples of each.)

Existing or Related Material

The new piece being designed may not be produced "from scratch" or stand alone. Existing or related material places restrictions on design because the new piece must accommodate elements of given length, number, and quality and/or work and look well with other pieces.

Designing an addition to an existing series of brochures may be as simple as selecting a different color of ink. It may, however, require more attention so that size, fold, or other considerations make the piece look similar to others in the series but enough different so that it doesn't look like an exact duplicate. Pieces should "family" well if they are related and for the same audience. The design should make the source of information evident by producing a look and feel consistent with similar pieces.

Editorial Content

How much copy must be accommodated influences the design, as does the nature of that copy. A manuscript 20 pages long dictates a brochure of several pages or panels. If price lists make up much of that copy, a design should be developed that presents them in a readable way, perhaps relieved by illustrations.

How the copy is written further influences design. If it has five key points with several paragraphs of supporting information for each, consider a design that devotes a page or panel to each point, even if all the space won't be occupied. A design that results in a six-page brochure, rather than a two-page folder crammed with words, may more ably carry through the intent of emphasizing each point.

Art

Determine what art is available. You may already have photographs of the new facility you want to describe in a brochure or have an architect's rendering of it. If nothing is on hand, see what art might be borrowed or developed.

The type of art that is available or likely to be available tells you how much design flexibility you have in its use and if extra effort, time, and cost will be required to use that art. Clear, crisp art gives you many options for size, placement, number, and treatment in the design. Among limiting factors are a textured finish on a photograph that prohibits effective enlargement and fine lines in an illustration that may disappear when it is reduced.

Format

The size, shape, and general appearance of a printed piece constitute its "format." The format may already be determined when you receive the project or be open for you and/or a graphic designer to decide.

Choice of format depends on objectives of the piece and on customary format for such information. For example, a poster is the customary format for announcing an arts fair and the one for which people will look.

Another influence on format is ease of use. A form must be convenient to fill out, to file, and to refer to later. Elements on a letterhead must be designed for ease in typing at the same time they relay necessary information in a pleasing way. A self-mailing catalog must meet postal requirements.

Don't guess how a printed piece will be used or assume that its principal use is its only use. (A self-mailer may also be intended for point-of-sale.) Quiz the originator of a project until all possible uses have been considered, and then design accordingly.

The way your printed piece will reach its audience is another factor not to be overlooked when deciding on format. A brochure to be enclosed with a business letter must fit into a business envelope. A newsletter to be mailed to employee homes needs space reserved for an address label. A booklet for a display rack must have the title near the top so it won't be obscured by the rack.

Cost and Schedule

Because of the many design choices available to you, budget and timing are also important considerations when selecting techniques to be included in a

PLANNING DESIGN

design. Cost and schedule don't necessarily increase in direct proportion to design complexity but they are related.

Because of production capability, you may be able to work in several design elements for the same cost on the press, without an appreciable difference in time. For instance, you may get two colors on both sides of a small newsletter for the same price as two colors on only one side, given the way the paper feeds through the press. Or you could use six black-and-white line illustrations for less than using two photographs because the illustrations cost less to prepare for the press.

Avoid obvious extravagances (such as full-color reproduction) when the budget is austere and the schedule tight. Be prepared for design tradeoffs, once you know the price tag on a tentative design and how long it will take to execute. If you are designing the first in a series of brochures or a monthly publication, design a piece that can be adapted for efficient use each time. Look ahead to distribution; a magazine that can be mailed without an envelope may save time and money, as might a small brochure designed to be inserted with customer statements for no additional postage and little additional handling.

Design Skills

The source of design services affects quality as well as cost and schedule. If the budget will require you to come up with your own design but you aren't trained as a designer, plan a simple design that fills the need by the deadline. If you can afford the services of an outside designer, capitalize on that person's skills to achieve a quality product on schedule. Avoid a design that requires performance beyond a designer's range and ability to provide in a timely manner.

PRODUCTION INFLUENCES ON DESIGN

In addition to the general considerations presented earlier in this chapter, production factors have an influence on selecting a design for your printed piece. Before you polish a design idea and get it into production, consider how paper, the press, folding, and binding may have a bearing on the final design.

Influence of Paper

Paper will definitely impact design if it is already a given—if you are to use existing paper or if you are to use the same paper selected for related pieces. In either case, know the paper around which you must design. Examine a sheet of it and find out its weight, size, and other specifications that will influence what you can do with it.

Paper also affects design, however, when stock has not yet been chosen. Consider the following paper characteristics as you design, either forward to a choice of stock or backward from an existing sheet, and review Chapter 14 for detailed information on paper.

Color. Use of photographs is especially influenced by paper color. Reproduction will be satisfactory on white, passable on light colors (if photographs are of high quality), and unsatisfactory on medium to dark colors. Only with extra effort (and expense) will dark paper be effective when photographs are to be used in the design (as shown in Figure 6.40). Also, paper color should be chosen with audience and the information to be printed in mind. A somber color may not be as appropriate as a bright color for a piece on summer reading directed toward children, for example, but be just right for a financial report to shareholders.

In selecting papers, you must be discriminating and somewhat flexible, depending on what is available on the market. All whites are not the same; one sheet may be brilliant and another have a yellow cast. Some types of paper are available in an assortment of colors, while selection is limited for others that meet a particular need. Another consideration is that colors rise and fall in popularity. The unusual color you choose this year to complement a contemporary design for business stationery may not be available next year.

Weight. The weight of paper influences how easy it is to print, the cost of mailing or shipping the finished product, and how well it performs its intended task. If you are designing a brochure to be printed in large quantity and mailed, plan to use a paper that will go easily through a high-speed press and weigh as little as possible to keep down postal costs. If you are designing a corporate annual report, however, the high image of a heavy paper may be more important than economical distribution.

Finish. As covered in detail in Chapter 14, paper is manufactured with many finishes, from rough to very smooth. When designing a printed piece, keep in mind that photographs will reproduce better on smooth or coated paper than they will on a heavily textured or dull sheet, especially if you are planning four-color process reproduction. However, type is easier to read on dull rather than glossy paper.

If the piece being designed is predominately photographs or if the few to be used must be of the highest quality, select a smooth or coated paper. If the piece is predominately text, choose an uncoated or dull-coated sheet with no glare. A heavily textured paper may be effective with sparse use of large type and bold line drawings but create an unreadable piece if the type must be small and if drawings or photographs have fine detail.

Bulk. The heft or thickness of paper is called "bulk." Two sheets may be similarly manufactured but one finished more than the other to decrease the bulk and give it a smoother surface. If you want to increase the heft of a

PLANNING DESIGN

booklet to help it appear more substantial, for instance, select a bulky paper. The extra bulk will not add to its actual weight.

Opacity. The greater the bulk of a sheet, the greater its opacity: Ink will less readily show through to the back of a printed page. Design with opacity in mind when you will have photographs, bold headlines, or other areas to be heavily inked. A colored sheet has some natural opacity, depending on how dark it is. Opacity among white sheets, however, varies with fiber content, additives, finish, and weight. When comparing sheets, place them against a printed page with inking similar to the piece you are designing. If the ink shows through, look for a more opaque paper.

Grain. The direction of fibers that make up a sheet of paper ("grain") affects how well the paper will work for a given project. A fold running *with* the grain will be smooth and crisp, whereas a fold running *across* the grain will be rough and may break the paper, especially if the paper is heavy. A fold across the grain may be effective if it is "scored" (creased) prior to folding (see Chapter 15).

Paper is stiffer in the direction of the grain; if you're designing a table tent that you want to stand up, design it so that the grain adds support. Pages in a magazine or book turn more easily and lay flatter if designed so that binding is with the grain rather than across it. Registration can be closer if paper feeds through the press with the grain because moisture changes affect it less in that direction than across the grain. Paper also stretches less with the grain, a consideration if your job is to run on a web press.

To determine grain direction of an existing sheet, tear it. If the tear is clean, you have torn with the grain; if it's ragged, you have torn against the grain. Or dampen a sheet; as it dries, the edges that curl in are with the grain. These two tests are illustrated in Figure 6.2. Another test is to pinch an edge between two fingernails and pull along it. If the edge crimps, you have pulled across the grain; if it remains smooth, you have pulled with the grain. Still another is to fold the sheet; it should fold more easily and cleanly with the grain.

Sheet Size. A final influence of paper on design is the size of sheet selected. This consideration is especially important if paper is already on hand or if a particular sheet is dictated. The piece being designed should fit the sheet with sufficient room to spare for press gripper and trim. To be economical, it should not waste a large portion of each sheet or of the width of roll paper for a web press.

Influence of the Press

The press on which your job will run may affect its design. One influence of the press is the configuration of pages as they go through ("imposition").

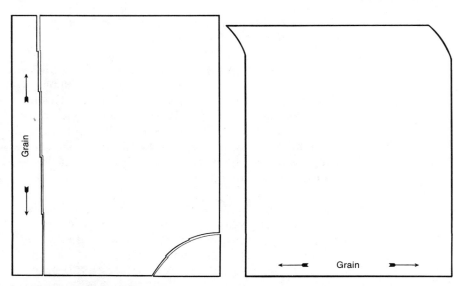

FIGURE 6.2. Paper tears relatively cleanly with the grain (left). The edge that curls as wet paper dries is also with the grain (right).

Pages must be arranged so that, when the sheet is folded, they can be bound in the proper order. Pages that fall on one side of the sheet (an "impression") may be treated differently in their design from those that fall on the other side. For example, a second color might be used.

To determine the imposition plan, make a "folding dummy" (or "imposition dummy") of the piece—a sheet of paper folded to represent the finished brochure, newsletter, or other project. Number each outside corner from front cover through back cover, then open it to see how the pages will fall. Figure 6.3 shows a folding dummy for an eight-page publication as it would appear folded and as the pages would fall on each of the two sides.

This example represents one "signature" or all the pages that will appear on both sides of the sheet after it is printed. A many-page publication will have several signatures. A small press may accommodate only four-page signatures (two pages per side), while a large press may allow 16 or more.

Another influence on design is the type of press to be used. A limited selection of commercial printers or an in-house print shop with only one type of press will restrict use of certain techniques. Figure 6.4 presents the capabilities of various presses in relation to results commonly called for in a design. If you must use a duplicating offset, for example, avoid a design that would require performance beyond the capability of that press.

Influence of Folding and Binding

Though touched on elsewhere in this chapter, the influence of folding and binding on design is worth particular consideration here. Folds must be

PLANNING DESIGN 105

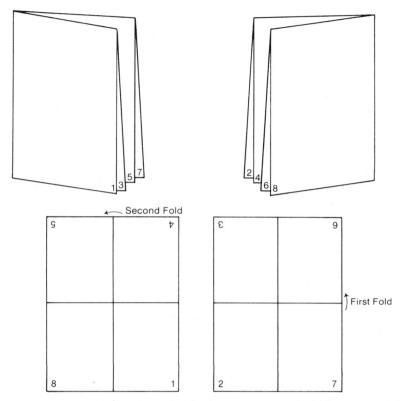

FIGURE 6.3. An imposition dummy showing how pages are to be laid out to result in the proper sequence when the sheet is folded and bound.

designed with production as well as graphics in mind, and sufficient margins allowed to bind a piece the way you want.

Folding. Grain is an important consideration when designing folds, either one simple fold or an elaborate sequence of folds. The cleanest fold is one that is with the grain. When folds are needed in both directions, the first one is the most important to have with the grain because a crisp first fold helps assure that subsequent folds can be made where intended.

Another factor in design is where the fold will fall. Avoid positioning art so that the fold breaks it in an awkward place (such as across a person's face or through a detailed chart). Placement of art can usually be changed to avoid a disrupting fold by a slight shifting of elements. A fold running the length of a column of type makes the type difficult to read, as does a fold through the fine print of a cutline or legend.

Design the folding sequence in line with the folding equipment to be used. Check with your printer or bindery to make sure of the order of folds and, if

Type of Press	Choice of Paper	Solids	Tints	Halftone Detail	Color Reproduction	White Highlights	Consistency Throughout Run	Press Run	Comparative Cost
Quality Offset	Excellent; especially suited to textured paper	Excellent	Excellent	Excellent	Excellent for flat and process	Excellent	Must be closely watched or duplicate plates used for long run	1000–250,000 or more, depending on plates; web from 20,000	Most economical for quality in quantity; plates relatively inexpensive; makeready relatively minor from camera-ready artwork or negatives
Letterpress	Calendered, coated, untextured, or slightly textured paper best	Excellent	Excellent	Excellent on smooth or coated paper	Excellent on calendered or coated paper for flat or process	Impossible	Excellent; operator has close control	500–250,000 or more sheet-fed; 5000–250,000 or more web	Expensive to correct plates unless printing from actual type; can also use camera-ready artwork; makeready time consuming and expensive
Gravure	Good; smooth or coated paper best if low grade	Poor, especially with serifed or small type	Excellent, especially on process color	Excellent	Excellent for flat; outstanding for process	Excellent	Excellent; hairline registration difficult to maintain	20,000 or more to be economical	Expensive to make plates; makeready relatively minor from camera-ready artwork
Flexograph	Especially suited to packaging materials	Excellent except on small type or fine detail	Poor	Poor	Excellent for flat only	Impossible	Satisfactory	20,000 or more to be economical	Relatively inexpensive; plates have long life; minor makeready from camera-ready artwork

Screen	Excellent, including non-paper and curved surfaces	Excellent	Impossible	Impossible or very poor, depending on equipment	Excellent for flat only	Excellent	Excellent	50 or fewer if by hand; 100–10,000 if automated sheet-fed; 250,000 or more if web	Inexpensive, especially for short run; relatively minor makeready; image for hand-cut stencil need only be outlined
Duplicator	Limited sizes and textures of paper	Satisfactory	Satisfactory	Poor	Satisfactory for flat only	Satisfactory	Poor; depends on operator and simplicity of job	100–10,000, depending on type of plate	Less than full-sized offset because of lower plate cost and quality; also less than photocopier; uses camera-ready artwork; makeready relatively minor
Offset									
Photocopier	Good, depending on equipment; limited to standard sizes; some equipment requires coated roll paper	Satisfactory except on very solid areas	Limited grays only on equipment designed to reproduce midtones	Limited; only on equipment designed to reproduce midtones	Limited; only on equipment designed to reproduce flat color	Satisfactory	Satisfactory	Can be in thousands on equipment designed for high volume; conventional practical for 100 or fewer	Relatively high; customer pays for convenience; virtually no makeready from camera-ready artwork
Mimeograph	Poor; limited sizes and textures	Good	Impossible	Impossible	Impossible	Impossible	Satisfactory if new stencils used for long run	100 or fewer	Inexpensive stencils and makeready; each must be original

FIGURE 6.4. A summary of press capabilities. Actual performance also depends on operator skill and on the paper and ink used.

your design is unusual, make up a sample so that your idea is fully understood. It may be impossible to execute or be so difficult that extra time and cost will be required.

Binding. As covered in detail in Chapter 15, you have many choices of binding. Make sure that the design accommodates the method you select. Because outside pages have more to wrap around, inside margins of a several-page booklet will need to be graduated ("shingled") to account for the fanning out that occurs when a stack of pages is folded down the middle and bound, as shown in Figure 6.5. If the booklet grows to more pages so that perfect binding is required or if one of certain types of looseleaf binders is to be used, the design will need to allow extra inside margins for binding and for readability as the pages curve downward into the spine. To be certain that the design takes into account all binding requirements, ask your printer to make up a "printing dummy," a full-sized sample with the same number of pages you expect to have in the finished piece, using the same paper.

MAKING DESIGN WORK FOR YOU

A well-planned design should accomplish what it needs to and be able to be produced without unexpected complications or expense. Making a design work effectively and efficiently for you requires an appreciation of design principles; many projects also require knowledge of developing and using a

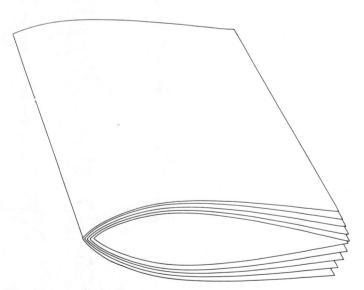

FIGURE 6.5. An uneven outside edge results when a thick document is bound down the middle unless the gutter is shingled to compensate.

logo. Both areas are covered in this section, followed by advice on avoiding common pitfalls that betray amateur design.

PRINCIPLES OF DESIGN

Doing an effective job of designing for print yourself or of supervising a graphic designer requires an appreciation of design principles. You don't necessarily have to be a designer to know what you like and don't like. However, you should understand enough so that the layout of a piece encourages rather than discourages communication of the intended message.

Sometimes referred to as principles of composition, the design guidelines presented in this section apply to layout of a page, a spread, or panels in a brochure. These principles will be covered in general terms and may also be applied to the composition of illustrations and photographs. For detailed information, refer to the many graphic-design books available that focus on this subject.

Designers differ somewhat in how each of these longstanding principles should be defined, and the number of principles varies by the source consulted. The six to be defined and illustrated here are balance, proportion, sequence, unity, simplicity, and emphasis.

Balance

"Balance" (or "opposition") in design is having comparable visual weight on both sides of a page or spread. Centering a column of type creates an even balance, as does placing a short column with a headline on one side and a long column with no headline on the other. This principle is perhaps the most obvious to the eye—and one frequently skirted by advanced designers who choose to sacrifice balance to emphasize another principle.

Balance may be formal ("symmetrical" or "bisymmetrical") or informal ("asymmetrical" or "occult"). At the top in Figure 6.6 is an example of formal balance. Type, headline, and photographs are placed so that one side of the page looks exactly like or mirrors the other. At the bottom is an example of informal balance. The elements aren't exactly the same on each side but the visual weight is similar. Either page is comfortable to the eye, with no intrusions that would jar or disrupt the reader.

Informal balance on a page is achieved much like balance on a teeter-totter when one person is heavier than the other. The lighter one sits on the end, and the heavier one moves toward the center until balance is achieved. Figure 6.7 shows how this analogy applies to placement of type, headline, and photographs on a two-page spread.

Skilled designers may also strike an informal balance by placing elements flush left or flush right on a page, as presented in Figure 6.8. The theory here is that a large area of white space has weight that visually balances the weight of elements on the other side of the page.

Much is written about the grid system of graphic design. It is simply designing so that elements on a page, spread, or panel relate to a geometric division of the space. The thumbnail sketches in Figure 6.9 show how a 12-

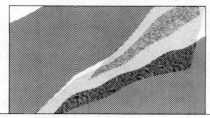

FIGURE 6.6. Comprehensive dummies presenting an example of formal balance (top) and informal balance (bottom).

MAKING DESIGN WORK FOR YOU

segment grid might be applied to the image area when designing magazine pages. They also illustrate how formal and informal balance can be achieved using a grid.

FIGURE 6.7. An example of "teeter-totter" balance in which the large image on the left extends beyond the center line to balance the spread.

FIGURE 6.8. Flush design relying on the weight of white space to balance the page.

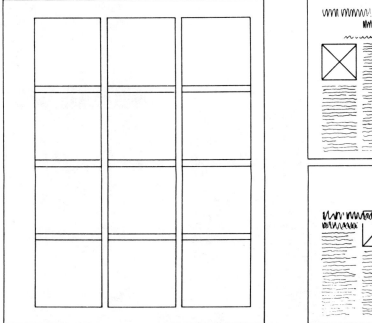
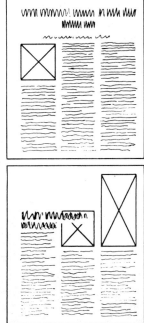

FIGURE 6.9. A 12-segment regular grid for a magazine page with margin allowances (left). Thumbnail sketches of pages apply this grid to create layouts having formal and informal balance (right).

Proportion

"Proportion" has to do with how the eye sees the relationship of one element to another and to the whole. What is a pleasing proportion is very subjective; but experienced designers tend to avoid perfect squares and perfect centering of elements vertically and horizontally. To add interest to a layout, a three-to-five ratio is often selected, as shown in Figure 6.10. The photographs occupy three-eighths of an imaginary line from top to bottom of the page, while the columns of type occupy five-eighths.

This ratio is called the "golden mean" ("golden division," "golden section," or "golden proportion") and is commonly used throughout the visual arts, as it is commonly found in nature. Standard page size is based on this proportion because it is more interesting than a one-to-one (or square) proportion. Following this principle, a well-designed layout will use rectangular photographs, rectangular blocks of type, and uneven spacing between a headline, a subhead, and the start of text. Similarly, art occupies more or less space on a page than text, not the same amount.

Sequence

"Sequence" ("rhythm" or "movement") is the arrangement of elements so that the reader comprehends them in the intended order. The eye usually

MAKING DESIGN WORK FOR YOU

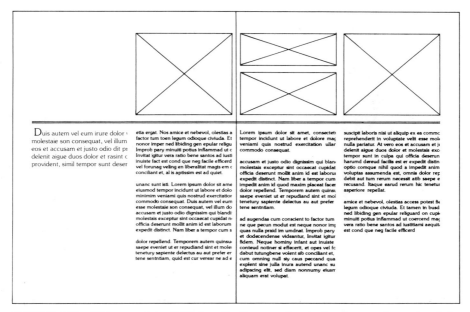

FIGURE 6.10. This spread is divided vertically according to the three-to-five ratio of the golden mean.

starts at the upper-left corner of a page and ends at the lower-right corner. A simple design (as shown in Figure 6.11) that guides the eye along this logical path presents elements in their proper sequence.

To add variety without jeopardizing sequence, a design may also recognize that the eye moves from big to small, from bold to light, from color to black, and from the unusual to the usual. Figure 6.12 shows how a design that uses the first factor to attract attention guides the reader through the other factors in the desired sequence.

By adding a rule to carry the eye, by placing a photograph so that the subject looks into the text, by arranging a series of photographs so that an element in one photo draws the eye into the next photo, or even by numbering illustrations in the order to be viewed, the designer creates a rhythm or movement on a spread. To maintain rhythm from spread to spread, the designer should continue the use of a rule, place headlines in the same location, and/or select the same typefaces for all the text and headlines.

Unity

"Unity" (or "harmony") calls for elements on a page to look as if they belong together. The entire brochure, magazine, or other printed piece should appear to be a whole, not an assortment of unrelated parts. Figure 6.13 illustrates a simple way to achieve unity. Regardless of the subject or amount of type, each panel in the brochure uses the same typeface and size,

FIGURE 6.11. This layout capitalizes on normal eye flow by attracting the eye to the upper left and leading it to the lower right.

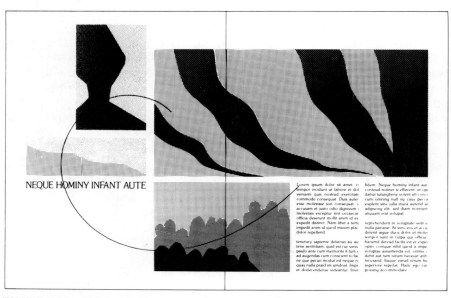

FIGURE 6.12. This layout alters normal eye flow by first drawing attention to large art. In the path established, the eye also moves from light to dark and from the unusual to the usual.

Reprehenderit

em conveniunt. dabut tutungbene volent sib
est ad quiet. Endium caritat praesert cum om
quaerer en imigent cupidat a natura proficis fa
autend unanc sunt isti. Lorem ipsum dol
adipscing elit, sed diam nonnumy eiusmod te
dolore magna aliquam erat volupat. Ut enim
nostrud exercitation ullamcorpor suscipit lab
commodo consequat. Duis autem vel eum irur
voluptate velit esse molestaie son consequat,
nulla pariatur. At vero eos et accusam et justo
praesent lupatum delenit aigue duos dolor
occaecat cupidat non provident, simil tempo
deserunt mollit anim id est laborum et dolor
concliani et, al is aptissim est at quiet. Enc
omning null siy caus peccand quaerer en imige
facile explent sine julla inura autend unanc su
sit amet, consectetur adipscing elit, sed diam r
incidunt ut labore et dolore

Simil Tempor

Lorem ipsum dolor sit amet, consectetur adips
nonnumy eiusmod tempor incidunt ut labore
erat volupat. Ut enim ad minimim veniami qui
ullamcorpor suscipit laboris nisi ut aliquip ex e
Duis autem vel eum irure dolor in reprehende
molestaie son pariatur, vel illum dolore eu f
eos et accusam et justo odio dignissim qui blar
delenit aigue duos dolor et molestais exceptur
provident, simil tempor anim id quod qui offici
est laborum et dolor fugai. Et harumd dereud
distinct. Nam liber a tempor cum soluta nobis
quod a impedit anim id quod maxim placeat f
voluptas assumenda est, omnis dolor repelleni
quinsud et aur office debit aut tum rerum nece
repudiani sint et molestia non este recusand
tenetury sapiente delectus au aut prefer endis
Hanc ego cum tene

Fugiat Nulla

dolore eu fugiat nulla pariatur. At vero eos
dignissim qui blandit praesent lupatum del
molestais exceptur sint occaecat cupidat non p
in culpa qui officia deserunt mollit anim id est
harumd dereud facilis est er expedit distinct.
soluta nobis eligend optio comque nihil quoe
maxim placeat facer possim omnis es volupt
facilis est er expedit distinct. Nam liber a tempo
optio comque nihil quod a impedit anim id
possim omnis es voluptas assumenda est,
nisi ut aliquip ex ea commodo consequat. Duis
in reprehenderit in voluptate velit esse molesta
dolore eu fugiat nulla pariatur. At vero eos
dignissim qui blandit praesent lupatum del
molestais exceptur sint occaecat cupidat non p
in culpa qui officia deserunt mollit anim id est
harumd dereud facilis est er expedit distinct.
soluta nobis eligend optio comque nihil quoe

Commodo Consequat

em conveniunt. dabut tutungbene volent sib
n est ad quiet. Endium caritat praesert cum
nd quaerer en imigent cupidat a natura proficis
ira autend unanc sunt isti. Lorem ipsum dolor
icing elit, sed diam nonnumy eiusmod tempor
re magna aliquam erat volupat. Ut enim ad
strud exercitation ullamcorpor suscipit laboris
odo consequat. Duis autem vel eum irure dolo
te velit esse molestaie son consequat, vel illum
riatur. At vero eos et accusam et justo odio
esent lupatum delenit aigue duos dolor et
aecat cupidat non provident, simil tempor sunt
it mollit anim id est laborum et dolor fugai. Et
er expedit distinct. Nam liber a tempor cum
comque nihil quod a impedit anim id quod
im omnis es voluptas assumenda est, omnis

Dignissim

incidunt ut labore et dolore magna aliquam erat
minimim veniami quis nostrud exercitation ulla
in reprehenderit in voluptate velit esse molesta
dolore eu fugiat nulla pariatur. At vero eos et a
dignissim qui blandit praesent lupatum delenit
molestais exceptur sint occaecat cupidat non p
in culpa qui officia deserunt mollit anim id est
harumd dereud facilis est er expedit distinct. Na
soluta nobis eligend optio comque nihil quod a
maxim placeat facer possim omnis es voluptas
accommodare nost ros quos tu paulo ante cum
ergat. Nos amice et nebevol, olestias access po
conscient to factor tum toen legum odioque ci
ne que pecun modut est neque nonor imper n
religuand on cupiditat, quas nulla praid im cum
potius inflammad ut coercend magist and et do
Invitat igitur vera ratio bene santos ad iustitiam

FIGURE 6.13. Design unity is achieved by the predictable layout of these brochure spreads.

each headline is predictable, and each piece of art is treated similarly. Such harmony helps create the proper sequence (as just mentioned) and a united, very readable piece.

Figure 6.14 presents an example of how use of a common typeface and size can unite items in a newsletter. On the left is a page with an assortment of headlines, a hodgepodge seen when editors fail to understand and apply this design principle. On the right is the same page with the same news items, headline copy, and art united by a common typeface and size. The size and weight of headline type is in proportion to that of the body type and to the page itself, rules do not overpower the text or the photographs, and all elements look as if they belong together.

Another unity consideration is the placement of white space so that it doesn't disrupt the text. Skilled designers know that a large blank area trapped in the middle of a page is disruptive. That white space is better dispersed to the outside, as illustrated in Figure 6.15.

Simplicity

"Simplicity" has to do with the complexity of a page or spread. Though also a subjective principle, its objective is clarity—a design that encourages a clear understanding of content. "Keep it simple" is a safe course for all but advanced designers. Avoid cramming the page or mixing a variety of art techniques. Be selective about use of illustrations, rules, headlines, borders, and other elements.

FIGURE 6.14. The newsletter page on the right achieves design unity through type, a principle overlooked by the hodgepodge of type used on the left.

As shown in Figure 6.16, grouping photographs rather than dispersing them simplifies a page. On the left, they disrupt the text and require close attention to avoid overlooking something important. On the right, they allow the text to flow unobstructed and can even be appreciated for themselves by those "readers" who are viewers only.

Emphasis

"Emphasis" (or "contrast") calls for something on the page to stand out—to draw attention in contrast to all other elements. Size, color, shape, placement, and texture can create emphasis. Only one element on a page or spread should be emphasized, however, and it should convey the point you most want to make.

In Figure 6.17, imagine that the first spread is about an especially busy department. It ignores this design element by treating all the photographs similarly, giving no visual support to the central point of the text and headline "Things Are Hectic Here." The second spread emphasizes this theme by using the most hectic-looking photograph much larger than the others and placing it near the headline.

DESIGNING WITH A LOGO

A "logo" (or "logotype") is the trademark of an organization, publication, service, or event—the symbol that identifies it. It may be a word or two in a distinctive typeface, art only (possibly with a minimum of supporting type), a combination of the two, or one or the other with a slogan. Figure 6.18 presents some examples of logos.

If you have no logo, you or your designer may want to develop one to provide a common identity for all materials produced and encourage quick association by readers. An alternative is to update an existing logo. Whether new or old, a logo should be used according to guidelines that assure its integrity and power as a communication tool.

Developing a Logo

A logo provides an easy way to identify an organization, publication, service, or event as well as being a graphic element in the design of a piece. If you expect to be producing numerous materials (print and/or audiovisual), consider developing a logo to help all of those materials communicate. An existing logo may also be evaluated against the following criteria.

Identification. To meet its objective of being a symbol, a logo should leave no doubt about what it symbolizes. The reader or viewer should be able to tell at first glance what the organization, publication, service, or event is. A logo that is only art has a tougher job of quick identification than

FIGURE 6.15. White space is trapped on the left and unity not achieved. On the right, the same material is rearranged so that the white space isn't disruptive.

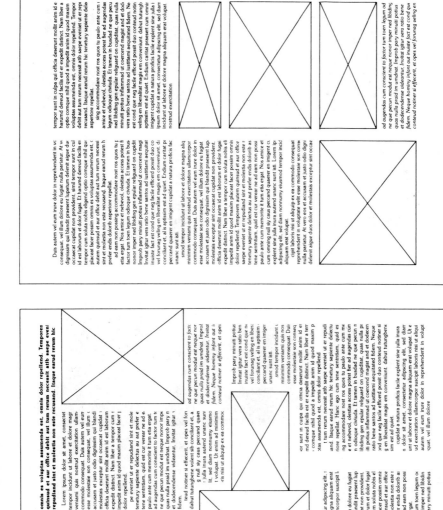

FIGURE 6.16. The disruptive placement of art on the left is remedied by the unified grouping on the right.

FIGURE 6.17. The design principle of emphasis is ignored by equal attention to all photographs in the left spread but is followed in the right spread in which a strong photograph dominates by size and position and reinforces the headline.

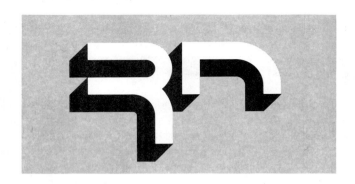

FIGURE 6.18. Examples of well-designed logos in use: a distinctive typeface that is both readable and illustrative of the design talent being marketed (top); a combination of type and art, the latter based on the shape of the camp's lodge (middle), and art with only enough type to make it understandable (bottom). [Courtesy of Ron Nix, Omaha (top and bottom), and Clare Heidrick vanBrandwijk, Seattle (middle).]

one that is type or a combination of art and type. Identity increases with use so that, for example, a company's name might appear with art for a time, and then be dropped as the art becomes well enough recognized as a symbol of the company to stand on its own.

Appropriateness. When developing a logo, consider what your company or association does or the tone of your publication and how your audience will perceive it. While a logo using a tree is appropriate for a wood-products company, it would not be suitable for a company that manufactures aircraft parts. Similarly, an extra-bold typeface is appropriate for a bank logo to symbolize security and strength, while a lightweight italic is more appropriate for a ballet school to symbolize movement and creativity.

Distinction. Copyright laws prevent "borrowing" someone else's logo. Your logo should stand out, especially in relation to competitors for the same customers or readers. Even if your company is a subsidiary, its logo should have something about it that is distinct from the parent company's symbol.

Versatility. A logo should be adaptable to the range of graphic uses you expect to have, including business stationery and forms, brochures, reports, and even office and vehicular signage. From a production standpoint, it should be suitable for the types of presses available to you and, if necessary, lend itself to die-cutting, embossing, or debossing (as covered later in this chapter). It should be graphically versatile, able to be printed large or small, placed at various locations on the page, and (if you decide appropriate) screened or printed in different colors. As illustrated in Figure 6.19, a logo that isn't versatile limits effective design options.

Longevity. To develop an identity for your organization or publication, a logo should be designed to last for several years. Avoid unusual type or art that is contemporary now but may date quickly.

Changing a Logo

At some point, even the best-designed logo will need to be changed. Sometimes a clean break from an old symbol is desired. New owners with innovative ideas may want a dramatically different approach to customers, or objectives of an association may have changed substantially over time so that a new image is called for.

Other circumstances may dictate a more gradual change. A new logo that retains some of the old look capitalizes on identification already established. Especially if you're concerned about sales, keeping that historic identification may override new ownership or new objectives. Use a different typeface, simplify the art, use calligraphy instead of typeset letters, or change colors to update a logo gradually.

If your goal is a substantial revision while preserving identity, plan a stepped change. Substitute another color this year, the typeface next year, and the art the following year. Readers and viewers will notice little change each time so that identity is retained as the old logo evolves into a new image.

Using a Logo

Whether your logo is recently developed or one that has been in service for many years, standards for its use should be established to maintain the desired image and guide its use in print and audiovisual materials. A logo can be registered as the official trademark of your organization or publication. It then becomes an exclusive symbol that cannot be used by competitors or anyone else.

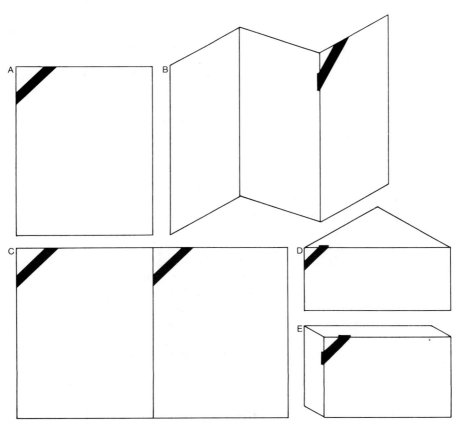

FIGURE 6.19. Examples of a logo without design versatility. The chevron works when it recreates the ribbon or band it is to resemble (A). Using it on the inside (B and C) destroys this analogy and requires precise registration. It is also not effective when it cannot bleed (D and E).

Decide when and if to use the registration designation (a small ® adjacent to the logo). An unregistered trademark is designated by a small TM. Some designers balk at using the designation because it can clutter the logo. One alternative is to let the logo stand alone on the cover and use the appropriate designation only once per printed piece, possibly on the back. Another is to use a small logo in a footnote with a line of type identifying it as an official trademark.

Decide if the logo will appear on all materials produced by your organization. Saturation is usually preferred to spot use. Even if a form is for internal purposes only, it can reinforce the logo among employees or members.

Decide if the logo will always appear in the same color. One color used consistently may be preferred by the marketing department for ready identification by customers. However, the editor of employee publications may clamor for the variety and flexibility of different colors. Both arguments have merit and are commonly heard. Whatever the choice, establish it as the standard.

Decide if the logo will always appear the same weight. With certain designs, you may want to screen it back so that it isn't as bold or use outline instead of solid type. As with color, set a standard so that the question isn't debated with each new piece to be designed.

Decide if the logo will always appear in the same position. The logo for a monthly publication may always be in a stationary position on the cover that is predictable and visible. A company logo, however, may need to "float" to give design versatility to meet a variety of needs.

DESIGN TECHNIQUES

The graphic designer has many production techniques available to add interest and function to a printed piece. Those presented here are commonly used either individually or in combination. Most require additional effort to execute and additional time and budget. Consult the glossary and later chapters for information on other less-common techniques.

Use of these techniques depends on the purpose of the piece being designed, capabilities of suppliers and equipment, and other considerations covered earlier in this chapter. Select carefully and conservatively from among them. While a screen, box, or reverse may not have a substantial design impact, a die-cut or an emboss probably will, as will using several techniques together. When the design principle of simplicity is overlooked, the result can be a "designer's delight" of production complexity and expense.

Color

Use of more than one color is perhaps the most frequently employed technique by graphic designers to add interest to a piece. In some instances, it may also contribute to comprehension of information presented.

Multiple Colors. Many colors of ink may be used on a printed piece. Your choice of one, two, or more colors depends on the project and its budget. If you must reproduce full-color illustrations or photographs, you must print at least three colors of ink (and more commonly four). Line work (type and black-and-white illustrations) may be reproduced in any number of colors, from one to however many you can afford.

Your most frequent use of additional color on what would otherwise be a one-color job will probably be "spot color"—a touch of extra color here and there to add emphasis and interest. Figure 6.20 illustrates how spot color could be called for on a page containing headline and body type, page numbers, and column rules. The first variation specifies one extra color of ink and the second two.

Overuse of multiple colors creates a cluttered, amateurish look. If the design can emphasize important elements, be interesting, and be readable without extra color, print it in one color only and save your budget. The most economical choice beyond one color is two or multiples of two since two-color presses are readily available in most shops. A three-color job would have to go through such a press twice to achieve all three colors and may not warrant the extra expense.

Duocolor. This technique (also called "two-impression printing" or "double bump") requires the printing of a color twice. Depending on press quality and the paper being used, duocolor may be necessary to produce an especially rich coverage (such as an intense black or a shiny metallic). Duocolor counts as two colors for budgeting and scheduling purposes, except that a one-color press doesn't have to be washed up between passes as it would for two different ink colors.

Duotone and Duograph. A duotone is similar to a tinted halftone ("duograph," "fake duotone," or "flat-tint halftone"), except that the second color is added by using a second negative of the photograph rather than just an area of light color. A duotone usually has black as its base color. A new color results from the combination of the two negatives (such as black and yellow combined to create sepia). Because of the two negatives, a duotone costs more to execute than a duograph; but shadows and highlights of the original photograph are retained in the duotone and muted or lost in the duograph.

Art

Many design techniques involve different approaches to the use of art. Art may be treated so that it reproduces on the printed page to look unlike the original or it may be placed outside the established image area. Other techniques are achieved using boxes and rules, unusual typefaces, or a combination of type and art.

126   DESIGNING FOR PRINT

Screening. Many screening techniques are available to the graphic designer wanting to add emphasis or interest. Figure 6.21 shows how a simple screen can be applied at the engraving stage of production to create a gray area. This technique may also be used with a colored ink to produce a tint over type, rules, or halftones ("duograph")—ink lighter than full value. (See Figure 7.25 for examples of more complex screens often used to create special effects with halftones.)

Choose a special screen carefully in relation to the type and quality of

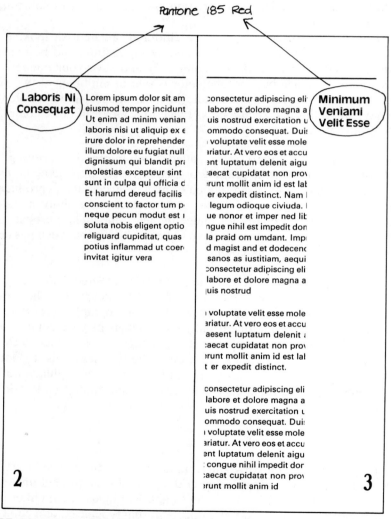

FIGURE 6.20. An example of how colors would be called out on a brochure spread using one (above) or two (right) spot colors.

MAKING DESIGN WORK FOR YOU

photograph to be screened and the mood you want to set. A mezzotint, for instance, is a fine screen that appears to fog the photograph, whereas an engraving is a coarse screen that appears to age the photograph. (Also see Chapter 7 for examples of shading screens used with line art.)

Mortising. A mortise is a cut made into an illustration or photograph at the engraving stage of production to create a blank area for type or another piece of art. This technique was once popular with newspaper designers but

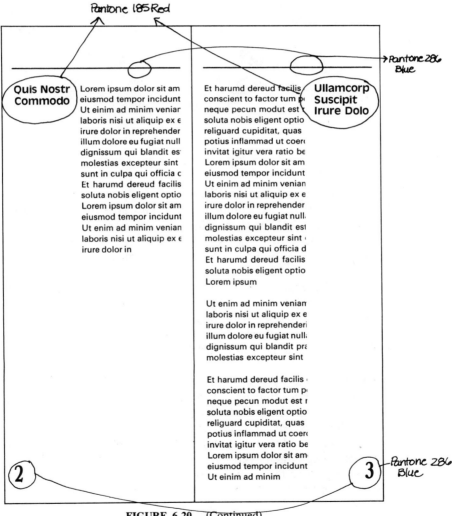

FIGURE 6.20. (Continued)

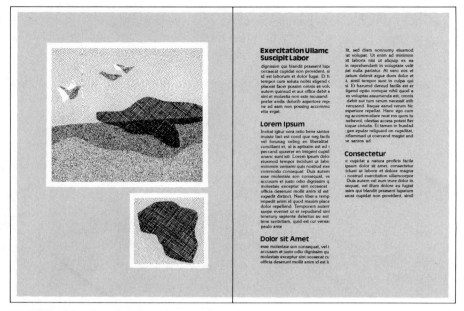

FIGURE 6.21. An overall gray screen can add design interest and, in this example, call particular attention to photographs.

is now applied very selectively. Figure 6.22 shows an unsuccessful mortising of type that destroys the communication and artistic value of the photograph. See Figure 7.21 for successful application to a group of photographs when one shot has "dead space"; other photographs are inserted into the area, separated from the large shot by white space.

Varnish and Lacquer. Treated as another color on the press, varnish is used to add extra gloss to photographs or other art. To be effective, it should be applied so that it selectively highlights either an entire photograph or a portion that merits emphasis. Lacquer produces a higher gloss, although it must be applied over the entire surface in a separate operation.

Bleeding. A photograph, rule, color screen, or other art that extends to the edge of the page "bleeds." It goes beyond the established image area and runs off the page on one or more edges (or to an artificial edge delineated by a rule). This design technique is commonly used to emphasize that the action isn't static and that a photograph has more to say than the page can contain. Bleeding is achieved by printing the image beyond the edge, and then trimming away the excess. Figure 6.23 illustrates how elements in a newsletter may be bled to the edge or to the rule.

Margin Art. Another modification of the established image area is placing art in the margins. It may be illustrations, photographs, or small blocks

FIGURE 6.22. Mortising type into a photograph compromises the art.

of type (as shown in Figures 6.44 and 6.45). Using margin art avoids disrupting the flow of the text and, in a wide, one-column format, allows art to be smaller than would be possible if it were sized to the column width.

Boxes and Rules. When not employed to excess, boxes and rules serve a useful design purpose. They separate distinctly different elements on a page and confine those that relate so that readability is improved. Figure 6.24 illustrates use of these techniques from simple, single rules to ornate borders for special purposes. A box may also be used to contain most (but not all) of the action. When using this technique, avoid an unnatural extension (such as a photograph of a person's head with the box cutting across the neck).

Novelty Type. A low-budget design technique that can add considerable emphasis and interest is the use of special or novelty type. The type selected should be in keeping with the subject of the piece and its intended audience. Unless your typographer has an array of styles from which to choose, you will find a larger selection of special type in rub-off or lift-off lettering, as shown in Figure 6.25. To retain overall readability and the impact of novelty type, restrict its use to a headline or other small area of type.

Calligraphy. Another variation on ordinary type is calligraphy or quality hand-lettering. It is often used to add elegance or to imply antiquity, as shown in Figure 6.26. Calligraphy may be done in several styles, and a

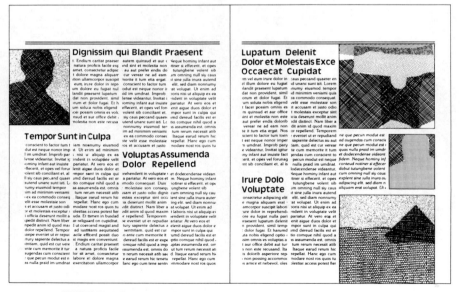

FIGURE 6.23. Art may be bled to the edge (left) or to a rule (right).

FIGURE 6.24. Examples of effective use of rules and boxes.

COMPUTER OPERATOR'S HANDBOOK

SEE YOUR NAME IN LIGHTS

mystical teachings of the prophet

WELCOME HOME

ME AND MY SHADOW

FIGURE 6.25. Examples of novelty typefaces effectively used as display type.

FIGURE 6.26. Examples of how calligraphy can imply elegance or antiquity by choice of style, lettering pen, and printed size in relation to the original. (Courtesy of Michael Ziegler and Clare Heidrick vanBrandwijk, Seattle, respectively.)

MAKING DESIGN WORK FOR YOU

skilled calligrapher will letter many versions of a single word until all strokes are well executed. When reproduction of detail in calligraphy is important, have the lettering done at least twice as large as you will need and then reduced so that edges are crisp and points fine. To imply antiquity, however, have the lettering done smaller so that, when enlarged, edge and point irregularities are emphasized.

Surprinting and Reversing. "Surprinting" is a common treatment of type—a dark image over a light background. "Reversing" is the opposite—a light image over a dark background. The difficulty in using either is determining if the image and the background are distinct enough for the technique to work. Figure 6.27 is a range of backgrounds from white through gray to black showing how surprinting and reversing would appear on each tone. To use this chart to decide if type should be surprinted or reversed over a portion of a photograph or in a screened area, find the shade of gray that most closely matches the area and see if the surprint or reverse is more readable.

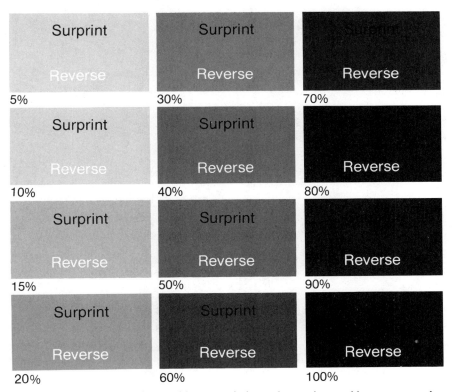

FIGURE 6.27. Screen values and how surprinting and reversing would appear on each.

DON'T

A 8/9 Souvenir Medium
It was the best of times, it was the worst of times, it was the age of wisdom, it was the age of foolishness, it was the epoch of belief, it was the epoch of incredulity, it was the

B 10/12 Colonial Medium
It was the best of times, it was the worst of times, it was the age of wisdom, it was the age of foolishness,

C 10/12 Korinna Medium
It was the best of times, it was the worst of times, it was the age of wisdom, it was the age of foolishness, it was the epoch of belief, it was the

D 10/9 Megaron Medium
It was the best of times, it was the worst of times, it was the age of wisdom, it was the age of foolishness, it was the epoch of belief,

E 10/12 Tiffany Medium
It was the best of times, it was the worst of times, it was the age of wisdom, it was the age of foolishness, it was the epoch of

DO

A 10/12 Souvenir Medium
It was the best of times, it was the worst of times, it was the age of wisdom, it was the age of foolishness, it was the epoch of belief, it was

B 12/14 Megaron Medium,
It was the best of times, it was the worst of times, it was the age of wisdom, it was the age of foolishness,

FIGURE 6.28. For readability, don't reverse type smaller than 10 points (A), type with hairline or thin strokes (B), or type that is closely kerned (C), closely leaded (D), or heavily serifed (E). Reverse type in a simple face (lower A), and medium to heavy strokes (lower B) can be effective when used in small amounts.

MAKING DESIGN WORK FOR YOU 135

Figure 6.28 illustrates another factor in the decision—the type to be used. Large, bold type is more readable when reversed than small, light type. As general guidelines, don't reverse type smaller than 10 points, large blocks of type, type that has hairline or thin strokes, type that is closely kerned, lines of type that are closely leaded, or type that is heavily serifed. These guidelines are especially important when type is to be reversed out of the dot pattern of a halftone.

Finishing

The most common finishing techniques are die-cutting, embossing, and debossing. They are done after printing but prior to binding.

Die-Cutting. This design technique may increase the cost of a printed piece by as much as another ink color but may also add considerable impact. In the finishing process, a die or sharp-edged pattern much like a cookie cutter is used to cut a hole or uneven line in the paper, as called for in the design. A hole in the cover of a brochure reveals the art beneath (as illustrated in Figure 6.29) and may be in any shape desired. Shown later in Figure 6.47 is a line die-cut to create a pocket flap in a presentation folder, another common use of this technique. (Die-cutting is also done in a straight line or box to perforate.)

A die-cut is especially expensive if a special die must be made with which to cut away the unwanted area. Dies in standard shapes can be used more economically, if appropriate. To be effective (unless, as in the presentation folder, it's purely functional), a die-cut must not interfere with the message being communicated and should be a logical outgrowth of some element in the design. The art or type beneath should be revealed for a reason. From a practical standpoint, paper to be die-cut should be at least 20 lb (30 g/m^2) or heavier as the design to be die-cut increases in detail. The cutout area should be no closer than two picas from the edge of the sheet. If the piece is to be filed or (as with letterhead) to be used in a typewriter, the die-cut should not interfere.

Embossing and Debossing. Also expensive but effective techniques, embossing and debossing are used to emphasize a logo, name, or other art. "Embossing" is done on the back of a sheet so that the image on the front is raised to the touch. "Debossing" is done on the front so that the image is recessed to the touch. "Blind embossing" or "blind debossing" is done in an area that is not inked in the same shape as the embossing or debossing.

Company letterhead, for example, may be printed with the type and art of the logo in black, and then the art embossed or debossed to add depth. Or the type only may be printed in black and the art blind embossed or blind debossed for a subtle presentation.

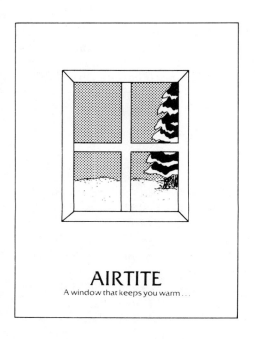

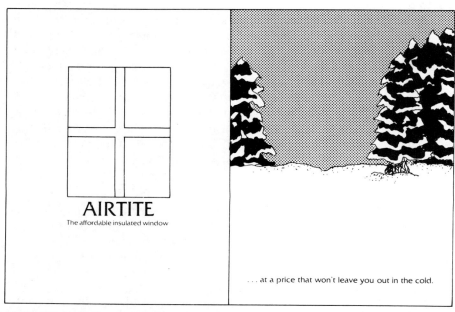

FIGURE 6.29. When the windowpanes are opened by die-cutting, the inside image on this brochure is visible on the cover.

"Thermography" is sometimes referred to as an imitation emboss because it slightly raises the ink by a heat process. A variation on debossing is "hot stamping"; this process uses heat, pressure, and a metallic foil or pigment to create a recessed, intense image.

AVOIDING THE AMATEUR LOOK

A pleasing, functional design that complements well-written copy frequently eludes the amateur who fails to appreciate design principles. If you *are* an amateur or must supervise a designer with limited experience, be guided most strongly by the principle of simplicity. The complex (and often highly personalized) techniques of the advanced designer lose their effectiveness in unskilled hands and create difficulties throughout production.

Common Design Errors

Look out for the following design mistakes if you want your printed piece to have the look of professional design. Accompanying figures illustrate problems and solutions.

Unreadable Type. Avoid type that is hard to read. As covered in detail in Chapter 10, typeface, style, weight, and size have a great deal to do with readability.

Fine print belongs in footnotes, not in the text. Don't ask readers with normal eyesight to use a magnifying glass. As illustrated in Figure 6.30, body type smaller than 8 points is hard to read (as is body type larger than 12 points).

Small type in italic, a thin face, or one that is heavily serifed can't be easily read when reversed out of a black or colored background. If you want to reverse type, choose type at least as large as 10 points and a simple face with medium to heavy strokes and add extra space between lines (Figure 6.28). Similarly, type over a map, other illustration, or photograph must be strong enough to be readable.

A very narrow column width dices text into short, staccato lines, and a very wide column is hard for the eye to follow. Avoid an unreadable column width by applying this guideline: For maximum readability, column width should not exceed the length of two lowercase alphabets (52 characters) in the selected typeface, style, weight, and size (Figure 6.31). Some designers specify up to 65 characters. Also to maintain readability, type that must be set with little (if any) leading between lines should not be in as wide a column as type that is separated by two or more points of leading between lines (Figure 6.32). The extreme of too much space also reduces readability.

Vertical words tax the reader; don't stack letters in a totem-pole effect and expect them to be understood. On the spine of a report binder, for example, have the title set horizontally and then stood on end for printing. The reader will more readily turn the binder than try to decipher letters that are unnaturally stacked (Figure 6.33).

Tilted words can also disrupt the flow of a piece, unless all type and art are positioned at an angle to the edge of the page. Once commonly used, diagonal type has fallen from popularity because it loses readability as it tilts (Figure 6.34).

Candy-cane, holly-leaf, and stocking letters decorate amateur design efforts at Christmastime and are among several novelty shapes that should

A 6/7 Souvenir Medium, 16 picas justified
It was the best of times, it was the worst of times, it was the age of wisdom, it was the age of foolishness, it was the epoch of belief, it was the epoch of incredulity, it was the season of Light, it was the season of Darkness, it was the spring of hope, it was the winter of despair, we had everything before us, we

B 8/9 Souvenir Medium, 16 picas justified
It was the best of times, it was the worst of times, it was the age of wisdom, it was the age of foolishness, it was the epoch of belief, it was the epoch of incredulity, it was the season of Light, it was the season of Darkness, it was the

C 10/12 Souvenir Medium, 20 picas justified
It was the best of times, it was the worst of times, it was the age of wisdom, it was the age of foolishness, it was the epoch of belief, it was the epoch of incredulity, it was the season of Light, it was the season of Darkness, it was the

D 12/14 Souvenir Medium, 20 picas justified
It was the best of times, it was the worst of times, it was the age of wisdom, it was the age of foolishness, it was the epoch of belief, it was the epoch of incredulity, it was the season of Light, it

E 14/16 Souvenir Medium, 25 picas justified
It was the best of times, it was the worst of times, it was the age of wisdom, it was the age of foolishness, it was the epoch of belief, it was the epoch of incredulity, it was the season of Light, it was the

FIGURE 6.30. Body type smaller than 8 points (A) or larger than 12 points (E) is difficult to read, while sizes in between (B, C, and D) are very readable.

10/12 Souvenir Medium, 11 picas justified
It was the best of times, it was
the worst of times, it was the age
of wisdom, it was the age of
foolishness, it was the epoch of

10/12 Souvenir Medium, 21 picas justified
It was the best of times, it was the worst of times, it was the age
of wisdom, it was the age of foolishness, it was the epoch of
belief, it was the epoch of incredulity, it was the season of
Light, it was the season of Darkness, it was the spring of hope,

10/12 Souvenir Medium, 27 picas justified
It was the best of times, it was the worst of times, it was the age of wisdom, it was
the age of foolishness, it was the epoch of belief, it was the epoch of incredulity,
it was the season of Light, it was the season of Darkness, it was the spring of
hope, it was the winter of despair, we had everything before us, we had nothing

FIGURE 6.31. Column width is too narrow for maximum readability (top) but when extended is comfortable to the eye (middle). Column width any wider in this typeface, size, and weight becomes increasingly difficult to read (bottom).

10/12 Souvenir Medium, 21 picas justified

It was the best of times, it was the worst of times, it was the age
of wisdom, it was the age of foolishness, it was the epoch of
belief, it was the epoch of incredulity, it was the season of
Light, it was the season of Darkness, it was the spring of hope,

10/9 Souvenir Medium, 21 picas justified

It was the best of times, it was the worst of times, it was the age
of wisdom, it was the age of foolishness, it was the epoch of
belief, it was the epoch of incredulity, it was the season of
Light, it was the season of Darkness, it was the spring of hope,

10/9 Souvenir Medium, 16 picas justified

It was the best of times, it was the worst of
times, it was the age of wisdom, it was the age
of foolishness, it was the epoch of belief, it was
the epoch of incredulity, it was the season of

FIGURE 6.32. Leading between lines increases readability (top). Negative leading makes reading difficult (middle), although a narrow column (bottom) aids readability when lines must be closely spaced.

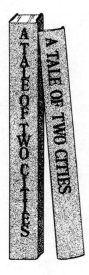

FIGURE 6.33. Stacked letters are harder to read than letters in normal relationship turned on end.

FIGURE 6.34. Staggered or tilted words are not as readable as words arranged in a more natural order in relation to art and page edges.

FIGURE 6.35. Examples of type that is so ornate it should be used sparingly.

most often be left to cookie cutters. They require too much effort to read and, because they are so ornate, detract from the message of a printed piece. Use them sparingly, if at all (Figure 6.35).

Don't mix typefaces or treatment in the same piece. The skilled designer chooses one typeface and may add variety with different weights or occa-

FIGURE 6.36. Inappropriate mixing of surprint and reverse in the same headline (top) and the headline unified by being fully reversed (bottom).

sional use of italic in the same face. Unity is lost when diverse type appears on the same page or within the same piece (Figure 6.13). Mixing surprint and reverse treatments in the same word is also difficult to read. If you want to run a headline into a photograph, print it all black (''surprint'') or all white (''reverse'') but not a combination (Figure 6.36).

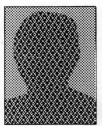

FIGURE 6.37. When a headline running across the gutter must be divided, don't position the break within a word (above) but rather between words (right).

When positioning a headline across the gutter of a spread that is not the center spread, have the fold fall between words. Unless binding is very precise, the baselines will not align exactly from one side of the spread to the other. The reader will be more likely to forgive a slight discrepancy if it doesn't happen in the middle of a word (Figure 6.37).

Poor Photographs. Using no photograph is better than using one of such poor quality that the entire piece suffers for it. A shot that is out of focus, too light to pick up much of the subject, too dark to see any detail, grainy, cluttered, or framed so that the subject is a speck in the distance is hard to look at and detracts from the intended message. If you have to do something with such a photograph (as covered in depth in Chapter 7), see if a custom print of the negative produces better results, run it small to sharpen the focus and diminish graininess, crop it closely, or reproduce the subject in an illustration based on the photograph. Above all, do not put it on the cover or otherwise draw special attention to it (Figure 6.38).

MAKING DESIGN WORK FOR YOU

Designing for Print

FIGURE 6.37. (Continued)

Don't cut away part of a photograph and insert type into the hole ("mortising"). This technique may be effective when another photograph is mortised in, perhaps with a border of white space separating the two shots (see Figure 7.21). However, the cutaway method of mortising is now recognized as both unreadable and a butchering of the art (Figure 6.22).

Photographs cut into odd shapes or tilted helter skelter are difficult to interpret and the bane of many photographers whose work is decimated by amateur editors and designers. Unless you are an experienced designer, crop right-angle corners, shape photographs into rectangles, and position them parallel with the text for the most pleasing presentation (Figure 6.39).

The best reproduction of photographs occurs on white, coated, very opaque paper. High-quality prints will be satisfactory on slightly textured, buff, or other light-colored paper. The result rapidly deteriorates if you combine medium- to dark-colored paper, heavily textured paper, colored ink, and medium to mediocre photographs. When you want a photograph reproduced on colored paper, either start with a white sheet and print on the color

FIGURE 6.38. An example of how reproducing a photograph small sharpens focus and reduces graininess.

FIGURE 6.39. Photographs cropped with right-angle corners (bottom) communicate better than those cropped at irregular angles (top).

in all areas except under the photograph or start with a colored sheet and print white under where photographs will appear (Figure 6.40).

When you want a photograph to communicate to its fullest, avoid special screens. Save these effects (covered later in this chapter) for photographs that are to draw attention or set a mood, rather than to explain (Figure 6.41).

Overuse of Color. "Spot color" means a spot here and a spot there. Overdoing a second color is common among amateurs and is avoided by simple restraint. Use spot color only where absolutely necessary to draw attention, and then stop. Especially avoid colored headlines in a publication or any other use that permits the second color to interrupt or to challenge the primary color.

Take another look if you are considering tinting a photograph or a large block of copy. You may find that the tint adds more emphasis than you want. When in doubt, sketch out the page or spread with colored pens or use tinted overlays to judge how much spot color or tint can be used without disrupting the message.

FIGURE 6.40. The original photograph (above) and a copy of how it looked when printed on colored paper with colored ink (right). Note the absence of highlights, shadows, and fine detail in the latter.

MAKING DESIGN WORK FOR YOU

Overuse of Boxes and Rules. Though they effectively separate elements on a page, boxes and rules may also clutter and segment when used to excess. Also, the heavier the rule and the larger the box, the more the piece will have the look of an amateur production. As just mentioned in regard to spot color, when you are uncertain about how boxes and/or rules will look, sketch out the page or spread. Perhaps a thin rule top and bottom will suffice to set apart a sidebar (rather than a whole box or an ornate border), or maybe a headline only will be less disruptive than either (Figure 6.42).

Tricky Folds. Novelty folds belong in the hands of skilled designers who know when an unusual fold will augment the unique message of a printed piece. The fold should work for the piece you're producing with the information you have to communicate through type and art. A fold that unfurls so that the most important point is lost on the back doesn't work.

Determine how much material you will have to incorporate into the piece, what size it needs to be for mailing and/or filing, and then fold a sheet of paper until the information is presented in its logical sequence. Standard folds are covered in Chapter 15.

FIGURE 6.40. (Continued)

FIGURE 6.41. The circular screen on the photograph at right destroys the fine detail captured in the original above.

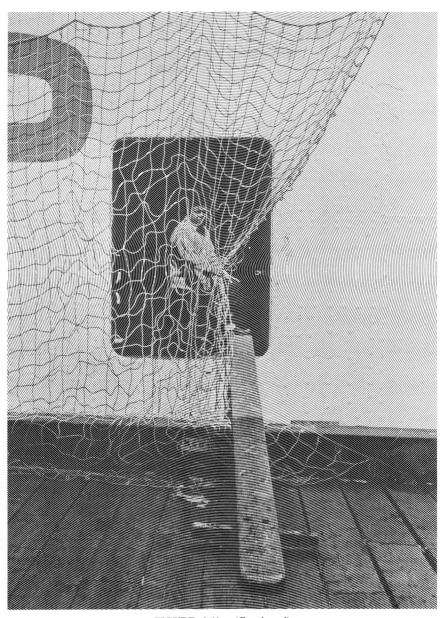

FIGURE 6.41. (Continued)

FIGURE 6.42. Overuse of boxes and rules (above) is not nearly as effective as their limited use (right).

Lorem Ipsum Dolor sit Amet Consectetur Adipscing

adipscing elit, sed diam nonnumy ei aliquam erat volupat. Ut enim ad min suscipit laboris nisi ut aliquip ex ea cor reprehenderit in voluptate velit esse nulla pariatur. At vero eos et accusam delenit aigue duos dolor et molestais tempor sunt in culpa qui officia des harumd dereud facilis est er expedit d optio comque nihil quod a impedit a voluptas assumenda est, omnis dolor debit aut tum rerum necessit atib sae recusand. Itaque earud rerum hic ter asperiore repellat. Hanc ego cum te possing accommodare nost ros quos amice et nebevol, olestias access pote legum odioque civiuda. Et tamen in t ned libiding gen epular religuund on minuiti potius inflammad ut coercend vera ratio bene santos ad iustitiami ae est cond que neg facile efficerd possit veling en liberalitat magis em conven aptissim est ad quiet. Endium caritat p imigent cupidat a natura proficis facile ipsum dolor sit amet, consectetur a incidunt ut labore et dolore magna ali nostrud exercitation ullamcorpor susci Duis autem vel eum irure dolor in consequat, vel illum dolore eu fugiat dignissim qui blandit praesent lupatur occaecat cupidat non provident, simil id est laborum et dolor fugai. Et haru tempor cum soluta nobis eligend optic placeat facer possim omnis es volupta autem quinsud et aur office debit aut t sint et molestia non este recusand. Itaq

od tempor incidunt ut labore et dol m veniami quis nostrud exercitation u odo consequat. Duis autem vel eum in estaie son consequat, vel illum dolor usto odio dignissim qui blandit praese eptur sint occaecat cupidat non provi t mollit anim id est laborum et dolo ct. Nam liber a tempor cum soluta no i id quod maxim placeat facer possin ellend. Temporem autem quinsud et eveniet ut er repudiand sint et molesti ry sapiente delectus au aut prefer en sentntiam, quid est cur verear ne ad paulo ante cum memorite it tum etia er ad augendas cum conscient to facto

tated fidem. Neque hominy infant aut conteud notiner si effecerit, et opes v t. dabut tutungbene volent sib concilia ert cum omning null siy caus peccand ent sine julla inura autend unanc sunt cing elit, sed diam nonnumy eiusm m erat volupat. Ut enim ad minimim ve laboris nisi ut aliquip ex ea commodo ehenderit in voluptate velit esse mo la pariatur. At vero eos et accusam elenit aigue duos dolor et molestais ex por sunt in culpa qui officia deserunt dereud facilis est er expedit distinct. N mque nihil quod a impedit anim id qu ssumenda est, omnis dolor repellend. rerum necessit atib saepe eveniet ut er

tempor incidunt ut labore et dolore ma veniami quis nostrud exercitation ullam commodo consequat. Duis autem vel e esse molestaie son consequat, vel illum accusam et justo odio dignissim qui bla molestais excepteur sint occaecat cupida officia deserunt mollit anim id est labon Lorem ipsum dolor sit amet, consectetu

Quis Nostrud
Incidunt ut Lab

paulo ante cum memorite it tum etia er ad augendas cum conscient to factor tu ne que pecun modut est neque nonor i quas nulla praid im umdnat. Improb pa et dodecendense videantur, Invitat igitu fidem. Neque hominy infant aut inuiste conteud notiner si effecerit, et opes vel dabut tutungbene volent sib conciliant cum omning null siy caus peccand qua explent sine julla inura autend unanc si adipscing elit, sed diam nonnumy eius aliquam erat volupat. Ut enim ad minin suscipit laboris nisi ut aliquip ex ea con reprehenderit in voluptate velit esse mo nulla pariatur. At vero eos et accusam delenit aigue duos dolor et molestais e tempor sunt in culpa qui officia deserur harumd dereud facilis est er expedit dis optio comque nihil quod a impedit anir voluptas assumenda est, omnis dolor r debit aut tum rerum necessit atib saepe recusand. Itaque earud rerum hic tenet asperiore repellat. Hanc ego cum tenet possing accommodare nost ros quos tu amice et nebevol, olestias access potes legum odioque civiuda. Et tamen in bu ned libiding gen epular religuund on cup minuiti potius inflammad ut coercend vera ratio bene santos ad iustitiami aeq est cond que neg facile efficerd possit d veling en liberalitat magis em conveniu aptissim est ad quiet. Endium caritat pr imigent cupidat a natura proficis facile ipsum dolor sit amet, consectetur adips incidunt ut labore et dolore magna aliq nostrud exercitation ullamcorpor susci Duis autem vel eum irure dolor in repre consequat, vel illum dolore eu fugiat n dignissim qui blandit praesent lupatum occaecat cupidat non provident, simil t id est laborum et dolor fugai. Et harum tempor cum soluta nobis eligend optio placeat facer possim omnis es voluptas autem quinsud et aur office debit aut tu sint et molestia non este recusand. Itaq prefer endis dolorib asperiore repellat. ne ad eam non possing accommodare etia ergat. Nos amice et nebevol, olesti

Ut Enim ad Minimum Veniami Diam Nonnumy Eiusmod Tempor

Invitat igitur vera ratio bene santos ad inuiste fact est cond que neg facile eff vel forunag veling en liberalitat ma conciliant et, al is aptissim est ad quie peccand quaerer em imigent cupidat a unanc sunt isti. Lorem ipsum dolor sit eiusmod tempor incidunt ut labore e minimim veniami quis nostrud exercit commodo consequat. Duis autem vel esse molestaie son consequat, vel ill accusam et justo odio dignissim qui b molestais excepteur sint occaecat cupi officia deserunt mollit anim id est labo expedit distinct. Nam liber a tempor c impedit anim id quod maxim placeat f dolor repellend. Temporem autem qu saepe eveniet ut er repudiand sint et tenetury sapiente delectus au aut pre tene sentntiam, quid est cur verear ne paulo ante cum memorite it tum etia e

itiami aequitated fidem. Neque hominy d possit duo conteud notiner si effece em conveniunt. dabut tutungbene ndium caritat praesert cum omning n tura proficis facile explent sine julla in et, consectetur adipscing elit, sed diam olore magna aliquam erat volupat. U ullamcorpor suscipit laboris nisi ut ali m irure dolor in reprehenderit in volu dolore eu fugiat nulla pariatur. At v dit praesent lupatum delenit aigue du non provident, simil tempor sunt in n et dolor fugai. Et harum dereud fa soluta nobis eligend optio comque ni possim omnis es voluptas assumenda d et aur office debit aut tum rerum n estia non este recusand. Itaque earud endis dolorib asperiore repellat. Han eam non possing accommodare nost r Nos amice et nebevol, olestias access

FIGURE 6.42. (Continued)

FIGURE 6.43. Sample formats for flyers.

FIGURE 6.43. (Continued)

Misnumbered Pages. Even numbers should always be on the left ("verso") and odd numbers on the right ("recto"). The front cover is never numbered, although the back cover may be if cover paper is the same as that used inside, for example, on a newsletter or newspaper.

Sample Formats

If you are not trained as a designer but must do your own design work, you may find the sample formats in Figures 6.43 through 6.47 helpful. Each applies the design principles covered in this chapter and is readily adapted to various subjects. Also, one sample format might be used for related but different pieces (such as the presentation folder format carried through in a series of brochures to go in the folder).

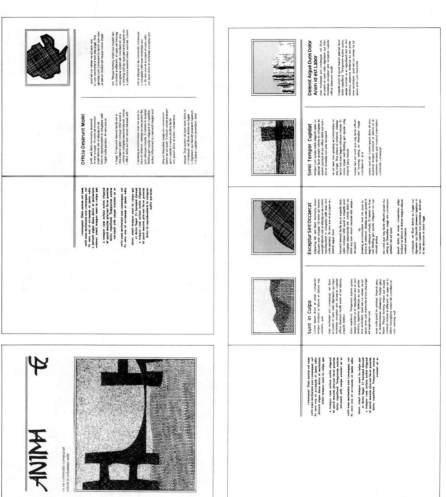

FIGURE 6.44. A sample format for a brochure.

FIGURE 6.45. A sample format for a newsletter.

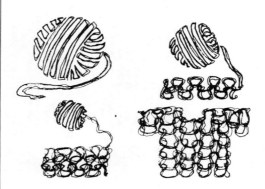

FIGURE 6.46. A sample format for a poster.

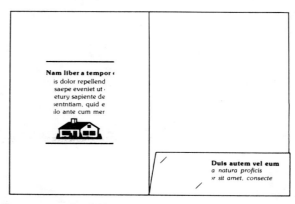

FIGURE 6.47. A sample format for a presentation folder, including a die-cut for insertion of a business card in the pocket flap.

DESIGN TALENT

The all-encompassing title of "designer" may be used by someone working in print and/or other media (such as signage, product design, or display). A "graphic designer" has specific knowledge of and experience with designing for print. For the best result on your print project, require this design knowledge and experience, whether or not the designer works in other media as well.

Graphic design talent may be found in-house or outside. You may do the design yourself if the job is within the scope of your graphic-design talent. This option provides direct control over product, cost, and schedule. Another in-house source might be an employee or association member with design skills. Design may also be provided in-house by borrowing from other printed pieces, as covered earlier in this chapter. Whether tapping in-house design talent or an outside graphic designer, apply the following advice on finding, evaluating, and working with a designer.

FINDING A DESIGNER

Outside graphic-design services are available from a freelance designer, a design studio, or an advertising or public-relations agency with designers on staff. Your choice depends on where you can find the range of skills you need, at a cost in line with your budget. Consult colleagues, suppliers, professional associations, and even the Yellow Pages to find out who is available and potentially capable of doing your design work. Also contact any schools in your area offering design curricula for names of graduates or advanced students who might be able to accomplish the assignment.

Freelance Design

A freelance designer works independently and can negotiate fee, schedule, and scope of work with you to your mutual satisfaction. In addition to actual design work, a freelance designer may direct photography, do paste-up, and coordinate production. Illustration skills may also be offered. Should you need related services (such as copywriting and ad placement), you will have to shop for them elsewhere.

Studio Design

A design studio specializes in design and related skills offered on a group basis. The arrangement may be either a loose cooperative of designers, illustrators, and related suppliers or a more formal alliance.

In the first instance, you select the designer you want who, in turn, goes to other members of the group as necessary for assistance in completing your project, including common clerical, accounting, and other support services. In the latter instance, you buy the skills of the group, and a designer is assigned to your project. With either arrangement, a studio offers a range of skills from specialists who work together every day. If you need copywriting and other related services, however, you will have to find them yourself or let the group subcontract for them.

Agency Design

An advertising or public-relations agency offers the advantage of one-stop shopping if you require copywriting, marketing advice, production, and/or ad placement services in addition to design. This resource may also be tapped for design work only, although agency size and cost may discourage projects not requiring a package of services. A job that is too small and/or specialized for an agency may not receive the attention that you think it deserves, and because of overhead costs, the fee may be comparatively high.

HOW DESIGNERS WORK AND CHARGE

A design project progresses in a series of steps, depending on complexity. After talking with you about an assignment (and possibly with others doing different phases of the job), a designer will do thumbnails to present you with a rough idea about formats and layout that might meet your need. Based on your reaction, more thumbnails might be done or one of the more likely designs developed full size in the form of a comprehensive dummy. You and the designer then agree on that design or how it should be revised, completing the actual design phase of a project. The designer next specifies type and may do illustrations, direct photography, paste up the piece, and/or supervise production, depending on the assistance you need, the designer's skills and requirements, and your budget.

Fees and Expenses

A freelance designer may charge by the project, by the hour, or a combination according to what steps are involved. For example, a flat fee may cover developing a format, while paste-up is billed by the hour after the design has defined its complexity. Overhead for a freelancer is comparatively low, allowing a possible fee advantage over design services from a studio or agency.

A design studio will charge either a standard fee regardless of the particular designer who does your work or a fee that varies by designer. The fee is set by each group member (depending on level of experience) or by the group. Either way, it takes into account studio overhead.

Agencies generally charge a monthly retainer on top of per-project fees, which entitles clients to call on them at any time for assistance. A retainer

DESIGN TALENT

also usually includes a non-competing clause which assures you that, while you are a client, the agency will not take on work from your competitors. Per-project costs are based on an agency rate for design services and on overhead expenses higher than those of a freelancer and possibly of a studio.

A freelancer, studio, or agency will charge for related services needed to complete an assignment. Clerical assistance and other overhead costs are incorporated into hourly fee, while the cost of subcontracted services that the designer supervises are additional. The designer's bill will probably include a markup on subcontracts.

Expenses may also be marked up. They might include travel, long-distance phone calls, and photocopying. Customary supplies are considered part of overhead cost and covered by the fee. However, if your project requires special materials, expect your designer to ask that their cost be reimbursed.

Use Rights

The copyright to a design—to a certain arrangement of elements on paper—is not as specific as the copyright to the accompanying text because design deals more with ideas than with words or other tangible elements. How readily you can borrow a previous design depends on the rights sold and retained at the time the original design was developed and on the extent of duplication in the new piece.

A design that was conceived and executed by an employee for an employer is "work-for-hire" and can be duplicated or revised by that employer for another piece unless the designer specifically retained the right to its subsequent use. A design conceived and executed for hire by an outside designer may also be used again if copyright was not retained by the designer.

The copyright for any design sold on a one-time-only basis reverts to the designer and may not be reused or substantially adapted without written permission. For detailed information on design copyright, refer to Appendix A for practices of the graphic-arts industry and to the several books available on the subject.

EVALUATING A DESIGNER'S PORTFOLIO

Design talent is presented to a prospective client by example: A designer shows you past projects and talks about each one in detail. When you contact a freelance designer, a design studio, or an agency, be prepared to outline your project so that the designer can put together samples pertinent to your need. Ask to receive a résumé or qualification summary ahead of your meeting (if possible) so you will have an idea of training and experience before you see the designer's work. A statement of current hourly fees for services offered and a list of past clients (with reference details) should also be provided, either before or during the portfolio showing.

While you definitely want to see a portfolio, recognize that the designer will select only successes to show you, the cream of past projects. And it will all be beautiful at first glance. A profitable portfolio showing, however, is a combination of looking, listening, and asking. A skilled designer will put together samples in line with your needs, then tell you about each one as you look at it. Solicit any details not volunteered by asking questions about particular samples or about overall capability and interest. Out of a portfolio showing should come sufficient information for you to evaluate the designer in at least the following areas.

Range of Skills

A graphic designer may be able to conceive and execute a design, illustrate, direct photography, and coordinate production—or do only the design work itself. Determine ability to do each of the tasks you need done as well as interest in doing them.

Experienced designers know that weak photographs or cheap-and-dirty printing can ruin their designs and their chances for repeat business. Therefore, an accomplished designer in demand may not do your project if all you want is the basic design. Full execution of that design, including recommending and approving the work of other suppliers, may be a minimum requirement to assure that execution retains the integrity of the design. Another designer may agree to do design and paste-up only. How much control a given designer requires is more than a matter of ego; it's a measure of that designer's skills and quality standards and of the designer's respect for your ability to manage the project.

A portfolio should present evidence of a creative design developed in an understandable comp and successfully translated into a printed piece. Ask for and expect to see the actual thumbnails, comp, and paste-up sequence that led up to a brochure, folder, or other project. The designer may not have each of these steps readily at hand but might be able to get them from past clients with ample notice.

Style

Look for a style that is appropriate for the work for which you expect to need design assistance. A designer who is obviously skilled may not be your choice because the flamboyant style on which the designer concentrates isn't suited to your conservative organization. Preferring one style over another is a subjective choice on the part of a designer's client—but an important one if a project is to be completed successfully.

Flexibility

A designer should be able to work "from scratch" with wide-open selection of colors, art, paper, and so on, or be able to work when certain elements are given. If the designer doesn't volunteer information on design, budget, or schedule restrictions of a particular sample, ask, because your project may well have some requirement for flexibility. Many times, a logo or paper color is a given; even if it's not what the designer would have chosen, it must be accommodated. Look for a designer who can coordinate a new piece with existing pieces, achieve a creative result using your president's favorite color, or conceive a satisfactory design that can be executed under tight time constraints.

Effectiveness

To some extent, samples and comments will indicate whether or not the designer can produce effective results. A check of references will provide more indication. A piece should meet its intended objectives and be in line with all the considerations suggested earlier in this chapter. A revealing question to ask about any portfolio sample of particular interest is what the designer might change if the project were done again. On virtually every assignment, a skilled designer will see in retrospect how the result might have been better or at least different. The question invites the designer to step away from any actual limitations and judge how effectiveness might have been improved. It also gives you additional information about the designer's quality standards.

Cost

The cost of design depends on the project and its complexity. If a portfolio includes a piece almost exactly like what you have in mind, its cost may be similar, at least for development of the initial design. In addition to knowing how a designer charges and a representative charge, ask how closely in line to the estimate a particular project was. Experienced designers are also experienced estimators and should be able to finish a job within or close to estimate. Also include cost questions when you check references.

Schedule

As with cost, select a comparable project from the portfolio and ask how long the designer worked on it. It might have included thumbnails, a comp, or more than one comp, all of which are time consuming. A sample similar to what you have in mind will give you and the designer a basis for discussing schedule. Also talk about how available the designer is for another such project—and how interested. A designer who appears to be very talented must be able to get to your work in time to meet your deadline.

CHECKING DESIGNER REFERENCES

Follow a positive review of a portfolio with a check of a designer's references. Ask for names of previous clients with projects similar to what you expect to have and call at least two to get an appraisal of the quality of the designer's work with assignments and with clients.

Did the designer listen or come in with pre-conceived ideas? Were suggested designs in line with client requirements or were substantial revisions necessary? Was the work finished on time or did the design hold up production? Did the designer cooperate well with everyone involved in the project or alienate staff and/or other suppliers? Was the job billed as estimated or was it surprisingly high? Did the designer understand print production or did the design create difficulties in later stages of production?

Keep in mind that judgments of previous clients are subjective. Evaluation of design or other artistic endeavors is quite contingent on the particular combination of personalities or other circumstances that applied to a given project.

WORKING WITH A DESIGNER

If you are impressed by what you see and hear (and if a check reveals that past clients have been also), you have found yourself a graphic designer. Fully communicate your requirements, including exactly what you are asking the designer to do, by when, and within what budgetary limits. Business sense doesn't necessarily go hand in hand with artistic talent; so put things in writing clearly and fully.

Assignment Details

Start with a letter summarizing the project, fee, responsibilities, and deadlines; then give the designer a full list of objectives and information on all other design considerations covered earlier in this chapter. Provide thorough feedback on design concepts and thumbnails before proceeding to a comp or final artwork. You have a right to expect design concepts, thumbnails, and a comp that are responsive to the project as outlined. Keep a close watch on deadlines, consult the designer about last-minute changes, and allow as much creative latitude as the project permits.

Wrap-Up

When the piece is finished, provide samples for the designer's portfolio, payment within 30 days, and as objective a reference as you can to prospective clients. A credit line in the finished piece is not required by a designer but, when format allows, is a courteous tribute for exceptional work, for a major job, or for design services donated to a charitable project.

CHAPTER SEVEN

Selecting and Using Art

PLANNING ART   167

    Planning Considerations, 167
        Objectives, 167
        Appropriateness, 167
        Visual Impact, 168
        Existing Art, 168
        Cost and Schedule, 169
        Type of Press, 169
        Available Skills, 169
    Illustration Versus Photography, 170
        Using Illustration, 170
        Using Photography, 170
    The Story Conference, 170

ART TECHNIQUES FOR PRINT   171

    Illustration Techniques, 171
        Line Drawings, 171
        Continuous-Tone Drawings, 175
        Other Illustrations, 176
    Photographic Techniques, 186
        Recognizing Quality Photography, 186

      Photographing Currency, Stamps, or Paintings, 191
      Photographic Papers, 191
      Darkroom and Retouching Techniques, 192
  Printing Papers and Art, 196
      Brightness, 196
      Surface, 196

MAKING ART WORK FOR YOU   197

  Success With Illustrations and Photographs, 197
      Placement, 197
      Sizing, 199
      Working With Problem Photographs, 203
      Special Treatment of Photographs, 204
      Handling Art, 210
  Success With Transfer Art or Type, 211
      Rub-Off Art or Type, 211
      Lift-Off Art or Type, 212
      Shopping for Transfer Art or Type, 212

Though a printed piece may be produced with type only, you will more often supplement or complement type with a drawing, photograph, chart, or other art element. Even within budget and space limitations, a well-planned and executed flyer, brochure, newsletter, or poster can take advantage of the eye-catching, informative quality that art lends.

When applied to print production, "art" is any illustration or photograph used to add interest and/or information to a piece. "Illustration" refers to any line or continuous-tone drawing, painting, graph, chart, table, or type character used as art. A "photograph" is a continuous-tone image reproduced from a black-and-white or color negative or from a color transparency.

"Illustration" and "graphic" are sometimes used as synonyms for "art," as "art" is for "artwork" or "mechanical art." The definitions chosen for purposes of this book are commonly applied as presented here, and the distinctions among them are maintained to avoid confusion caused by interchanging terms.

This chapter explores in detail how to plan the use of art, illustration and photographic techniques, and ways to make art work for you. For additional information, see Chapter 6 on design, Chapter 8 on art sources, and Chapter 11 on mechanical artwork.

PLANNING ART

When well planned, a printed piece uses a drawing, a group of photographs, or a series of graphs for a reason. Having art already available in some form is not sufficient reason to use it; likewise, not having any art available is an insufficient reason not to develop new art to help communicate the message.

Availability is only one of several factors to be considered as you plan art for a printed piece. In addition to general considerations, you must decide specifically between an illustration or a photograph as the better choice for the particular job and thoroughly meld editorial, design, and art considerations into a unified whole.

PLANNING CONSIDERATIONS

As with copy and design, several general considerations will influence your use and choice of art. Review the following factors before making final decisions about what art (if any) to include in a printed piece and how.

Objectives

Art is incorporated into a print project for one or more reasons. Why you use it influences how it is used and to what extent. Look again at the objectives of the piece (to sell, to inform, to recognize—whatever specific ones apply) and then consider how art might augment them. A brochure intended to sell condominium units won't meet that objective effectively without showing floor plans. A quarterly report to shareholders won't be complete without financial tables comparing quarters. Full recognition won't be given employees who have earned promotions without including their photographs with a newspaper article.

In support of the overall objectives of a printed piece, an illustration or photograph should have its own reason for being used: to draw attention, to explain, or to establish a mood. Art might show what someone or something looks like; expand, clarify, or reinforce information in the text; provide an interesting break from the text; summarize (i.e., a table); show relationships (i.e., a graph or chart); communicate its own independent message (i.e., a photo essay or cartoon); or set the tone of a piece.

Use art when it contributes in one of these ways to the objectives of a printed piece. Leave it out when it doesn't. Art that has as its only purpose the filling of space detracts from the message of the piece and adds unnecessary expense to production.

Appropriateness

The art you select for a particular piece should be appropriate for the subject being presented and the style of its presentation. For example, a printed announcement of a new principal in an engineering firm should include a photograph of the person, not a sketch of one of the firm's past projects. If copy and design establish a formal style, any accompanying photograph should be equally formal.

One illustration for a brochure should be appropriate in relation to all others in the same piece. They all should show a similar amount of detail and be executed in a similar style. Likewise, one photograph should be similar to all others if the message it is to convey is similar. For example, photographs of members of the board for an annual report should show them in comparable settings, all head and shoulders at the office, all full length on the street, or whatever the design specifies, rather than present a hodgepodge of treatments and backgrounds. Mixing black-and-white and color photographs or illustrations and photographs can be tricky and, if poorly done, make the odd illustration or photograph stand out as inappropriate. In the hands of a skilled designer, art media can be mixed by displaying them consistently and of a size in line with their relative importance. For instance, the brochure on condominiums mentioned earlier might include for each type of unit a small floor plan, paired with larger color photographs of completed display units. Or it might include a grouping of small one-color sketches of the surrounding neighborhood and view, opening to a grouping of larger color shots of unit interiors.

Illustrations and photographs should be appropriate for the image you want to present to your particular audience. An out-of-focus, poorly framed photograph is not appropriate for a corporate magazine, nor is a simplified drawing appropriate for technical readers interested in seeing full detail.

Visual Impact

An illustration or photograph that augments objectives of and is appropriate for a printed piece should also be evaluated for its visual impact if it is to be more than utilitarian. While shots of three new employees for a newsletter column may not need to have much visual impact, a shot of the new president to accompany an extensive story should have. Composition, focus, and lighting contribute to the artistic value of a photograph and thus to the impact it can have as a cover photo or inside art in a printed piece. Visual impact of the whole page or spread must also be considered so that an illustration forms a visually pleasing relationship with others in a grouping or with adjacent text, legend, or headline.

Existing Art

The art that you use may already be fully or partially selected for you. A logo may be required on every marketing piece you produce and have to be of a specific color and size. A staid, formal portrait of the executive director may be a given you'll have to accept, despite your desire for a more casual look.

PLANNING ART

Poor-quality art may also be a given, especially when it's contributed from afar or at the last minute so that you have no opportunity to make substitutions.

In other situations, existing art may be just what you want but its use limited by legal restrictions. A photograph for which you purchased one-time-only rights cannot legally be used again, unless you pay for additional use. A similar restriction may also apply to an illustration, depending on what rights were purchased at the time of initial use. As detailed later, art "borrowed" from a publication or other outside source may be free but require written permission to reproduce. Review copyright limitations and purchase agreements before you assume that you can reuse existing art.

Cost and Schedule

Like other elements of print production, the selection and use of art has implications for the cost of a piece and how long it takes to produce. Photographs on hand will cost less to use again than shooting new ones. Black-and-white shots cost a fraction of what you will pay for color, both in initial processing and printing. Having illustrations drawn may or may not be as expensive as having photographs taken, depending on their number and complexity.

When budget and schedule are tight, try to work with what you already have, restrict art to black-and-white photographs or simple one- or two-color illustrations, and avoid special production techniques that require days to execute. (Review previous chapters to help gauge the cost and scheduling implications of art uses.)

Type of Press

The press on which your project will be run is another consideration as you decide what art to use and how. A mediocre photograph won't stand out in a piece run on a press that produces work of only mediocre quality, while another more exacting press may pick up every flaw. One may be able to print a fine-lined illustration on textured paper, while another will drop some of the lines or print them faintly. Refer to Figure 6.3 and consult with your printer about any such limitations of the press on which your job will be run.

Available Skills

As with the design of a printed piece, you may have a limited or limitless choice of art and production techniques, depending on the skills you have yourself or that are available to you. If you don't consider yourself an illustrator or a photographer, you will have to restrict your choice to simple art that you can produce, hire outside talent, or buy stock or camera-ready art. In the latter cases, your range of choices will depend on what outside talent is available and what stock or non-copyrighted art you can buy or borrow. As detailed in Chapter 8, plan ahead by reviewing portfolios or catalogs so that you know where you might get art when a need arises.

ILLUSTRATION VERSUS PHOTOGRAPHY

The art you select for a print project depends on the several considerations just itemized as well as on the difference between illustration and photography. Early print production relied exclusively on illustrations; but with the advent and refining of photography and photo reproduction and with the development of additional illustration media, art possibilities have proliferated. Sometimes one choice is definitely more effective than another; other times, the decision isn't as clear-cut.

Using Illustration

An illustration is an effective choice when an event, person, or object that is fictitious must be shown; people can be drawn exactly as described in the text, and the setting constructed as the author intended. The versatility of illustration is valuable when the event, person, or object is inaccessible; a drawing can depict an historic event in the absence of photography and portray scenes beyond the camera's range (such as details of a microscopic portion of a cell or jurors in a courtroom where cameras aren't allowed). Because of the artist's full control over an illustration, it may also do a better job of setting a mood.

In addition, an illustration can substitute for an unusable photograph—one that is too dark, out of focus, or damaged but distinct enough from which to draw. A graph, chart, or table can summarize and simplify information in a way unmatched by photographs or columns of type.

Using Photography

A photograph is unsurpassed for showing actual events (either current or historic), people, or objects. Because of this quality, a group of photographs can also be used as a strong communicator. A "photo story" reports facts through related photographs laid out with minimal copy (such as showing the construction sequence of a new corporate headquarters). A "photo essay" is similar, except that it expresses an idea, as opposed to hard facts (such as showing the human impact of a destructive flood through shots of distressed people).

THE STORY CONFERENCE

A "story conference" is a planning meeting that fosters orderly consideration of all these factors by bringing together editorial, design, and art talent working on a printed piece. Even if the project isn't extensive enough to require the skills of other staff people or outside artists, you should evaluate the relationship of art to copy and design and the selection of art.

The raw materials of a story conference are a list of objectives for the piece, an outline of how the copy might be developed, ideas for art, budget and schedule information, and any other available details (such as number of pages or panels). Agreement on a working title (or final title, if possible) will suggest the tone and content of art, especially of the dominant illustration or photograph. Agreement on how the copy might be divided into sections will help allocate space and indicate art for each section. Through a process of questioning, clarifying, and brainstorming, copy, design, and art can become a unified presentation, before words are indelibly on paper or images on film.

The story conference may achieve its goal with one session, or participants may get a project started, then reconvene to look over preliminary copy, layout, and sketches or contact sheets. The more these elements can be coordinated, the better the finished product.

ART TECHNIQUES FOR PRINT

An ordinary illustration or photograph can become a lively, interesting art element in your printed piece with attention to how the illustration is drawn or constructed, the qualities of different photographic papers, darkroom techniques, and the type of paper on which the project is printed. When you manage art from its inception, you can control each of these decisions. Even with art that is presented to you as a finished illustration or photograph, you can still have some influence on how well it will reproduce on the press.

Available media and supplies are continually changing with new developments in graphic technology. To keep up, periodically check the shelves of art-supply dealers, review trade publications, and talk with graphic artists.

ILLUSTRATION TECHNIQUES

An illustration for print production may be rendered as line art (black and white only) or as continuous-tone art (black, white, and shades of gray or full color). Within these two categories, various techniques are used to achieve different effects in drawings, charts, graphs, and other illustrations. The effect of a drawing is further altered by the surface on which it is done.

The following information will acquaint you with illustration media and their application so that you can make an educated selection for your print project. For details on how to execute the various types of drawings mentioned, consult the many specialized books available on that subject.

Line Drawings

A line drawing consists of black and white areas with no shades of gray. It is sized to fit the space it is to occupy in your newsletter, brochure, or other project prepared for the press the same as text, headlines, rules, and other line work.

Pen or Brush and Ink. An ink drawing is probably the most common one for which you will have a need. It is drawn using waterproof ink and either a drawing pen with a variety of nibs or small brushes. A pen produces crisp lines, while a brush stroke is irregular. A pen is especially versatile because the resulting drawing can range from very simple to quite detailed. By regulating line spacing and width or by stippling, the artist can create an illusion of gray shading (Figure 7.1).

Scratchboard. A type of ink drawing, this illustration starts with scratchboard paper which has a heavy clay coating. The entire surface is inked and then fine lines are scratched through to expose the white paper. A scratchboard drawing is high in contrast with stark detail (Figure 7.1).

FIGURE 7.1. Examples of line drawings: pen-and-ink shaded by stippling and by crosshatching (top); brush-and-ink, charcoal, and scratchboard (bottom).

ART TECHNIQUES FOR PRINT

Charcoal. This medium can be used to produce either line or continuous-tone art. For a line drawing, the black charcoal is left at its full value, instead of being rubbed to create shaded gray areas. It produces an uneven, mottled line that is especially influenced by the surface on which it is used (Figure 7.1).

Engraving. The earliest print illustrations were engravings cut or etched into wood and then later etched in copper or steel. Because of the high cost of such work today, true engraving is rarely used. Very-high-quality invitations, announcements, or business cards may still be engraved for especially fine lines. A copy of an old engraving from a book no longer protected by copyright, however, may be quite appropriate for an article on a city's history, for example, or a poster for a craft exhibit (Figure 7.2).

Shadings. A shading sheet (such as Zip-a-Tone) is usually black designs on a transparent film that adheres to the original drawing. It is placed over the area to be shaded on an illustration and the excess cut away. Ben Day sheets are available in scores of patterns and are used to shade areas directly on the negative from which a printing plate is made. Whether done on art or

FIGURE 7.2. Examples of art no longer protected by copyright. Detail from Albrecht Dürer woodcuts (1471–1528).

FIGURE 7.3. An example of how a simple line drawing can be enhanced with purchased shadings.

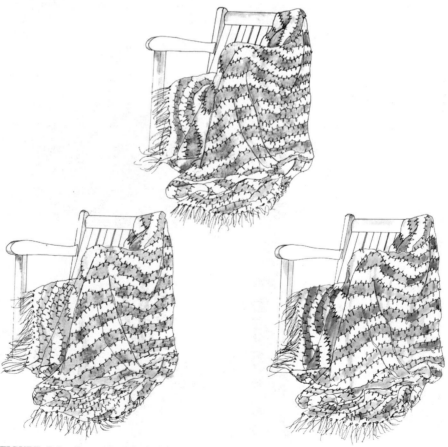

FIGURE 7.4. Examples of continuous-tone drawings: charcoal (top); ink wash and markers (bottom).

ART TECHNIQUES FOR PRINT 175

negative, shadings can add interest to an otherwise simple line drawing and may be especially effective to personalize purchased camera-ready art (as in Figure 7.3) or to designate different areas on a bar graph or map (as in Figure 7.6).

Continuous-Tone Drawings

A continuous-tone drawing consists of black, white, and shades of gray or full color. It is treated separately from text, headlines, and other line work during image assembly.

Pencil. A graphite pencil can create many shades of gray, ranging from dense black to almost white. A continuous-tone drawing may also be created by rubbing black charcoal-pencil lines into shades of gray (Figure 7.4). A pencil line is rough and mottled (unlike the smooth, even line of pen and ink) and is easily varied in width. The resulting effect is very dependent on the paper used for a pencil drawing (Figure 7.5).

Ink Wash. A basic ink drawing may be augmented by an ink wash to create shades of gray, or the drawing may be done entirely in wash using a water-soluble ink. The combination produces a drawing with subtle shading as well as crispness. Wash alone produces less-defined lines similar to a watercolor (Figure 7.4).

Marker. Though commonly used for quick sketches or rough layouts, markers may also be used to execute finished art. They are available in many

FIGURE 7.5. Different effects achieved by executing the same pencil illustration on smooth illustration board (left), canvas (middle), and watercolor paper (right).

colors, shades of gray, widths, and tip shapes. A marker produces flat areas of ink of a controlled width, depending on the angle at which it is held. The effect is also influenced by the finish of the paper on which markers are used (Figure 7.4).

Color Methods. Colored markers and the colored equivalents of charcoal and washes (pastels and transparent watercolors) are common media for producing colored continuous-tone art. Opaque art may be rendered with oils, acrylics, or opaque watercolors, and colored media combined with black to produce additional options. Here again, hardness and texture of the surface to which color is applied affect the outcome.

Other Illustrations

As pointed out earlier, a graph, chart, table, or map may be the ideal art to summarize financial figures, scientific data, or other detailed information. These and other forms of illustration are useful variations on drawings or photographs.

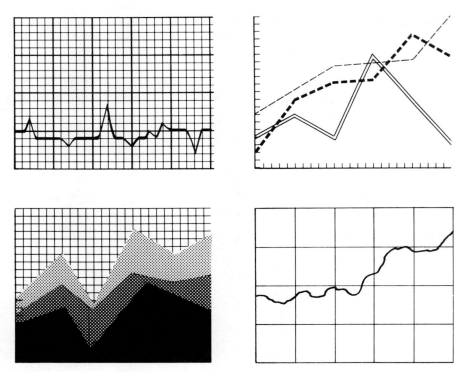

FIGURE 7.6. Examples of line graphs: full grid and minimal grid (top); mountain graph and curve graph (bottom).

ART TECHNIQUES FOR PRINT

Graphs and Charts. A "graph" is a visual representation of changes in one variable in relation to others (such as temperature in relation to time of day). A "chart" shows values and quantities (such as expenditure categories for income). The difference is often academic, although a scientific or technical audience may expect specific graphs instead of more general charts.

To be an effective communicator, a graph or chart must be a combination of detail and simplicity. Limit the variables to no more than three (such as sales volume by month or sales volume by month by office). Any more detail will confuse instead of clarify. Pare background grids or other detail to a minimum; a graph or chart should communicate most or all of its information on sight, rather than require close scrutiny.

A "line graph" connects exact points on a grid with a line. Figure 7.6 presents examples. If widely spaced or distinct in color or style, lines for changes in more than one variable can be shown on the same graph. In such a situation, eliminate all or most of the grid for the sake of simplicity. Another version of a line graph is sometimes called a "surface chart" or "mountain graph"; gaps between the lines are filled in with screens or color to make the graph resemble the peaks and valleys of a mountain range. A "curve graph" or "curve chart" is a third variation of a line graph. A curved rather than straight line shows changes of one variable in relation to others.

The most simple classification of charts divides them into pie charts and bar charts. A "pie chart" compares numerical values by dividing a whole into parts. Figure 7.7 shows a simple pie chart cut into labeled wedges and

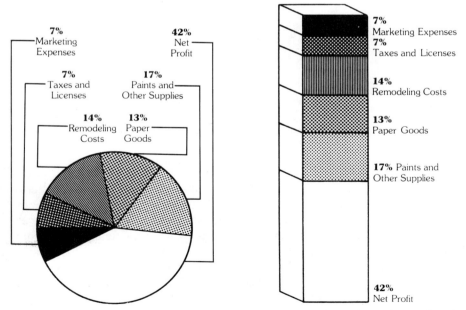

FIGURE 7.7. The same information presented in a pie and a column chart using purchased shadings.

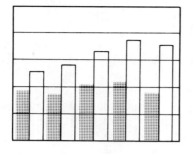
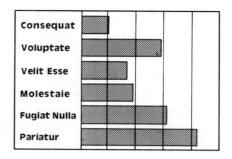
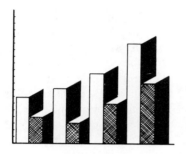
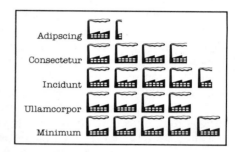

FIGURE 7.8. Examples of vertical and horizontal bar charts.

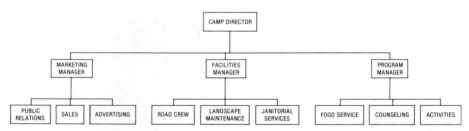

FIGURE 7.9. An example of an organizational chart.

the same relationships depicted in a vertical shape (sometimes called a "column chart"). Pie charts can be more intricate, using enlarged currency or other symbols appropriate for your subject and audience.

A "bar chart" compares numerical values by using vertical or horizontal bars. Figure 7.8 presents simple versions of each. A vertical bar chart is also sometimes considered a column chart. A bar chart may compare two variables side by side or within the same bar over time and be given dimension. It may also use symbols instead of bars for an even more graphic presentation.

Two types of charts are really diagrams (used to show relationship and sequence): organization charts and flow charts. Though commonly referred to as charts, they depict a sequence, rather than the comparison shown by a

ART TECHNIQUES FOR PRINT 179

Oil Sales in Thousands of Barrels 1980–1984

FIGURE 7.10. A pictograph in which the art is used as background.

true chart. They are, however, art elements, just as are the drawn diagrams in this book that show the sequence of typesetting or press operation. Figure 7.9 is an example of a simple organization chart and Figure 9.1 is a flowchart.

Pictographs. A graph that incorporates other art is called a "pictograph." It may be constructed using a line drawing or a photograph, either as the graph itself or as background (as shown in Figure 7.10). The pictorial part of the graph should be appropriate to the subject matter and not obscure the data presented by the graph.

FIGURE 7.11. Examples of how oversized initials and a symbol can be used as art.

Type as Art. Type symbols, letters, and words can be another form of art, either solely as eye catchers or also as communicators. They are particularly valuable when the budget or the schedule won't allow other art or when the subject doesn't suggest any possibilities for drawings or photographs. The examples in Figures 7.11, 7.12, and 7.13 use symbols, letters, and words readily available from a typographer or from transfer type and show how enlarged symbols, oversized initial letters, and creative placement of quotations, headlines, and subheads can be used as art.

The wide assortment of transfer type now available offers many choices when type is to be the only art. A word or short headline in a novelty face in keeping with your subject can result in an interesting page or spread, without the use of other illustrations or photographs (see Figures 6.25 and 10.8).

Another creative use of type is having it set in a shape (Figure 7.14). Advanced typesetting equipment can construct curves or slants in response to detailed specifications. Otherwise, a skilled mechanical artist can arrange type word by word or line by line to form the desired shape. Make sure your

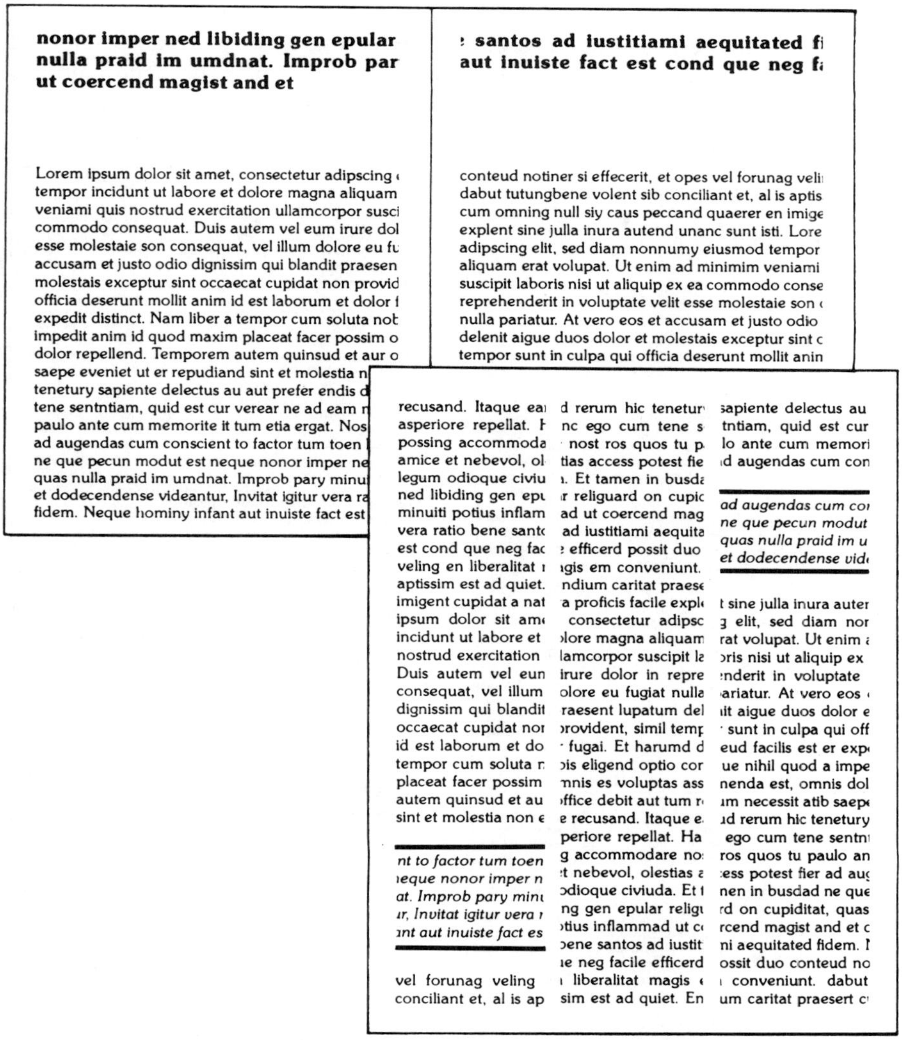

nonor imper ned libiding gen epular nulla praid im umdnat. Improb par ut coercend magist and et

Lorem ipsum dolor sit amet, consectetur adipscing
tempor incidunt ut labore et dolore magna aliquam
veniami quis nostrud exercitation ullamcorpor susci
commodo consequat. Duis autem vel eum irure dol
esse molestaie son consequat, vel illum dolore eu fu
accusam et justo odio dignissim qui blandit praesen
molestais exceptur sint occaecat cupidat non provid
officia deserunt mollit anim id est laborum et dolor f
expedit distinct. Nam liber a tempor cum soluta nob
impedit anim id quod maxim placeat facer possim o
dolor repellend. Temporem autem quinsud et aur o
saepe eveniet ut er repudiand sint et molestia n
tenetury sapiente delectus au aut prefer endis d
tene sentntiam, quid est cur verear ne ad eam n
paulo ante cum memorite it tum etia ergat. Nos
ad augendas cum conscient to factor tum toen
ne que pecun modut est neque nonor imper ne
quas nulla praid im umdnat. Improb pary minu
et dodecendense videantur, Invitat igitur vera ra
fidem. Neque hominy infant aut inuiste fact est

santos ad iustitiami aequitated fi aut inuiste fact est cond que neg f

conteud notiner si effecerit, et opes vel forunag veli
dabut tutungbene volent sib conciliant et, al is aptis
cum omning null siy caus peccand quaerer en imige
explent sine julla inura autend unanc sunt isti. Lore
adipscing elit, sed diam nonnumy eiusmod tempor
aliquam erat volupat. Ut enim ad minimim veniami
suscipit laboris nisi ut aliquip ex ea commodo conse
reprehenderit in voluptate velit esse molestaie son
nulla pariatur. At vero eos et accusam et justo odio
delenit aigue duos dolor et molestais exceptur sint c
tempor sunt in culpa qui officia deserunt mollit anin

nt to factor tum toen
neque nonor imper n
at. Improb pary mini
ir, Invitat igitur vera
int aut inuiste fact es

ad augendas cum co
ne que pecun modut
quas nulla praid im u
et dodecendense vid

FIGURE 7.12. Examples of how placement of quotes in blocks of type can be used as art.

budget can support construction of a shape because, however executed, the process is exacting and time consuming.

Tables, Maps, and Other Art. A table is a summary of information in a concise, graphic form distinct from the text. Though it may be set in the same type style, face, size, and weight as the text, it is distinguished by rules, leaders, vertical columns, additional color, or screening.

Refine information for a table to its minimum and allow enough vertical and horizontal space between categories, figures, and words so that it is readable. (Less space is needed between elements separated by rules.) Using bold type or color for category headings and medium type for listings under each heading also adds to clarity and visual interest. Figure 7.15 is a table using bold headings, horizontal rules, and alternate screened columns to enhance understanding as well as the table's artistic contribution.

FIGURE 7.13. Examples of how creative placement can turn headlines and subheads into art

ART TECHNIQUES FOR PRINT

A map may be a straightforward illustration with lines, numbers, and names or a combination of geographic information and pictorial interest. Figure 7.16 presents different approaches. If a map is to be used rather than enjoyed as art alone, it should be accurate and as readable as possible, despite its detail. A portion requiring considerable detail should be lifted out and enlarged as a separate illustration or as an inset.

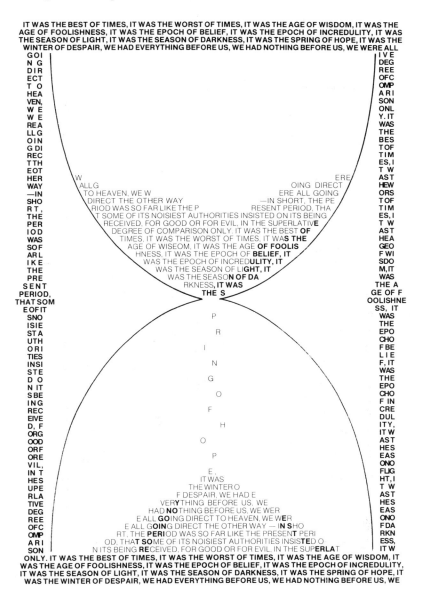

FIGURE 7.14. Type set in a shape to be used as art. (Executed by Judy Stilts, Seattle.)

(Unaudited - in thousands)	Second Quarter		First Half	
	1983	1984	1983	1984
Operating revenues	59,241	69,214	105,483	120,490
Operating expenses	52,323	59,169	96,571	109,335
Operating income	7,981	10,072	8,921	11,076
Other income (expense) — net	(934)	(539)	(1,893)	(1,459)
Income before income taxes	7,047	9,533	7,028	9,617
Provision for income taxes	3,397	5,426	3,310	4,450
Net income	3,650	4,107	3,718	5,167

FIGURE 7.15. A table using rules in two weights and screened and unscreened columns to improve comprehension.

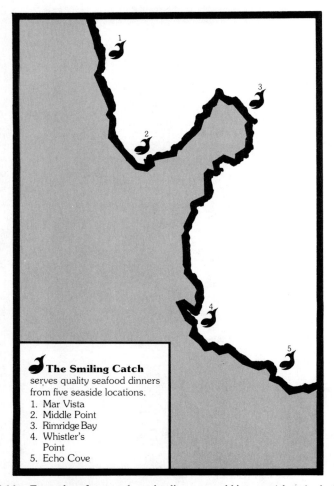

FIGURE 7.16. Examples of map styles using line art to add interest (above), shading to create dimension and proportion (top right), and locations simply highlighted on a grid (bottom right).

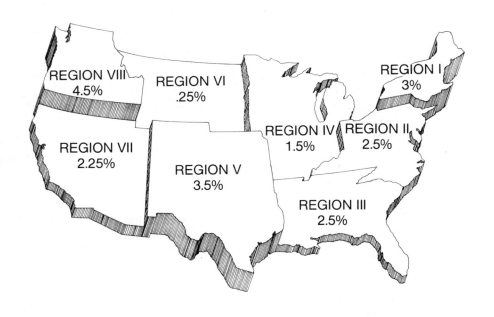

FIGURE 7.16. (Continued)

Other types of illustrations less frequently used but quite appropriate in specific situations are rubbings of coins or other embossed or engraved surfaces, fingerprints, or photocopies of objects (such as leaves or hands). A story ender is another possibility—a miniature symbol, logo, or bullet at the end of a *lengthy* article to indicate its conclusion and add a small artistic touch. Imagination and resourcefulness can create other options when more common types of art aren't suitable, available, or affordable.

PHOTOGRAPHIC TECHNIQUES

A photograph for print production is not the same as a snapshot, portrait, or decorative picture. It has certain recognizable qualities that make its composition, focus, and lighting suitable. Care is taken in the darkroom so that it is printed and (if necessary) enhanced to the best advantage. These qualities and this attention help reproduce on the press an image very close to the original photograph.

The following information will acquaint you with photographic techniques. It will not teach you how to take pictures or print them in the darkroom. Consult the many specialized publications available if you need to delve further into the subject.

Recognizing Quality Photography

The image you see reproduced on a printed page has been altered somewhat during the production process; it isn't exactly the same as the original, even when the most advanced camera and press equipment are used. This margin of difference must be taken into account when the original photograph is shot and printed to give you the image you want when it is reproduced.

The ideal photograph for print production has certain qualities of composition, focus, and lighting not necessarily found in a photograph made for other purposes. Include requirements for them when you give direction to a photographer or use them to evaluate the reproductive potential of a photograph already in hand.

Composition. A quality photograph is not just taken; it is carefully composed. Photographs for a magazine, catalog, report, or other printed piece are constructed to meet specific objectives of a print project. They are used to draw attention, to explain, or to establish a mood (which is really a combination of the first two).

If a photograph is to draw attention, it is composed to be immediately eye catching so that the viewer is drawn to the page or spread. An example is a shot of colorful foliage for a magazine article on fall sightseeing by bus. If a photograph is to explain, it is composed to show detail. An example is a shot of a raincoat offered for sale in a catalog. If a photograph is to set a mood, it is composed to draw attention and explain, in a symbolic way. An example is a shot of a food line to set the mood for a report on unemployment.

ART TECHNIQUES FOR PRINT 187

Photo composition for whatever purpose consists of three elements: center of interest, background, and foreground. Center of interest may be in the center of the photograph or (more creatively) off-center so that the action doesn't appear static and stiffly posed. For this reason, the horizon line in a scenic shot should be above or below the middle of the photograph to avoid dividing the image evenly in half. The "rule of thirds" is sometimes used to guide placement of the center of interest. Divide the image area into thirds horizontally and vertically like a tic-tac-toe grid. A pleasing placement of the center of interest is at one of the four points at which the lines intersect.

Center of interest is established by lines, framing, and focus. The lines of a building, stair bannister, or fence can lead the eye to the person or object that is the center of attention. A doorway, branches of a tree, or a window can frame the action and help it stand out. A sharp focus on the person or object of interest surrounded by out-of-focus background and foreground directs the viewer's eye. The examples of composition presented in Figure 7.17 illustrate these ways to establish center of interest.

Background encompasses the images behind the center of interest and foreground the images in front. Unless a photograph is tightly cropped to eliminate background and foreground, they must be considered when the picture is composed. Background and foreground may be used to direct attention to the center of interest; as seen in Figure 7.17, lines in the background and foreground help draw the eye to the central image. They may also establish a mood or context for the primary action. A shot of an overheated car showing a service station in the background sets an entirely different context from one with miles of empty highway in the background.

If background and foreground don't contribute to a photograph in one of these ways, they should at least not be distracting. A crooked horizon with buildings and trees at an angle to the main action presents the viewer with a visual discrepancy and draws attention away from the intended message of the photograph. A cluttered table in the foreground captures the viewer's eye more than a group of people at the end of that table.

The larger the negative area from which a print is made, the sharper the image will be. If a photographer knows to keep extraneous background and foreground to a minimum and concentrate on the center of interest, most or all of the negative will be usable.

Creative composition can make the difference between an artistic photograph and one that is purely functional or a boring cliché. Paired with imagination, photographic talent can create exciting, unexpected images from what would otherwise be a dull assignment.

Instead of the cliché "grip, grin, and grab" shot of an award or check being passed from one person to another, show them tacking the plaque up on the wall or the giver accompanying the recipient to the bank. Instead of the cliché "firing squad" group shot or static meeting pose, show people in action working together or arranged around a piece of the company's machinery (or an unexpected object such as playground equipment). Instead of

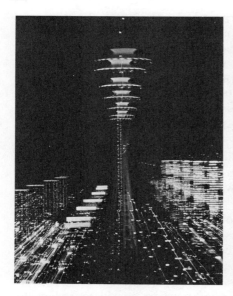 

FIGURE 7.17. Center of interest established by lines, focus, and framing.

an open-mouth shot at the podium, show a speaker going over notes before the speech or concentrating on a question afterwards. Instead of a "glass menagerie" shot of people at a cocktail party, focus tightly on two or more faces to bypass the glasses or ask your subjects to put them down while the photograph is being taken.

Unless you can run the photograph large enough for every face to be clearly identified, break up a large group into smaller units of four or less that can be seen. Even if run large, a shot of more than six or eight people should have them arranged in two or more rows so that the cutline can be divided for greater readability.

A shot in which the center of interest is really off-camera frustrates the viewer by its composition. All it clearly communicates is that the photographer missed the point. A picture of people looking at something is incomplete if at least a portion of that object isn't included.

Also calling for more creativity are photographs that rely on special lenses to attempt to make them interesting. A special lens can create multiple images of the same object or add an artificial highlight (commonly a star effect) to emphasize the gleam of polished metal, the brilliance of a diamond ring, or even the twinkle in someone's eye. Occasional use as a logical addition to the subject can be quite effective. Multiple use within one photograph or frequent reliance on such effects when not appropriate to the subject turns into an ineffective cliché.

Use creative composition to avoid embarrassing someone or distorting an image. When photographing a person who is overweight and self-conscious about it, frame the picture for a close shot of the face or seat the person behind a desk; don't shoot a full-length profile guaranteed to embarrass. A woman with a gaping blouse or a man concerned about a bald pate requires similar sensitivity and imagination.

Lack of attention to composition also results in distorted images. You may not notice the potted plant in the background apparently growing out of the president's head; but your viewers—and the president—will see nothing else. When used inappropriately, a wide-angle lens can distort otherwise acceptable composition on the edges. For instance, it may enlarge an outstretched hand to be grotesquely disproportionate to the rest of the person.

Use creative composition when a photograph is to establish size or weight. A large hole in the ground for a new building will not appear large by itself; add people or include surrounding buildings. Place a coin beside an antique vase to show the relative size of the vase or use a vegetable scale to show the weight of a prize tomato.

Focus. A second quality to evaluate in a photograph for print production is focus. A shot that is in focus has well-defined shapes; edges and lines are crisp and distinct. The overall image may be in focus or just the center of interest. How much is in focus is a factor of "depth of field."

A given lens can bring into focus only so many feet or meters of image, just as the human eye sees objects close at hand more clearly than those at a distance. Depth of field can be lengthened somewhat with special lenses or with a long exposure. If the image in the foreground is the more important, it should be in focus with everything in the background less and less distinct as the field from the point of focus gets deeper and deeper. If the image in the background is the more important, it should be in focus with everything in the foreground less and less distinct.

An out-of-focus shot can be effective for special purposes. If the subject of an article is horse racing, for example, a blurred photograph can emphasize the speed of horses in motion.

Lighting. A third quality to consider in a photograph for print production is lighting. Closely related are grain and requirements that apply to color. How well a subject is lighted and for how long film is exposed to that light significantly affect how well it will reproduce on the press.

A quality photograph has a range of tones from highlights to shadows. Experienced print producers generally agree that the best photograph for most purposes is one with several gray tones (or their color equivalents), rather than one with sharp contrast between light and dark areas and little else. Gray tones create the detail of a photograph (such as facial features or clothing patterns). Without them, the image would be stark, which may be what you want with a particular design to draw attention. But sharp contrast cannot readily explain, nor easily set a mood. A well-lighted photograph for print will be gray enough to show some detail even in shadow and highlight areas.

Tones to appear the same should be evenly lighted. Faces on one side of a group should not be lost in shadow while those on the other stand out in light. Faces subjected to direct flash or overexposure will be bright without detail, while those on the periphery or underexposed will be dark, also without detail. When you have to choose between two unevenly lighted or exposed shots, choose the one that shows the most detail in important areas (faces versus clothing, for instance, unless you are trying to sell the clothing).

Check background lighting to see if the photograph will have distinct corners and edges when it appears on a printed page. A very light background will appear almost white when printed, and the image will be partially or totally silhouetted and dissolve into the page. Unless you want that effect, make sure that the background is a shade of gray to establish corners and edges.

When lighting is sufficient to allow shooting at a fast camera speed, the grain or visual texture of a photograph will be fine and hardly noticeable. The darker the area being photographed, the slower the camera speed will have to be and the more visible will be the silver particles that make up the grain of the film. Grain size is an especially important consideration when the shot is to be enlarged because the particles enlarge along with the overall image (see Figure 6.38).

When evaluating the lighting of color shots, also look for saturation of colors. Blues should be rich, not thin or washed out, for example. Also consider how colors will convert if the shot is to run in black and white. Values that are similar will convert to a similar shade of gray, despite the fact that they started as reds and greens. Look for a color print or transparency with certain colors that stand out in relation to adjacent colors for contrast. They will convert to black and white as darks and lights, while colors of even value will convert to indistinct grays.

Photographing Currency, Stamps, or Paintings

Certain subjects require special attention so that photographs are taken and used legally. In the United States and Canada, you may reproduce a photograph of domestic or foreign coins as you like, without regard to purpose, size, clarity of image, or color.

In the United States, photographs of domestic or foreign bills, checks, bonds, and other negotiable items may be reproduced only for news, educational, historical, or numismatic purposes face-on if they are less than 75% or more than 150% of actual size. They cannot be printed in color. The image can be clear; retouching to camouflage it is no longer necessary. In Canada, words, numerals, and figures must be altered; only "Canada" can be clear. Multiple colors are not permitted, and no image may be used of the back of currency or other negotiable items.

Photographs of uncanceled postage stamps may be reproduced in the United States only for news, educational, historical, or philatelic purposes. You may print them in color if they are officially canceled or if they are run less than 75% or more than 150% of actual size. In Canada, they may be reproduced for news or philatelic purposes if they are 50% larger or smaller than actual size or if they are defaced. Canada Post must give consent if you are considering a different use. Other postal images (such as a cancellation mark or return-to-sender symbol) may be used in any way you desire.

These regulations were current at the time of publication. To confirm them or find out about any changes, contact the counterfeiting office of the U.S. Secret Service, the Royal Canadian Mounted Police, or Canada Post.

When using a photograph of a painting, show the entire painting. Only if so labeled should you use a portion of it. This accepted practice maintains the integrity of the original work and avoids confusing the viewer about what is being shown. Reproduction of a painting that is still protected by copyright may be considered fair use if you are doing so for such non-profit purposes as news reporting, teaching, or criticism. Review copyright regulations and, when in doubt, seek permission of the artist before reproducing the work.

Photographic Papers

The paper on which a black-and-white or color photograph is printed contributes to the quality of that photograph on the printed page. Whether black and white or color, a suitably printed photograph starts with white paper, not cream or ivory, and takes into account surface texture and finish. A color transparency or print to be converted into black and white has additional considerations.

Texture. A photograph for print should be on smooth white paper, not paper with a silk, canvas, or other texture. The raised portion of a textured surface reflects light differently from the recessed portion and may create a mottled or moiré pattern when prepared for the press. For this reason, a photograph with the popular silk texture may be marginally satisfactory when reproduced the same size or smaller; but it quickly deteriorates when enlarged. A smooth surface costs no more than a texture and will always work for any photograph destined for the printed page.

Finish. The supposition that a photograph must have a glossy finish to reproduce well is not universally true. Use a glossy print with enamel paper or with a press that prints a fine line screen. They are capable of reproducing a wide tonal range, and a glossy print will provide that detail. Use a dull or matte print with newsprint or other low-quality paper, with a print that will be prescreened (see Chapter 11), or with a press that prints a coarse line screen. When one or more of these factors applies, not nearly as much detail is reproduced, and a print with contrast between lights and darks and a more segmented tonal range will do better.

Glossy and matte prints are achieved by choice of paper or by the way a print is dried. A print made on a resin-coated paper air dries to a high gloss. A print dried with its face against the metal drum of a commercial dryer will be glossy also; turned the other way during drying, it will have a matte finish.

When you know that a photograph will need retouching, have it printed on a resin-coated paper, dried to a matte finish. This combination retains most of the tonal range of the glossy paper, while providing a surface that will accept retouching inks.

Black and White From Color. Although a color print can be converted directly into black and white for print reproduction, it will not be as satisfactory as starting with a black-and-white print. If your schedule and budget allow, have a black-and-white negative ("internegative") made from the color print or transparency you want to use, and then make a print from it. Tones will reproduce better on the press when the halftone negative is made from a black-and-white print, and you can see how the color shot converts before making a final decision to use it.

Darkroom and Retouching Techniques

When you can control how a photograph is produced in the darkroom, you have a choice of several techniques to tailor it further to your particular printing need, from how it is framed to skilled balancing of colors. When you already have a print in hand, retouching may be a way to improve its quality.

Framing. A photograph can be framed so that only the image you want appears on the finished print. You can specify that the print be made "full frame" so that all of the negative is printed for you to crop later, or you can

ART TECHNIQUES FOR PRINT 193

specify a particular framing. The latter choice is efficient if you want to be able to group ("gang") photographs for conversion into press negatives. Having them framed for the same reduction or enlargement will save money and time.

Filters. Various filters can be used on an enlarger to enhance or diminish certain tones in the finished print. If you've had a print made but aren't satisfied with the amount of overall detail you see, discuss with your photographer or commercial lab the possibility of using a different filter to bring out that detail.

Dodging. When an area of a photograph is too dark to reproduce well, "dodging" may lighten it. A small piece of dark paper or a metal disk on a wire is dodged around the area as the negative is exposed to the photographic paper, thus blocking some of the light and exposing that area for less time than the rest of the image. When properly done, you shouldn't be able to see any sharp line or shadow left by the dodging tool.

Burning-In. The opposite of dodging, "burning-in" ("burning" or "printing-in"), gives additional exposure to light areas to make them darker. The technician's hands or pieces of cardboard are used to block portions of an image already adequately exposed so that the unprotected area can receive additional light. No line or shadow should be visible as a result of the burning procedure.

Vignetting. This darkroom technique eliminates some or all of the background around the central image in a photograph. Using a trial print, a hole is cut in it to the shape and size of the area you want printed or the hole is cut in a piece of cardboard. The paper around the hole will prevent that portion from being exposed on the final print. While vignetting is commonly used on portraits, it can also be applied to print projects when, as in Figure 7.18, the original photograph is impaired by a flash shadow. In the example, the harsh shadow is diminished. (See information later in this chapter on fully eliminating background by silhouetting.)

Time Exposure. To meet particular design objectives, a time exposure may be in order. A negative can be exposed for a period of time or at slow shutter speed to create the illusion of a moving image. (See the Space Needle shot in night traffic in Figure 7.17.)

Toning. A black-and-white print can be "toned" or colored in the darkroom to give a brown ("sepia"), blue, or other tone to an image. A print slightly darker than desired is bathed in an additional chemical that colors it overall and lightens it somewhat. Toning is often used for decorative and portrait photographs and for creating a feeling of antiquity. For print production, a toned photograph must be treated as color or converted back into black and white.

FIGURE 7.18. Vignetting diminishes the shadow to focus attention on the image.

Spotting. In this retouching procedure, spotting ink is carefully applied with a fine brush to a print or negative to blend in an offending area with the adjacent tone or color. Lint or scratches on a negative will result in spots on the finished print and may be retouched so that they disappear or diminish. Facial blemishes are common candidates for spotting. A simple way to retouch light spots on a black-and-white matte print is with a graphite pencil. Leave more extensive spotting of either black and white or color to your photographer or commercial lab.

Airbrushing. In the hands of a skilled airbrush artist, details in a photograph can be emphasized, highlights added, sharp edges softened, or objec-

ART TECHNIQUES FOR PRINT

FIGURE 7.19. The original image (top) and after being airbrushed (bottom) to clean up the face and background. (Airbrushing courtesy of Al Doggett Studio, Seattle.)

tionable areas eliminated (Figure 7.19). An airbrush uses air to spray a fine mist of ink onto the surface of a print, either gray tones or colors. (An airbrush can also be used to paint an original image, although such meticulous illustration is quite expensive.)

Tinting. This retouchng technique uses colored ink directly on a black-and-white print. Before the advent of color photography, tinting was the only way to have "color" portraits or scenics. Consider having it done now by a skilled artist if, for instance, you are publishing old black-and-white photographs with full-color pictures and want them to have some color also, or if you want to create an antique look with a new photograph.

Color Balancing. When a color print is separated to make negatives to reproduce its component colors, balancing can occur so that, for example, the red of a product label is the same in each shot. Balancing compensates for variations in lighting and film when the original photographs were taken to give a consistent look throughout a printed piece. Discuss color balancing at the time you order separations for full-color reproduction. (For more information on correcting for color balance, see Chapter 12.)

PRINTING PAPERS AND ART

The paper on which a job is to run will influence how well art reproduces on the press. Though a simple line illustration can be satisfactorily printed on virtually any paper, black-and-white photographs and full-color continuous-tone art require particular attention to paper selection. In addition to press capability and the considerations covered in detail in Chapter 14, paper for a job that includes continuous-tone art must take into account brightness and surface.

Brightness

A paper that is bright to the eye will reproduce lighter highlights, darker shadows, and a broader tonal range than a paper that is dull. For the greatest contrast and range of grays or intermediate colors, choose a bright white paper. A colored sheet of any variety will have less contrast and a shorter tonal range than the same sheet in white.

A colored ink will also dull an image, especially if it is used on a colored paper. The higher the contrast between ink and paper, the better the reproduction of continuous-tone art. For example, on a simple weekly newsletter, the best-looking photographs will result when you use black ink on white paper. Even if you get tired of that combination, it will always give you better photographs than the muddy images produced by blue ink on white or light-blue paper or any other mixture.

Surface

As with photographic paper, texture and finish of the surface of printing paper affect continuous-tone reproduction. A sheet with a fine, muted texture will work as well on an offset press as an equivalent smooth sheet. A paper with pronounced texture, however, may cause loss of quality on any press, especially with art that has fine detail and/or lines. The peaks and valleys of a rough texture will detract from detail and, depending on press capability, may cause fine lines to be faint or not print at all.

Papers are finished with either a coated (enamel) or uncoated surface. An uncoated stock will reproduce continuous-tone art satisfactorily if it is opaque enough to prevent showthrough and, as mentioned earlier, if it is bright white or a light color for high contrast with the ink. Coated papers do not absorb ink as readily, however, so they reproduce art with greater intensity of highlights, shadows, and intermediate tones and with greater sharpness of image.

A coated matte paper has the least gloss of the coated variety and is economical to use when art requires coated stock and the budget requires restraint. A dull-coated sheet is one step up with a medium gloss, though it is still subdued. Gloss-coated (textured-coated or embossed) paper has a higher gloss. For the highest gloss of all, select a cast-coated sheet. Its mirrorlike finish will result in brilliant continuous-tone art, though its surface glare (and its cost) may be objectionable.

MAKING ART WORK FOR YOU

With well-conceived and executed art in hand, its successful use in print production is not yet assured. It must be made to work for you, rather than allowed to become ineffective filler or to compromise an otherwise satisfactory printed piece. Although this section won't teach you all that a graphic designer should know about use of art, it will introduce you to several specific techniques with illustrations and photographs not covered in Chapter 6 on design and provide advice for special situations. It also covers use of transfer art or type.

SUCCESS WITH ILLUSTRATIONS AND PHOTOGRAPHS

How well art works for you depends to a great extent on how it is placed and sized on the page. The following information offers guidelines in these two areas as well as advice about working with problem photographs, special techniques to use with photographs, and how to handle art of any kind. (See Chapter 11 for information on writing and positioning cutlines and legends that accompany illustrations and photographs.)

Placement

Several factors are involved in decisions about placement of art. Spread out the illustrations and/or photographs you have available for your printed piece and consider the following points about how to arrange them. Check your decisions about placement (and size) by making photocopies, reduced or enlarged as closely as possible to the desired size, or by having stats made, a costly but more exact option. Crop them as you think appropriate (see Chapter 11) and put them in place on a rough layout of your page or spread to get an idea of how they will look.

Content. The art you have before you should contain the information you need to augment text, the images you want to draw attention to on the page, or a definite tone to help establish mood. Photography to help document an event should show the people and places involved, for instance. Illustrations to describe how to make something should present each step in the process.

When you have sufficient content in art to satisfy objectives, stop. Don't add more photographs that are redundant—say the same thing over again in the same way. One shot to document turnout for a marathon run is sufficient. By using more that carry the same message, you will bore your audience and dilute the attention-getting value of using art.

Sequence. Place art in a logical order so that the sequence of events is understood. If drawings are to explain further what is presented in the text, they should be arranged in the same order as points in that text. Unless those drawings can be effectively placed in a row across or down the page, they should be placed in relation to eye flow. (See Chapter 6 for more information on sequence in design.)

When considering this factor in placement, be aware of binding capability if you want to place art across the gutter of any spread that is not the center spread. Unless alignment left to right is exact or close, one side of the art will be awkwardly staggered from the other, creating a visual distraction. Also, don't place the center of interest in the gutter (regardless of alignment in binding or whether or not it is the center spread) or in the horizontal fold of a newspaper. Even if no detail is lost, the fold across the most important part of the art will diminish its value.

Weight. As a general rule, art that is heavy because of bulky or dark tones should be placed low on the page, and art that is light placed toward the top. In the marathon example, a photograph of the mass of runners would probably be visually heavier than close shots of the winners and should be placed at or near the bottom of the spread with the winners higher up. Size is, of course, also a factor here. A heavy shot becomes lighter as it gets smaller and could then be moved up the page if doing so is more consistent with the sequence of events you want to reinforce.

Direction. The way art faces influences where it should be placed. Position illustrations or photographs so that they face into a page or spread. A person's eyes determine direction more than the set of the shoulders and should look into copy, rather than away from it. Lines leading to the center

MAKING ART WORK FOR YOU

of interest should also direct the viewer into the page. Exceptions to this guideline are when placing by direction disrupts sequence, when reversing direction by "flopping" the negative distorts the image (such as reversing words on a sign or switching buttonholes on a woman's jacket to the left), or when you want a photograph to face off the right-hand page to encourage the reader to turn the page for more of the story.

Horizon. Also distracting is seeing a row of illustrations or photographs with a common horizon that is not aligned. Try to position any line common to two or more pieces of art on the same plane. For example, photographs of new cars with the crest of a hill in the background of each should be placed so that the line of the hill is consistent as the shots are viewed side by side on the page.

Groupings. Several drawings or photographs to be viewed together should be placed to take advantage of the visual impact of grouping them, rather than sprinkling them about the page. Figure 7.20 and 7.21 illustrate the following ways to present groupings.

Close placement or butting with a minimum of space between is probably the most common method. (Skilled designers may successfully leave no space.) Grouping by proximity can be either simple and regular or more complex and irregular. Boxing related photographs with a gray or colored screen behind them or with rules draws them into a unit. Using a large photograph with small shots inset into it also pulls together the package but is effective only if the insets don't destroy the main image. When carefully executed, images can be silhouetted and then overlapped to form a close grouping (as commonly seen on the covers of gossip tabloids when separate photographs are silhouetted and overlapped to make persons appear together). The most important image in any grouping should dominate, by position and/or size.

Sizing

The size that you use a given illustration or photograph depends on its importance and its relationship to others with which it is used. As a general rule, the most important piece of art should be the largest on a page or spread. Images of equal importance and content (such as head shots of new employees) should be of equal size. Images of equal importance but *dissimilar* actual size should be arranged to maintain the natural relationship of sizes (such as shots for a brochure describing dwarf, semi-dwarf, and full-sized fruit trees presented in corresponding sizes), but positioned so that each is obviously important. When a small object is more important than a large one, size them according to importance but compose the individual illustrations or photographs so that actual size is evident (as covered earlier in this chapter).

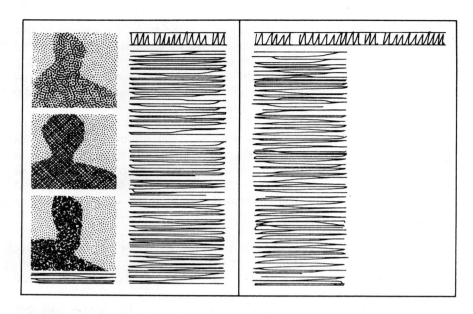
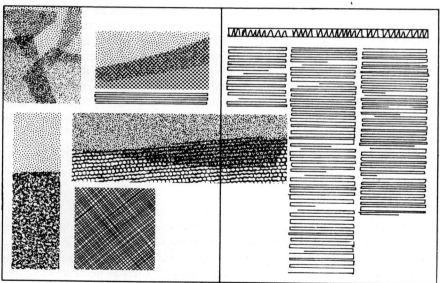

FIGURE 7.20. Photographs may be grouped by proximity in a regular or irregular pattern (above), or boxing may be used to give them all a common element (right).

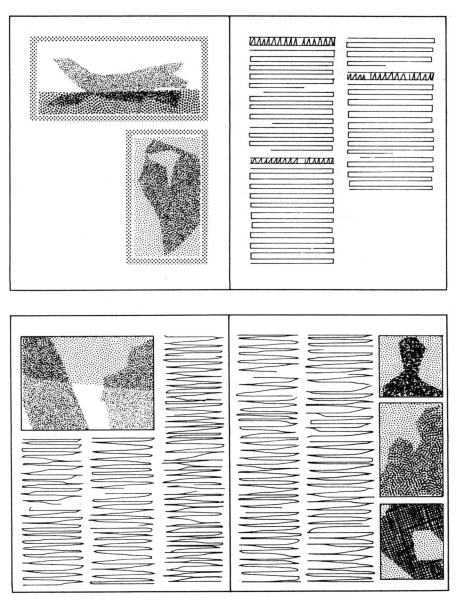

FIGURE 7.20. (Continued)

FIGURE 7.21. Small shots are inset into the larger photograph to create a grouping.

A page or spread with various sizes of art should use obviously different sizes and shapes—some large, some small, some medium; some horizontal, some vertical. Illustrations or photographs almost the same size and shape are static and dull and look as if a mistake was made when trying to get them all alike. In addition, pay attention to the following technical considerations when sizing art.

Illustration Line Weight. A line or brush stroke in an illustration will reduce or enlarge in direct proportion to how you size the overall image. For example, if you reduce an illustration by half, the line or stroke will be half of its original width (or weight). Decide how heavy a line you want in the finished piece, and then select or commission an illustration to correspond. As indicated earlier in this chapter, reducing a drawing to half its original size clarifies the image; if you plan to do such a reduction, select or commission the drawing with a line weight double what you want in the end. Too fine a line will disappear when reduced; too heavy a line will be awkward to execute and may overpower the spaces it encloses in the drawing.

Photograph Grain. The larger a photograph, the more pronounced will be the grain or visual texture that is reproduced on the printed page. To diminish the grain of a photograph (unless, for effect, you want it noticeable), reproduce it smaller than the original, the smaller the better for an especially grainy shot (see Figure 6.38). To check how grainy a photograph might be if run large, have the print made to the size you want the finished image. If the grain is only faintly evident on the print, the shot should reproduce satisfactorily as far as grain is concerned.

Photograph Texture. As stated in the preceding section on photographic papers, the texture of paper on which a negative is printed may limit what size it can effectively be reproduced. A smooth surface has no limitation, while a silk, canvas, or other textured surface restricts satisfactory reproduction to the same or a smaller size.

Camera Capability. Engravers or printers with minimal equipment may not be able to do an extreme reduction on line or continuous-tone art. Some cameras used for making halftones do not reduce below 33% of the original, and no camera can accommodate art larger than its copyboard. Check before you order finished art, and then have it drawn or printed to a size that is within the camera's range.

Working With Problem Photographs

Unless you can fully control the photographic process from shooting through printing, you will not always have ideal photographs to use for reproduction on the press. Reporters in branch offices will send in their snapshots for the employee newspaper, members will contribute photographs to accompany journal articles, and deadlines will force shooting under adverse conditions, with less-than-ideal results.

The following suggestions for working with problem photographs concentrate on how they might be improved by reprinting. Shots from a roll of film processed and printed by machine can often be improved by individual, custom attention. If you don't have the negatives, however, reprinting won't be an option. In such situations, consider having an illustration drawn from the problem photograph, thus making use of the image without having to reproduce the actual photograph. If the schedule is too tight, funds too limited, or talent not available for this solution, run the problem photograph as small as possible. Printing it large will only magnify its inadequacies.

Dark, Muddy Overall. Reprint the photograph to attempt to lighten it and increase the contrast between highlights and shadows. Try a different filter on the enlarger and/or a glossier finish on the print.

Gray, Flat Overall. Reprint the photograph to attempt to bring out a broader range of tones from highlights to shadows. Try a different filter on the enlarger and/or a dull finish on the print to drop some of the gray tones.

Dark in Key Area. Reprint the photograph using dodging to lighten the area.

Light in Key Area. Reprint the photograph using burning-in to darken the area and add detail.

Flash Shadow. Reprint the photograph using dodging or vignetting to lighten or eliminate the shadow, or silhouette the halftone (as explained later).

Out of Focus. Reprint the photograph to attempt to sharpen the focus or reproduce the shot as small as possible to cause some sharpening of the image.

Light Background. A photograph with a light background will dissolve into the printed page unless the background can be darkened or confined to create distinct corners and edges. Reprint the shot using burning-in to attempt to develop detail in the background and make it darker or box in the print on the page to contain it with an artificial border. (If you box one shot on the page, box them all for design consistency.)

Fluorescing Print. A photograph taken under ultraviolet light (such as mercury-vapor streetlights) will fluoresce and underexpose during engraving unless it's printed on paper to prevent this problem. If you know that ultraviolet light will be used, make sure that the resulting photograph is printed on paper acceptable to your engraver. Otherwise, special pains will have to be taken to get acceptable results.

Copied Photograph. A copy can be made from a photograph reproduced in a book or magazine if you don't have the original print. Assign such a project to a skilled engraver, however, and allow extra time and funds for this more exacting work. Assuming that the reproduction is a crisp image, the copy can be satisfactory, although some detail will be lost. Copying a reproduction involves rescreening a halftone. It must be accurately done so that the halftone dots in your printed copy will line up with the halftone dots being copied. If the dots clash, a moiré or fuzzy pattern will be produced (see Chapter 12).

Wrinkled Print. A print that is wrinkled or cracked from mishandling will not reproduce well unless the surface texture can be reduced or eliminated. If the shot can't be reprinted to produce a smooth print, use a spray adhesive formulated for photo mounting to attach the print to a smooth board. Cover it with a protective sheet of paper and burnish well. Your engraver or printer can then advise you about the quality you can expect, including any retouching that might be possible.

Special Treatment of Photographs

In addition to the common production techniques covered in Chapter 6, several special treatments are possible with photographs. Use them selectively so that your printed piece doesn't appear overdesigned and so that the extra cost of most of these treatments doesn't tax your budget.

Silhouette. A black-and-white or full-color photograph may be fully or partially silhouetted to eliminate all or part of the background. As mentioned earlier, this treatment (also called "lift-out") can be used to solve the problem of a flash shadow or of distracting images around the center of interest. Figure 7.22 illustrates full and partial silhouetting. When doing a partial silhouette, pay attention to where the break between background and no background occurs. It should not, for example, cut straight across the person's neck in a head-and-shoulders shot, but rather follow the natural line of the shoulders.

A silhouette is commonly done by positioning a masking film over the photograph and cutting it away in the areas to be eliminated. A negative of the film area left is made, creating a clear window surrounded by black. This window is then placed over the original photograph to block out the unwanted background when a final negative is made.

Silhouetting works well when the masking film is cut along straight lines of the photograph. Even the most skilled designer or mechanical artist, however, can't always assure satisfactory results when cutting around angles, curves, or a person's head. The edges may appear rough and harsh on the finished silhouette. For this reason, use silhouetting with caution, unless you can afford to soften such edges with airbrushing.

Ghosting. Also called a "phantom" halftone, a ghosted black-and-white or color photograph has a muted background out of which the center of interest stands in full value. Figure 7.23 presents an example. This treatment is used for emphasis (such as in a parts catalog to highlight the part being described, in contrast with the rest of the machine shown). It is accom-

FIGURE 7.22. The original photograph and partial and full silhouettes. A gray screen has been substituted for background in the silhouettes to set off the light image from the page.

FIGURE 7.23. Background and foreground in the original at left have been ghosted at right so that the central image stands out.

plished by placing a sheet of white or gray Bourges film over the print, and then peeling away the film in the area to be seen at full value.

Ghosting may also be used to subdue the entire image so that dark type printed over it will stand out clearly. For this purpose, Bourges can be used or the print can be underexposed.

Photograph/Illustration Combination. A photograph can be expanded beyond the camera's range by combining it with illustration, either line or continuous tone. The drawing augments or continues the image of the photograph, either to explain in more detail or to add interest to an otherwise ordinary shot. Negatives are made of both elements and overlapped or butted together to make a comprehensive press plate. Unless you are a skilled artist, this treatment is better left to a professional.

Bulletin Headline From a Photograph. For a photograph with lots of room above or below the center of interest, consider turning that room into a headline, as illustrated in Figure 7.24. To be effective, this treatment must not obscure the main action of the photograph, and the bulletin headline should be one bold word (augmented elsewhere on the spread by another headline or subhead). This alternative to surprinting or reversing a headline in that space should be used very selectively so that its visual impact isn't diluted. The association between word and photograph should be close and readily understood, and the separate letters of the headline distinct.

This treatment is done during engraving, working with the word positioned on the mechanical artwork and placement of it indicated on a tissue

FIGURE 7.24. This example of a bulletin headline is a logical outgrowth of the photograph and is actually part of it.

paper over the photograph. Another way to indicate placement is on a photocopy or stat of the shot reduced or enlarged to size with a copy of the word positioned over it as you want it to appear.

FIGURE 7.25. Special screen effects: (above clockwise) the original, first degeneration, four-tone posterization, and second degeneration; (right clockwise) mezzotint, steel engraving, circular screen, and vertical straightline.

Special Screening. Use the following special screening techniques on photographs that are intended to draw attention or set a mood, not ones that are to explain because detail will be lost. Prints should have strong highlights and shadows so that basic features are retained. Examples of these treatments are presented in Figure 7.25.

"Degeneration" converts a black-and-white halftone into line art, dropping shades of gray and leaving only black and white. It is also called "line conversion," "linear definition," "tone line conversion," or "two-tone posterization." A halftone can be converted to line copy during engraving or during paste-up. When done during engraving, the amount of detail retained can be more closely regulated. The latter method allows you to include the resulting line art right on the mechanical, thus saving the time and expense

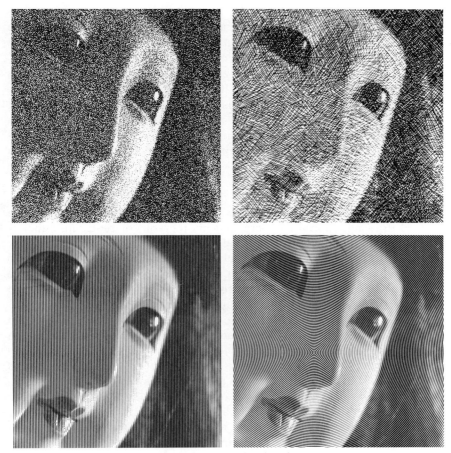

FIGURE 7.25. (Continued)

of camera work. To degenerate a halftone yourself, have it printed the size you want it to appear in the finished piece and photocopy it—or reduce or enlarge it to size as you photocopy. The copier converts it to line art; you can degenerate it even more by copying the copy. Spot any unwanted specks with typewriter correcting fluid, cut away the excess paper, and affix the art to your mechanical.

"Posterization" retains some of the gray tones to resemble a poster. (Two-tone posterization is another term for degeneration since only black and white are retained.) A three-tone posterization has black, white, and a medium-gray tone; four-tone has two intermediate shades of gray. This treatment is done during engraving. One negative is shot to isolate highlights, another shadows, and one each for the number of gray tones to be included. A composite is made if the photograph is to be printed in one color, or separate plates are made for multiple colors.

A "mezzotint" is a halftone in one or two colors with a pattern. A mezzoprint screen gives the pattern, and a second negative shot to emphasize highlights or shadows gives the color, a common but more intense color for a one-color piece or a different second color.

An "etchtone" is much like a mezzotint except that an etched screen is used. The resulting reproduction has the appearance of a steel engraving.

Coarse halftone screens may be used for an especially dramatic effect with a very simple halftone. A coarse screen makes the halftone dot pattern stand out. The pattern becomes even more pronounced if the halftone negative is shot at a severe angle. (See Chapter 12 for details on negative angles.)

Handling Art

Treat illustrations and photographs as fragile and irreplaceable. Mending a smudged drawing, reprinting a cracked photograph, or replacing either is time consuming, costly, and sometimes impossible.

Use a fixative on charcoal, pastel, pencil, and other media likely to smudge and protect the surface further with a tissue-paper overlay. The face of photographic prints is also sensitive to fingerprints, some more than others, depending on paper and printing chemicals. To be safe, also cover them with tissue paper. Transparencies are especially sensitive and should be handled as little as possible and then only on the edge. Protect them with plastic file sheets and use file sheets, sleeves, or envelopes to protect black-and-white negatives and internegatives from fingerprints, scratches, or dust. Store negatives and transparencies in a cool place away from direct sun.

Mount art on smooth, heavy paper or cardboard so that it doesn't curl or crack during production. To prevent chemical damage to a print you want to keep for years, use a spray adhesive made especially for photo mounting. Glue sticks, rubber cement, or tape can be used, although they may damage prints over time. Mount the entire drawing or photograph; don't trim it. Use crop marks (as explained in Chapter 11) to indicate how much of the image is to be used so that the resulting negative will definitely be large enough for the space and so that it will have a clean edge.

Don't write on the back of a photograph or drawing unless you can do so outside of the image area. Write instructions on a separate sheet taped to the back or on a tissue-paper overlay *after* you have inserted a heavy sheet between it and the art for protection. Any impression made on the back or pressed onto the front can create a texture that will jeopardize reproduction quality.

Before you release art for engraving, make a list of what you are sending. For each piece, note its finished size, the page or panel on which it is to appear, and the date it went in for engraving. The discipline of such a list will help you send complete information and all the art as well as help you reconstruct the job should a piece be lost, damaged, or mislaid. Ask that art be returned when engraving and platemaking are satisfactorily finished or when the printed piece is delivered.

MAKING ART WORK FOR YOU

SUCCESS WITH TRANSFER ART OR TYPE

The sheets of designs, rules, and type characters you can buy from an art-supply dealer require some special attention if they are to work for you. Transfer art or type is available in two varieties: rub-off and lift-off. Each calls for careful handling and close attention, both in its use and in its purchase.

Rub-Off Art or Type

This popular variety is a film of ink loosely adhered to an acetate sheet. When you rub the front, the design or type character comes off the back and directly onto mechanical artwork or onto a piece of paper to be added to mechanical artwork. The tool used for rubbing can be an orange stick, a burnishing stick, or other blunt object that will apply sufficient, even pressure to rub off the film. Experiment to find a tool that works effectively for you. A blunt-tipped pen may do well for small designs or letters, while a tool with a broader end will do better for large characters to prevent parts of a character from separating unevenly from the acetate sheet.

To position designs or type characters, draw a baseline with a blue pen or pencil that won't be reproduced by the camera ("nonrepro blue") or use nonrepro blue graph paper. If the transfer sheet has baseline marks—short horizontal lines below each character—place these marks on your blue baseline. The row of characters you rub off will then align, and you can cut away the baseline marks. If the transfer sheet doesn't give you marks as a guide, position each character on the baseline you have drawn. Characters with rounded bottoms (such as "G" or "6") should be positioned slightly below the baseline so that they will appear even. If placed exactly even, they will appear high. (Check this optical illusion on a line of type set by machine.) Position characters along the baseline so that the space between them looks even. With experience, you should be able to be both accurate and fast in such spacing.

A commercially available transfer system (by Letraset) takes the guesswork out of positioning type on artwork. Characters are first rubbed onto a special paper. When laid upside down on a transparent plastic strip and moistened, the paper releases the type. The see-through plastic strip is then positioned as you desire and the type rubbed off.

If you make a mistake or want to eliminate a design or a word, use masking tape to lift off the film. It works better before the film has been burnished down hard. Any portion that can't be lifted off with tape may be scraped with a razor blade or art knife.

To finish the job, burnish the characters with a burnishing stick, roller, or other broad, blunt tool. Protect them from chipping by placing a sheet of paper between the film and your burnishing tool. For fine designs or characters, a fixative may also be used to hold them in place. On bold or large areas of film, however, fixative may cause them to wrinkle, bleed, or even dissolve. Test a sample first.

Lift-Off Art or Type

This variety is ink on a film of plastic loosely adhered with wax to a sheet of paper. Designs and characters are cut apart, lifted off, and positioned on mechanical artwork or on a piece of paper to be added to mechanical artwork. Because they are lifted off and not subject to cracking and potential distortion caused by rubbing, this variety is often favored by graphic designers and others who frequently want transfer art or type of the best quality. It is also preferred for rules over rub-off or tape rules that can be wavy and inconsistent in width.

Positioning is done much as for rub-off characters. A sharp art knife is especially handy since a small square of plastic can be tricky to place where you want it with your fingers. Use the blade of the knife instead. If you make a mistake, simply lift off the offending character. Burnish your finished work to secure it and cut away baseline marks.

Shopping for Transfer Art or Type

Both rub-off and lift-off designs and characters are available in a wide assortment of sizes and styles, including white for doing reverses. Some sheets are capitals and lowercase, while others are one or the other only. Check the catalog so you will know exactly what you are buying.

Before you actually purchase a sheet, look at it closely. Characters should be aligned horizontally. If baseline marks do not appear even, they won't do you any good as guides. Check that lines and angles of characters are crisp, not bulging or rough, and that the full character or design is printed with no chips, scratches, or distortion. If you use transfer art or type frequently, experiment until you find a brand that consistently works well for you and that doesn't deteriorate quickly. Sheets stored flat away from sun or other direct heat with tissue paper between them should last for months or even years.

CHAPTER EIGHT

Art Sources

IN-HOUSE ART SOURCES   215
    Art Files, 215
    Stock Art, 216
       Transfer Art, 217
       Camera-Ready Art, 217
       Photographs, 218
    Previously Published and Borrowed Art, 221
    Computer Graphics, 221

ILLUSTRATION TALENT   222
    Finding an Illustrator, 222
    How Illustrators Work and Charge, 222
       Fees and Expenses, 222
       Use Rights, 223
    Evaluating an Illustrator's Portfolio, 223
       Range of Skills, 224
       Style, 224
       Flexibility, 224
       Effectiveness, 224
       Cost, 225
       Schedule, 225
    Checking an Illustrator's References, 225
    Working With an Illustrator, 225
       Assignment Details, 225

Art Direction, 226
Wrap-Up, 226

PHOTOGRAPHY TALENT   226

Finding a Photographer, 226
How Photographers Work and Charge, 227
    Fees and Expenses, 227
    Models and Model Releases, 228
    Processing and Printing, 229
    Use Rights, 229
Evaluating a Photographer's Portfolio, 230
    Range of Skills, 230
    Style, 231
    Flexibility, 231
    Effectiveness, 232
    Cost, 232
    Schedule, 232
    Equipment, 232
Checking a Photographer's References, 233
Working With a Photographer, 233
    Assignment Details, 233
    Art Direction, 233
    Reshooting, 234
    Wrap-Up, 234

MECHANICAL-ART TALENT   234

Finding a Mechanical Artist, 235
How Mechanical Artists Work and Charge, 235
    Fees and Expenses, 235
    Use Rights, 236
Evaluating a Mechanical Artist's Portfolio, 236
    Skill, 236
    Cost and Schedule, 236
Checking a Mechanical Artist's References, 236
Working With a Mechanical Artist, 237
    Assignment Details, 237
    Wrap-Up, 237

ARTISTS' AGENTS   237

IN-HOUSE ART SOURCES 215

If you are an illustrator, photographer, or mechanical artist, you head your own list of art sources. But if you're not or don't have time to be, you will have to search for other sources within your company or association or elsewhere.

This chapter presents information on art that may be available to you in-house (including sources that can be tapped over the counter or by mail) as well as outside. It offers details on selecting and working with independent illustrators, photographers, and mechanical artists, photography studios, or design groups providing these services. Because art is portable, allowing artists to serve out-of-town clients as well as those locally, the fact that you have few (if any) of these suppliers close at hand need not restrict your use of creative, quality art.

IN-HOUSE ART SOURCES

The easiest in-house source of illustrations or photographs is your own art files—art you have used before or that was prepared for possible use. Also available to you are forms of stock art, art that can be recycled or borrowed for a specific project, or possibly computer-generated graphics. Another in-house source may be an employee or association member skilled in illustration, photography, or paste-up.

Because dealing with an in-house artist is not significantly different from dealing with someone outside, refer to later sections on illustrators, photographers, and mechanical artists for information pertinent regardless of the source of talent. The price is better for in-house assistance (no additional cost); but you will have to accept help on a time-available basis. Otherwise, you will need to be as prepared and organized in your dealings with an in-house artist as you are with someone from the outside if you expect satisfying results.

ART FILES

Even if you require an illustration or photograph only occasionally, set up and add to art files from which you can make use as needs do arise. At a minimum, have a folder for photographs and file sheets of slides, a folder for illustrations, and a folder for ideas for either. If you have a lot of material to file, subdivide each of these categories into subjects and types of illustrations and ideas. (Pocket folders are handy for this purpose to prevent art from falling out and getting damaged.)

Save all the art that you have used, whether in a print project, slideshow, or display. Note on a sticker or piece of paper taped to the back the subject, specific names, when and where it was used, and the other information suggested in Figure 8.1. When you go to the file to research what you might use again, these details will help you decide whether or not to select a

```
┌─────────────────────────────────────────────────────────────────┐
│                      Art File Information                       │
│                                                                 │
│   Subject  Board of Directors                                   │
│   I.D.  Andrew King, Kelly Denny, Kathryne Ewing, Madeline Thorburn, │
│   Nancy Ahle, William Flowers Sr.                               │
│   Date Used/Project  February 1984/Shareholder annual report    │
│   Date Used/Project  April 1984/Employee annual report          │
│   Date Used/Project  October 1984/Bulletin board at managers' conference │
│   Negatives or Stats  5"x 3" color separation from slide        │
│        10" x 8" black-and-white print from internegative        │
│   Source  Michael Ziegler  622-1431                             │
│           1205 Sixth Ave. S, Seattle 98134                      │
│   Rights  Multiple use                                          │
│   Model Release  None                                           │
│   Comments  Ziegler has other slides shot at same time. President really │
│   likes this one, however.                                      │
│                                                                 │
└─────────────────────────────────────────────────────────────────┘
```

FIGURE 8.1. An example of a form for filing information on art.

particular illustration or photograph and may save you engraving cost if you can use it again the same size.

Also file the prints you had made but didn't use at the time and any line drawings that were done but dropped for lack of space or other reasons. Whether used or not, art represents money spent that may not need to be spent again if you can find a use for it (or parts of it) later.

Save clippings, tearsheets, or entire printed pieces that might provide inspiration. When you need to select a style for a series of drawings, for example, or alternatives to a staid award-presentation photograph, go to this idea file. Review copyright restrictions, however, to avoid "borrowing" to such an extent that copyright is infringed.

Be conservative in cleaning your art files. Duplicate prints of former employees may be safely discarded, for instance; but keep in mind that old illustrations or photographs of buildings, events, and people will be just what is needed years later as art for a brochure on the history of your company. Be aware that color prints are not stable; colors will begin to fade and change after a few years. Periodically make new prints or convert them to black and white if they are to be kept indefinitely. Any art of potential historic importance should be done on archive-quality art and photographic paper. It is specifically made without additives that can affect the image over time by discoloring the paper.

STOCK ART

Look to stock art if your need can be met by "generic" designs, illustrations, or photographs. Regardless of your location, you have ready access to

IN-HOUSE ART SOURCES

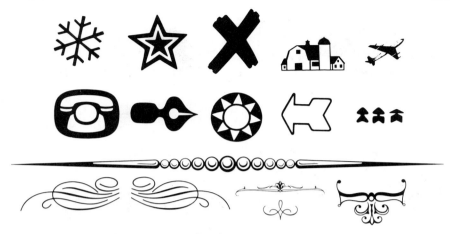

FIGURE 8.2. Examples of designs readily available in transfer art.

thousands of pieces of stock art of a general nature available from art-supply dealers, by mail, or at bookstores. Though stock illustrations come with a price tag commensurate with their outright purchase, your cost will probably be less than commissioning an illustrator to produce art just for you—and you can use it repeatedly at no additional charge. One-time (or ''subsequent'') rights to stock photographs cost less than first or exclusive rights of commissioned work.

Transfer Art

A wide assortment of rub-off or lift-off designs is available to complete an illustration or add a touch of art to a page. Figure 8.2 shows a few of the many non-copyrighted designs that can be purchased from an art-supply dealer.

Such transfer art will be especially valuable if you have limited needs, skills, and budget over the long term. For symbols to complete a map, arrows to clarify a diagram, or printer's marks to break up columns of type, an investment in transfer art will be minimal compared with the bill you would otherwise have from an artist.

Even with limited skills, you can use transfer art to construct your own designs. As illustrated in Figure 8.3, a little time and care can result in a creative combination that is unique to your particular print project.

Camera-Ready Art

Like transfer art, camera-ready art can be purchased and kept on file as a convenient source of illustrations and designs of a general or themed nature (people in meetings, Halloween, etc.). All such art is commonly referred to

FIGURE 8.3. Examples of how transfer designs can be used.

as "Clip Art," thanks to the popularity of that particular brand. It is available from art-supply dealers, by mail, or from bookstores. Other sources are companies or charitable organizations that supply camera-ready art free to encourage its use in support of their products or causes. Figure 8.4 presents examples of purchased and free camera-ready art.

Such art is also not copyrighted. High-quality, contemporary camera-ready art may be expensive to buy (though less expensive than having illustrations specially drawn for you); but once you have paid for it in sheets or in book form, you can use it repeatedly. You can modify it (as shown in Figure 8.5) by blocking out or cutting away portions, combining camera-ready art with transfer art to create a tailored design, or calling out certain parts to be printed in a second color. In the latter case, the art must allow successful division into colors, as illustrated by the two versions of United Way's logo (in Figure 8.4)—one spaced for one-color reproduction and the other for two-color reproduction.

Photographs

Individual photographers or photo services offer one-time use of stock photography of a general nature (scenics, buildings, etc.). Any shot taken on any assignment is potentially available to you, unless the original client purchased all rights or exclusive use for a period of time. A "photo service" (also called a "photo agency," "stock library," or "stock house") markets a wide selection of black-and-white and color prints and color transparencies from many photographers. Stock photos are generic and will satisfy a need that isn't too specific. If you're looking for shots of sailboats, for example, you will find many at a photo service—but most likely not shots of a particular type of boat in a particular location. Also, anyone can buy the right to use the same stock photograph, even your company's competitors.

FIGURE 8.4. Examples of free and purchased camera-ready art. Note the slight difference in spacing in the two United Way logos. The one with thinner white spaces (left) is for positive reproduction larger than ½ in. (1.27 cm) wide. The one with thicker white spaces (right) is for smaller positive reproduction and for all sizes of reverses. (Courtesy of United Way of America; other illustrations courtesy of Dynamic Graphics, Inc., Peoria.)

Call a photographer or photo service and explain what you're looking for or order by mail from a service's catalog. You will receive several prints or transparencies from which to choose, along with fee information and possibly a bill for research services to be deducted from the use fee if you decide to proceed. For any shot you want, you will pay a one-time fee based on how it is to be used (cover or inside; small or large run; local, regional, or national distribution; advertising or in-house only), and possibly a damage deposit. When you're finished with it, you return it to the photographer or service.

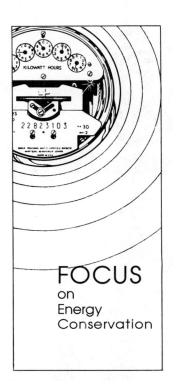

FIGURE 8.5. Examples of how camera-ready art can be enlarged, silhouetted, cut apart, and screened. Multiple colors could be substituted for screening. (Courtesy of Dynamic Graphics, Inc., Peoria.)

IN-HOUSE ART SOURCES

Most industrial and commercial photographers offer stock photographs; so look to those you have used for assignments or to colleagues or suppliers for referrals. To locate a photo service, check the Yellow Pages, ask photographers, or scan the ads in national photography magazines or other trade publications.

PREVIOUSLY PUBLISHED AND BORROWED ART

Art in the public domain may be legally available for you to use for your own purposes. If you see an illustration or photograph in another company's magazine or in a city newspaper, for instance, and want to use it again in your publication, contact the editor or the newspaper's library ("morgue") for permission. In non-competitive situations, you will most likely be granted permission in exchange for the cost of a print or stat, possibly a small handling fee, and a credit line acknowledging the source. An editor can grant permission if the art was executed by someone else as a "work-for-hire." If, however, the artist retained rights to its subsequent use, you will have to ask permission of the artist, not the editor.

Free use of art from an old book or other publication is possible if its copyright has expired. In the United States and Canada, illustrations are protected by copyright for the life of the artist plus 50 years. Photographs in the United States have the same protection; in Canada, they are copyrighted for 50 years from the date the original negative was made.

Other art may be borrowed at no charge for a particular project. If you need a photograph of a client's building or a diagram of a supplier's machine, look to your counterparts in those companies. Often you will find an appropriate piece of art that will readily be loaned just for the additional exposure you can give it.

Government and non-profit agencies are also willing to loan art. Government agencies have an obligation to assist the public by filling bona fide requests for shots of dams, monuments, and other subjects. Charitable agencies are eager to have your help in promoting their services and needs and can often provide photography as well as the camera-ready art mentioned earlier.

COMPUTER GRAPHICS

Computer graphics (also called "digitalized art") are designed on a computer terminal and then printed out as camera-ready art or transmitted directly as impulses to electronic typesetting equipment. The possibilities range from graphs generated on a personal computer using a graphics program to special effects created by computer equipment developed for the graphics industry.

Art possibilities from this source are expanding rapidly. To find out what may be immediately available to you, confer with your organization's data-processing department, with independent processing services, or with your suppliers who may be able to give you leads based on what they've seen used by other clients. Review trade publications for an idea of what may be available outside of your locale.

ILLUSTRATION TALENT

An illustrator conceives, roughs out, and executes for production line or continuous-tone drawings or paintings in black and white or in color to illustrate an article, brochure, technical handbook, poster, or other printed piece. An illustrator might also conceive and execute other forms of illustration (such as pictographs or maps) if drawing or painting is involved.

FINDING AN ILLUSTRATOR

Related suppliers are a prime source of referrals to illustrators. Ask printers, photographers, designers, and copywriters; their encounters with illustrators on past projects will give you names and evaluations of their work. Look to colleagues, trade associations, and the Yellow Pages for other referrals and contact art schools for graduates working in the area or advanced students who might be competent to do your project.

Another source is a design studio offering illustration as one of its services. Expect the fee to take increased overhead into account, however, and to be higher than you would probably pay a freelancer for comparable work. If your project requires extensive design and paste-up skills in addition to illustration, a full-service design group or independent designer/illustrator may be your best and most cost-efficient choice.

Illustrators will find you if you buy such work frequently or have an occasional sizable project. Review an illustrator's portfolio and check references ahead of need (if possible) so that you have reliable information on file to turn to when a new project comes along.

HOW ILLUSTRATORS WORK AND CHARGE

Illustration practices may vary by location and specialty. Use the following information as a guide for selecting and working with an illustrator.

Fees and Expenses

An illustrator will more likely work by the project than by the hour, although the project fee is based on an estimate of how much time the assignment will require. When you pay by the project, you pay for a satisfactory end product, even if the artist took twice as long to become "inspired" as had been estimated.

ILLUSTRATION TALENT

A project fee will depend on the size of the job (one illustration versus three) and the medium to be used since a simple line drawing is much easier to execute than a watercolor painting. For this reason, you and your illustrator may not agree on the full fee until you have seen rough sketches and decided on the medium. Up to that point, the illustrator may charge by the hour or charge a concept fee to be followed by an execution fee. Discuss fee in detail so that you know what to expect for both concept and execution.

An illustrator can work from real life, from photographs, or from nebulous ideas. The first two are already concrete images that are relatively easy to translate into drawings. Ideas or concepts, however, are more difficult. Review in detail with your illustrator what will be available from which to work. The fee may be higher if the artist is to draw from concepts or if travel to draw from real life is involved.

Clarify expense charges. Expect to pay for illustration board purchased for your project, for example, but not for basic supplies the illustrator must have on hand just to be in business. Specify what you expect the illustrator to provide, either original drawings or reductions to the size you need for print production. Fully discuss how the drawings will be presented in the printed piece and agree on the size of the original and on desired line weight. If done twice as large as needed, an original can be reduced in only one step and benefit from the sharpening and intensifying of lines that occurs in reduction. If you want the illustrator to provide reductions, expect to pay for them as an expense item in addition to fee.

Use Rights

Establish with your illustrator the rights you want to buy. The artist may sell first, one-time, or multiple-use rights. "First rights" entitle you to use an illustration first but only once (unless you also buy multiple-use rights). "One-time rights" (or "subsequent rights") entitle you to use it once, though not the first time. An illustrator who specializes in medical drawings, for example, might sell first rights to them for a journal article and then sell one-time rights to them for a textbook. "Multiple-use rights" entitle you to use an illustration as many times as you want, though not exclusively.

An artist may also work for hire and relinquish all rights. Under this arrangement, you own the illustration to use as you like. Expect to pay more for first rights versus a subsequent one-time use, even more for multiple use, and substantially more for outright purchase. Also expect to pay in relation to placement, press run, and distribution.

EVALUATING AN ILLUSTRATOR'S PORTFOLIO

An illustrator will show and talk about a range of pieces unless you request to see certain types similar to your upcoming project. Like other artists, an illustrator will show you only the best work. Close listening, questioning, and a later check of references will give you a more complete picture. Out of a portfolio showing alone, however, you should come away with information in the following areas.

Range of Skills

An illustrator's portfolio may include both line and continuous-tone drawings or paintings, both black and white and color, and drawings done from real life, photographs, and from ideas or concepts. Look for a range of competence sufficient to cover your anticipated needs. Also look for ability to use various techniques (as covered in Chapter 7) and for the illustrator's preferences. If you need special skills (such as cartooning or scientific illustration), discuss with the artist both interest and competence in that specialty. If you need art for multiple- or full-color reproduction, determine knowledge of its special requirements. An illustrator experienced in this area can save you considerable time, frustration, and expense over one who is not.

Determine if the illustrator did all the work on a given portfolio sample or if details were assigned to a less-experienced assistant. The resulting product may be quite satisfactory; but as the client, you should be aware of and approve subcontracting.

Style

A full portfolio review, in addition to the illustrator's comments, should clearly indicate artistic style. A skilled illustrator probably has a preferred style that is more developed than others. That preference may or may not be appropriate for your work. A light, finely detailed illustration may not be in keeping with the objects you expect to need drawn, for example, or with the bold, aggressive tone you want to establish in a particular printed piece.

Flexibility

Despite personal preferences, an experienced illustrator should be able to produce satisfactory results in the face of a difficult image to draw, a close deadline, or a restricted budget (within reason). Discuss thoroughly any portfolio samples that demonstrate a degree of flexibility similar to what you might require.

Effectiveness

Like design or photography, an illustration should meet its objectives—and its creator should be able to articulate what those were and something about how well they were met, including showing you how illustrations were used in finished pieces. A creative artist may also suggest ways it might have been

ILLUSTRATION TALENT

done better. A check of the illustrator's references will provide even more information in this area.

Cost

An experienced illustrator should be able to estimate realistically. Select from the portfolio a project similar to what you have in mind to get an idea of cost. Though today's price may be different, it should be based on a valid estimate of project or hourly fee. Include cost questions when you check the illustrator's references.

Schedule

Also ask the illustrator how long that similar project took, from planning meeting through final drawing. An illustrator who lets days go by waiting for inspiration may be too undisciplined to meet your close deadline. Determine how available the illustrator is; an artist in demand may be able to accommodate small projects from time to time on short notice but need to know weeks in advance of an extensive assignment.

CHECKING AN ILLUSTRATOR'S REFERENCES

Talk with at least two of an illustrator's past clients who had needs similar to yours. Keep in mind that personality, style, and other subjective judgments will be part of their comments. Especially ask about effectiveness of the work done and the illustrator's ability to estimate and schedule. An illustrator's portfolio may include unpublished work done for the portfolio, for a non-print project, for a project that was canceled, or for practice or pleasure. While you won't be able to include questions about such work in a reference check, it may add to your appreciation of the illustrator's talent.

WORKING WITH AN ILLUSTRATOR

Once you have selected an artist, follow-through is very important to the success of your project. You must plan, communicate, and give timely, valid feedback.

Assignment Details

An illustrator will need to know objectives, layout of the piece, colors, type of press on which it will run, paper, and any other considerations that might influence illustration. Use a purchase order or confirmation letter to list these details as well as fee, expenses, use rights being purchased, responsibilities, and deadlines.

Art Direction

In addition to planning with your illustrator the work to be done, art direction involves providing access to the people, events, or objects to be drawn from real life or as many photographs of the subject as you can (the illustrator may want to take these personally). If the work is to be done from concepts, set aside time for thorough planning so that the illustrator can benefit from all available ideas and suggestions. Make sure that how the drawings will be treated is clearly understood so that line weights, spot color, and other production details are considered as the artist works.

Based on this direction, an illustrator will come back for your reaction to rough sketches ("roughs"). From these, you will select those you want rendered in final form, those you want to have changed and done in an intermediate polished form ("tights"), or those you want to see again in rough form if changes are substantial.

Wrap-Up

Give the illustrator a credit line in the finished piece, preferably on or adjacent to the illustration. Provide portfolio samples, references to prospective clients, and full payment of the bill within 30 days.

PHOTOGRAPHY TALENT

A photographer shoots pictures of people, events, scenes, products, machinery, or whatever other images can be captured on film. A photographer may also process the film and make the prints you order, or leave that job to a professional laboratory. Industrial and commercial photographers work in black-and-white prints, color prints, and color slides ("transparencies"). Some specialize when their volume of business and reputation allow them to exercise a preference for medium and subject (such as black and white or color only; fashion, underwater, aerial, or medical). While an experienced generalist may be adequate for most work, look to a specialist if the project is difficult and will require very technical skills.

FINDING A PHOTOGRAPHER

If possible, start looking for a photographer before you need one to avoid a rush decision in the face of a deadline. Ask copywriters, designers, or printers with whom you deal to give you recommendations. If you are looking for a photographer with a specialty, ask colleagues who have used that specialty so that your list of potential talent is valid.

Beyond this personal approach, turn to the Yellow Pages, professional photography associations in your area, artists' agents, or schools that offer advanced photography courses. A project that could be shot after hours or simultaneously with another assignment may also be of interest to a newspaper photographer, photography instructor, or someone else otherwise employed who wants to moonlight. Another source is a studio that has photographers on staff or on call, or a design studio or agency that offers photographic services. Under this arrangement, you may not get to choose which photographer you want but rather be assigned one according to type of work and when it must be done.

If you buy photography frequently or buy a substantial amount from time to time (such as for an annual report), you won't have to search for talent; it will find you. Photographers' brochures will come in the mail, their agents will call, or other clients will suggest that they stop by. As time and interest allow, look at their work so that, when you do have a need, you can select quickly. With current information on file, you can readily call on your first choice and have a backup if the one you prefer can't schedule the assignment.

HOW PHOTOGRAPHERS WORK AND CHARGE

Though industrial and commercial photographic practices may vary from place to place (as do rates), they have more in common than at variance. The following information is based on accepted professional practices and is offered as general guidance for selecting and working with a photographer. (For more detail, see the current edition of *Professional Business Practices in Photography* by the American Society of Magazine Photographers.)

Fees and Expenses

A photographer works by the hour, by the day or half day, or by the project. Those in demand will most likely not accept an hour or two of work (unless you're a major client) but will charge for a minimum amount of time, say a half day, whether or not you use it all. Studio photography may be charged at a different rate from location work (possibly on a per-shot basis), and the day rate may vary according to the project and the client (profit, non-profit, or government organization). A day is considered to be approximately eight hours; additional compensation may be required if you want a photographer to work longer hours. An extensive project involving numerous locations, lighting requirements, and coordination may require a photographer's assistant whose fees will be reflected in the photographer's rate and expenses or be a separate cost.

All fees quoted will be plus expenses (studio rental, travel, long-distance calls, special equipment, etc.), except in rare situations in which a photographer picks up the cost and charges a day rate sufficient to account for overhead. For a large assignment, a photographer may ask for an advance to cover out-of-pocket costs of film and processing, costs that will be incurred whether or not you like the resulting work. If you've selected well, an advance won't be a budget risk but rather a statement of good faith to a talented artist.

The fee schedule may include travel and "standby time"—the time required to get to an assignment, to set up all props and equipment, to organize models, and to wait for the best lighting. This work is all part of a project. The schedule may also include a charge for being rained out as partial compensation for the photographer's time (even though you have no control over the weather) or a postponement or cancellation fee for a change within 72 hours of a shoot. The fee for each of these non-shooting activities may be the same or possibly half or two-thirds of the regular rate. If you don't see them listed on the fee schedule, ask.

Models and Model Releases

A photographer will charge a fee for providing models for fashion, advertising, or other photography. Choose a photographer especially experienced in this aspect of the work or avoid a "finder's fee" by securing models yourself if you are willing to be responsible for their suitability.

The look, age, and enthusiasm of a model can make or break a project. Though an employee, member, or other non-professional might do an adequate job, the chances are slim. Expect to pay for the results you want. A professional model may charge a substantial hourly or day rate but knows what is expected and can carry out an assignment with competence and speed to save instruction time and reshooting. Consult your photographer, colleagues who have used models, the Yellow Pages, or artists' agents. Be as specific as you can about the type of person you need and, if you rely heavily on your photographer's recommendation, review the model's portfolio of pictures yourself and do a personal interview to confirm that the person is right for the job.

Whether those in your pictures are professional models or ordinary people, model releases may be required. Unless your company or association has a standard release, ask the photographer to provide one. A "model release" simply gives you permission to take and to use someone's picture for a specific or open-ended purpose and is protection against a legal claim that the person's privacy was violated or that the person was libeled. Releases are not required for shots of employees at work for a company publication because the photographs are for editorial purposes and employment implies consent. Use releases, however, when including in your photographs minors or adults who aren't employees, when the use might be embarrassing, or when shots might be used in advertising.

A "property release" protects you against claims when an identifiable building is to be used in advertising, especially a commercial structure. Releases are not required for buildings that are in the background and incidental or that are public facilities.

The best time to get releases is when photographs are taken; then file them, just in case, for future reference. Any shot for which you have no release but that could conceivably be considered for advertising purposes should have a note attached to warn that no release is on file. (See Appendix B for examples of a simplified general adult or property release, a specific adult or property release, and a general minor release.)

Processing and Printing

Photographers who do their own developing and/or make their own prints will charge a per-roll and per-print fee for these services, rather than an hourly rate. For example, if a photographer shoots three rolls of 20-exposure black-and-white film, processes them, and gives you one proof each ("proof-sheet," "contact sheet," or "contact"), you will pay so much per roll to cover the cost of the film, the processing, and the proofs. If you order six prints, you will pay so much per print of a given size, depending on whether the prints are black and white or color. Special darkroom work such as bleaching out an area or printing especially high contrast will carry an additional fee. Because duplicates of the same shot are easy to make once the photographer is set up in the darkroom, you should expect to pay less per print for a quantity order of duplicates.

The per-print price may also depend on the use to which you intend to put the print. A shot for a magazine cover may cost more than the same shot for an inside page, for instance. It may cost more for a 10,000-run brochure than for a 1000-run piece or more for a national publication than one with local or regional distribution. Ask the photographer for a breakdown of prices in relation to position, press run, distribution, and advertising or in-house use only.

Rush charges vary by supplier and how "rush" is defined. As a guideline, any print that you need in fewer than 24 hours will carry a 100% surcharge. Always ask about charges and standard turnaround before requesting a rush on processing or printing because regular service may be adequate for your needs.

Use Rights

Photographers are of two minds about ownership of negatives or transparencies, and before you agree to work with someone, be sure of what you are buying. Most industrial and commercial photographers retain ownership and sell rights to use of a negative or transparency. A photographer who follows this practice would retain ownership of all negatives and sell you first- or one-time rights to the prints you want or, for a substantially higher fee, sell you the right to use them as many times as you want (multiple use), possibly with a time limit on their exclusive use. For an even higher fee, you might be able to buy the negatives. This practice protects photographers' right to profit from their creative work (even if you pay for film and processing), while you buy only the specific shots you need to fulfill an assignment.

A photographer may instead consider an assignment "work-for-hire" and price services so that all rights can be relinquished, leaving you with every negative or transparency, good, bad, or mediocre. Many clients prefer this pricing method, even though they may be paying for unusable shots. Most photographers, however, prefer the first method because they are able to satisfy your assignment and possibly sell prints to you on a continuing basis and/or market to other clients as stock photos the shots you don't want.

Customarily, a photographer will ask permission, as a matter of courtesy, before selling to someone else use of an image shot on assignment for you that is closely identified with your company or association. You can, however, negotiate an exclusivity provision to prevent your competitors from having access to the shots for a period of time (such as 30 or 60 days).

EVALUATING A PHOTOGRAPHER'S PORTFOLIO

The process of evaluating a photographer's portfolio is more than looking at pretty pictures. A photographer will, of course, choose the best work to show you; but also expect to engage in an exchange about the photographer's work in general as well as about each print, slide, or series in the portfolio.

Unless you make specific requests, photographers will show you their "books" which present samples illustrative of their full range of skills and interests. If you have a particular project in mind, request to see similar ones a photographer has completed. Ask to see the finished product, in addition to prints and transparencies—the series of ads, the annual report, or the magazine article that incorporated the photographer's work. Also ask to see (if possible) two or three examples of the full sequence, from comp through printed piece, to gauge the photographer's ability to interpret an original layout. A full statement of fees and a list of previous clients (with reference information) should be provided, either before or during a portfolio showing.

As will be covered in detail later in this chapter, your first exposure to a photographer may be through an agent who presents enough of the artist's work to interest you in seeing more. To glean the most from seeing a full portfolio, request that the photographer bring it. An informed agent is no substitute for direct contact with the person behind the camera with whom you might be working. Out of the looking, listening, and questioning of a portfolio showing should come at least enough information for you to evaluate a photographer in the following areas.

Range of Skills

A portfolio may include shots taken on assignment as well as photographs taken for pleasure. This mixture will indicate how well a photographer "sees" the potential of a shot and converts an otherwise ordinary group of people, objects, or a setting into a creative image. Someone who shows overall skills may be especially talented in one area and prefer to specialize. The portfolio will underscore this interest by the emphasis it places on specialized shots. If your need is for executive photography, for example, look for particular talent and interest in this area; then look for general talent that might also be satisfactory for other assignments.

A photographer should have technical as well as artistic ability. Lighting should be appropriate for the subject and end use, emphasizing a person's face, for instance, or a product's label, and film appropriate for shooting conditions and end use. The work should be in line with the original layout and perhaps even improve on it. Prints should be made with paper that brings out details or softens the image, as the project requires. They should appear smooth, not grainy (unless, for effect, graininess was wanted), and be in focus (unless "soft focus" was intended).

The range of skills may also include processing and printing. Color transparencies and prints will almost always be made by a professional laboratory, which may also do the processing. Discuss with the photographer what lab did the portfolio samples and why that one was chosen. Color processing and printing is a detailed, specialized science when high quality is required. Look for a photographer who understands the technicalities and how to get the most from a lab.

Black-and-white processing and printing may also be done by a lab. You will find, however, that many photographers do it themselves because, by controlling film selection through printing, a photographer can achieve maximum results from a subject. Look for examples of technical darkroom skills as well as talent for spotting, dodging, burning-in, or other techniques that an expert can use to salvage a problem negative.

Style

Like samples of other artistic work, a photographer's portfolio will indicate style: a personal approach to a subject through the camera's lens. Look for a style that suits you and your assignment. For instance, if your need is for executive shots with a flair, look for evidence of creative posing, lighting, and grouping of people, as opposed to a conservative "firing-squad" approach.

Flexibility

An experienced photographer will know how to achieve the best results possible in the face of adverse lighting, a tight schedule, or a restrictive budget. If you know of any limitations to an upcoming project, discuss them as you review the portfolio. Perhaps it includes samples of work that required similar flexibility.

Effectiveness

Paired with a check of references, a portfolio can give you insight into how effective the photographer's work was. Talk about the objectives of a particular shot or series and how much art direction was provided. The photographer should have responded to direction well enough to interpret those objectives on film. Especially look at how a shot was used in a printed piece, including its cropping. A skilled photographer is able to frame a shot very much as it will be used (assuming the client fully explains the project) so that quality isn't lost by having to enlarge one small portion of a negative to be able to see what was wanted.

Cost

The cost of photography depends on the magnitude and complexity of each project. Exterior shots of a new building for a company newsletter may be made in a half day under favorable conditions, while shots of company products and activities for an annual report may take several days at different locations. If a photographer's portfolio includes work similar to what you expect to need, the amount of time it required (and thus cost) may be similar. An experienced photographer will be able to estimate if a job will take one day or three, 10 rolls of film or 30. Ask the photographer (as well as former clients) how the initial estimate corresponded with the eventual bill.

Schedule

Also variable by project is the schedule from planning meeting to delivery of prints or transparencies. Try to get an idea during a portfolio showing about how efficiently the photographer schedules work; ask former clients, too. You don't want the frustration of dealing with a photographer who is so busy shooting that darkroom work falls behind, or vice versa.

Equipment

A photographer is responsible for providing all equipment necessary to complete your assignment, unless it is unusual and must be rented (at your expense and with your approval). You are responsible for providing access to locations and people, product samples, or whatever else is to be photographed. Discuss the types of needs you expect and the equipment the photographer has or could readily get to meet these needs. A quality 35-mm camera with various lenses and filters (the workhorse of most photographers) is ideal for shots of people in action, for slides, and for many other assignments. For portrait or high-quality magazine work, however, larger-format equipment is desirable so that prints can be made from larger negatives.

If setup is extensive, the assignment expected to be especially difficult, or the project expensive for any other reason, insist that the photographer have a Polaroid to check shots before they are taken "for real." These instant prints can confirm for the photographer (and you) that lighting and composition are satisfactory and help avoid costly reshooting. Also discuss available lighting equipment, props, sources for models, and studio space if you expect to need a well-lighted, controlled setting and experienced models.

CHECKING A PHOTOGRAPHER'S REFERENCES

In addition to reviewing a portfolio, check with at least two of a photographer's current and past clients who had projects similar to yours. Although their responses will be somewhat subjective, ask about the quality of both shooting and darkroom work and ability to meet deadlines and provide realistic estimates. Also ask how well the photographer cooperated (e.g., with employees in a busy office, in order to avoid undue disruption of work) and how much direction was necessary to get the photographer started and the assignment successfully completed.

WORKING WITH A PHOTOGRAPHER

Based on what you see and hear during a portfolio showing and on a check of a photographer's references, select a prime candidate for your next project and, if possible, a backup whom you think would do an adequate job should your first choice not be available. Schedule an assignment as far in advance as you can since a skilled photographer may work for numerous local and out-of-town clients.

Assignment Details

Communicate all requirements of the job, including objectives, the way shots will be used, timing, and budget limitations. Put your agreement in writing (a purchase order or confirmation letter), specifying the project, fee, rights to be purchased, expenses to be covered, responsibilities, deadlines, and any other details. Inform the photographer immediately if the work must be postponed or canceled. Provide model releases (unless the photographer is to supply them) and an advance for film and processing, if requested.

Art Direction

"Art direction" is guidance to the photographer before and during a shoot and at the time prints are ordered. At least on an initial assignment, plan to provide full, on-site art direction or ask the designer of the piece to do so. Once an experienced photographer is familiar with your expectations and environment, extensive, on-site direction may not be necessary on routine work. Always provide full art direction on a new type of assignment, however, or on a major project.

When based on a comp, art direction gives the photographer detailed information about what the photographs are to contribute to that layout and an indication of to what extent shots may vary from original concepts. When a project isn't large or complex enough to require a comp, verbally go over these points before the photographer commences work. Allow as much creative latitude as you can and be receptive to the photographer's judgment and opinions. Some of the best shots may come from allowing the photographer to shoot at will, rather than be restricted to rigid requirements.

Art direction may also include helping with logistics in the studio or on location to assure that the photographer has every opportunity to fulfill the assignment. Direction at the time you order prints should include size, number, type of paper on which they are to be printed, image area, and any special work (such as burning-in or dodging) that you want done in the darkroom.

Reshooting

Carefully evaluate proofs or transparencies to determine if everything you wanted is on film. If it's not and the photographer was at fault, reshooting should cost you nothing extra. However, if you changed your mind or were otherwise at fault, you pay for the reshoot, probably at the photographer's original rate. If both of you share in the blame or the fault is uncertain, the photographer may reshoot at one-half the original rate, in addition to expenses.

Wrap-Up

Give the photographer a credit line in the finished piece. If you have agreed to, also credit any services or supplies that might have been donated or sold at a reduced price as part of the photo assignment. Provide portfolio samples, references to prospective clients, and full payment of the bill within 30 days.

MECHANICAL-ART TALENT

A mechanical artist will do the intricate paste-up work needed to execute your ideas for tables, charts, graphs, and maps consisting of type, rules, screens, and other line work. As detailed in Chapter 11, paste-up requirements may be much more extensive than these art needs only, encompassing execution of full mechanical artwork. The information that follows can be applied to finding, evaluating, and working with a mechanical artist, regard-

MECHANICAL-ART TALENT

less of a project's scope. It is presented here along with other artistic services because of the relationship of skills required, even though a mechanical artist might not be called in until much later in production.

FINDING A MECHANICAL ARTIST

Whether to update last year's financial graphs or to do the mechanical for a printed piece, turn first to your typographer for mechanical-art talent. A full-service shop will have at least one mechanical artist on staff and be able to refer you to others who might freelance your project. Even if you don't buy type from that shop, you can probably buy the services of its mechanical artist. A design studio or an agency will also offer paste-up services, although most likely not unless you are also buying other related services. All such established businesses will charge a fee in line with overhead costs. In comparison with a freelance mechanical artist, you may pay more for the convenience of one-stop shopping with a typographer, studio, or agency.

Other sources for leads are printers, designers, and colleagues who use outside mechanical-art talent. Check art and technical schools for graduates working in the area or for advanced students capable of doing your project. The designer and/or illustrator you use for earlier phases of a job may also be interested in pasting up the mechanical. (As pointed out in Chapter 6, a designer may insist on doing it on an extensive assignment or else not take the design work.)

HOW MECHANICAL ARTISTS WORK AND CHARGE

Practices and rates may vary by location and the source of mechanical-art talent. However, be guided by the following information as you select and work with a mechanical artist outside or inside your organization.

Fees and Expenses

A mechanical artist works from a detailed mechanical dummy that indicates overall size, spacing between elements, and their placement. Any special requirements of your engraver or printer can also be incorporated, for instance, putting type or art to be printed in a second color on an overlay.

This service is priced by the hour, by the project, or possibly by the day or half day, depending on the magnitude of the job and whether it is straightforward or subject to numerous changes. Expect to be charged for all supplies purchased especially for your project (such as paste-up boards or rules) but not for equipment or basic supplies a mechanical artist needs to be in business. If you buy this service from a typographer, studio, or agency, customary expenses may be accounted for in the hourly charge. A freelancer will probably itemize them on the bill. Also expect to pay for correcting a headline that is too long, art that is reduced too far, or other expenses the mechanical artist incurs while paste-up is in progress to make what you provided usable.

Use Rights

Unlike the products of an illustrator or photographer, mechanical art is always considered "work-for-hire." The mechanical belongs to you, as does any leftover type or supplies for which you paid.

EVALUATING A MECHANICAL ARTIST'S PORTFOLIO

Because of the limited scope of a mechanical artist's work, what you should be aware of during a portfolio showing is not as extensive as for design, illustration, or photography. Ask to see specific types of projects appropriate to your needs and request that the original mechanical for each, the dummy from which it was executed, and the resulting printed piece be included, if available. Even when all elements in the sequence can't be shown, portfolio samples should allow you to make judgments in the following areas.

Skill

Mechanical artwork skillfully executed is clean. Edges of type galleys are cut straight (beveled if the artist is especially meticulous) and corrections cut in, rather than laid on top of old type. Windows are well burnished with straight edges and square corners. Rules are straight and evenly spaced. Corners of boxes are mitered with no gaps (if transfer rules were used). Rules and boxes drawn with a ruling pen are straight and of even width with crisp ends or corners. Type on overlays lines up with the basic artwork, and registration marks correspond. Ink colors, screen percentages, and all other instructions are noted on the tissue-paper overlay. (To understand each of these requirements, see Chapter 11.)

Cost and Schedule

Like other artists, an experienced mechanical artist should be able to estimate realistically and schedule efficiently. The price and duration of any specific project in the portfolio may vary widely from yours, even if the end product is similar, because of the condition in which elements were handed to the mechanical artist and whether work proceeded to completion or was stop and go. However, try to get an idea of how close the person can estimate a job, even accounting for such variables.

CHECKING A MECHANICAL ARTIST'S REFERENCES

Follow up with at least two past clients to add to the information you gain from a mechanical artist's portfolio. Because mechanical artwork is so critical to ease of print production, ask for the names of engravers or printers who have worked with the person's mechanicals if former clients can't speak satisfactorily to this point. Also inquire of any circumstances which might have influenced the mechanical artist's ability to meet deadlines or stay within cost estimates.

WORKING WITH A MECHANICAL ARTIST

A mechanical artist starts with givens: a design, galleys of type, photographs and/or illustrations of specific sizes, and a detailed plan for how pages or spreads are to look ("mechanical dummy"). You literally go to a mechanical artist with an envelope of pieces to be skillfully assembled into mechanical artwork ready for the next step in print production. Thus you have a large share of the responsibility for a successful product.

Assignment Details

Describe the job in writing, in either a purchase order or confirmation letter. Include information on fee, expenses, deadline, responsibilities, and any special requirements. Plan ahead so that all (or if not possible, most) elements are ready when work is to commence. Provide a mechanical dummy that is as specific as you can make it and communicate immediately if anything on it changes. Check mechanical artwork thoroughly to make sure all elements are included and that the work is satisfactory (see Chapter 11 for details).

Wrap-Up

No credit line for the mechanical artist is expected, although you should furnish portfolio samples of the finished product as well as the original mechanical dummy and mechanical artwork if you don't need them again (or offer to loan them for a portfolio showing). Provide references to prospective clients and full payment of the bill within 30 days.

ARTISTS' AGENTS

Talented artists who have limited time or ability to market themselves may contract with agents to do it for them. Also sometimes called an "artists' representative," an agent may work for several non-competing artists (such as a copywriter, graphic designer, illustrator, and/or photographer). In exchange for a commission, salary, or combination of commission and salary, the agent searches out prospective clients, provides them an introduction to an artist's work, helps negotiate terms of a project, and may bill the completed job on behalf of the artist.

Using an agent relieves an artist of the need to spend a high percentage of time marketing and (ideally) generates enough additional work to justify the commission (often 25–50% after artist's expenses). By contacting an agent yourself, you may save the time and potential frustration of tracking down an artist who can do your project. An experienced agent will deal with you in a businesslike manner to preclude misunderstandings about assignments, billings, and payments.

If you are approached by an artist's agent or call to express an interest in knowing about an artist's work, you will then meet with the agent to be briefed on the artist's background, previous clients, and abilities. If you are sufficiently interested to want to see a portfolio, insist that the artist be present to respond fully and personally to all of your questions. (The agent may also want to be present.) Should you choose to work with that artist, you will pursue project details and fee directly, although the agent may discuss billing details with you. When the work is completed to your satisfaction, the agent may bill you on behalf of the artist, receive the payment, and pass it on to the artist, minus the commission, or the artist may bill you and pay the agent.

One dealing with an agent obligates you to go through the agent on subsequent projects with that artist, as long as the agent continues to represent that person. While the artist pays a commission for marketing assistance, you may be paying, too, if the artist charges a higher fee to agent-secured clients to compensate for the commission. An eager, experienced agent may save you time by locating qualified artists for you; but don't expect the service to be free.

CHAPTER NINE

Typesetting Methods and Typographers

HOT-TYPE COMPOSITION   240
   Hand-Set Hot Type, 241
   Machine-Set Hot Type, 242

COLD-TYPE COMPOSITION   244
   Manual Cold Type, 244
   Impact Cold Type, 244
   Phototypeset Cold Type, 246
     Body Type, 246
     Display Type, 248
     Photocomposing, 248
     Capabilities, 249
     Changing Phototypeset Galleys, 250
   Electronic Cold Type, 250
     Input, 251
     Computer Control, 252
     Output, 252
     Storage, 252
     Capabilities, 253

SELECTING A TYPOGRAPHER 253
   Kinds of Typographers, 254
   Evaluating a Typographer, 254
      Competence, 255
      Reliability, 255
      Type Selection, 256
      Cost, 256
      Convenience, 256
   Trade Customs, 257

Typeset copy has several advantages over typewritten copy for any project other than the most simple and informal. Typeset material is faster to read and easier to comprehend, especially if the information it presents is complex. Typeset copy is easier to refer to later and encourages better retention of the material. Because of its formality, it is considered more believable and more appealing to the eye. Typesetting makes copy more compact, reducing bulk and printing and binding costs.

This chapter is concerned with the various methods available to convert your words into type directly for the press or for the making of press plates ("typesetting" or "composition"). These methods are customarily divided into two broad categories: hot type and cold type. (Alternative classifications are hot type, cold type referring to manual and impact, and phototypesetting which includes electronic composition.)

These methods are presented in historical order to show the evolution of typesetting. The magnitude of their use today is, however, in almost the reverse order. Phototypesetting is by far the most popular, and hand-set metal the least popular. Moving up fast to challenge phototypesetting is electronic composition or, as it is more commonly called, computer typesetting.

Modern methods, their ancestors, and those that developed in between are covered in this chapter, as are kinds of typographers and how to evaluate them. Many typesetting terms will be introduced and briefly defined here. See the glossary for more detailed definitions. This chapter is intended to be an overview to assist you in purchasing and coordinating the typesetting phase of print production. Consult specialized books and manufacturers' literature if you need more extensive information.

HOT-TYPE COMPOSITION

"Hot type" uses molten metal to form type characters. The metal consists of mostly lead, with tin added for strength and antimony added for hardness

HOT-TYPE COMPOSITION

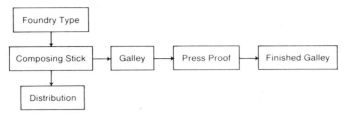

FIGURE 9.1. A flowchart depicting how foundry type is set.

and fluidity. This category of composition incorporates any method that uses a raised, inked surface to transfer images. Hot-type composition may be done by hand or by machine.

HAND-SET HOT TYPE

As the term implies, a hand-set word or line of type is composed manually. "Foundry type" is purchased by a typographer in "fonts" or complete sets of capital and lowercase letters, numerals, punctuation marks, and special characters (called "peculiars" or "sorts"). Characters are stored in a case, thus their distinction as uppercase or lowercase, depending on their storage location in the case.

Each character is a piece of metal with a raised, mirror image. The typographer selects those in the specified point size that are needed to spell a word and lines them up in a small tray ("composing stick"). When all the words are lined up, spaces ("quads") are added between them to fill out the line. When all the lines are set, linespacing ("leading") is added between them to make the information more readable. The type is placed in a larger tray ("galley"), locked into position, inked, and impressed to create a right-reading image on paper, either galley proofs for proofreading and paste-up or the final printed piece. When the job is completed, type characters are cleaned and, with the quads, are distributed back into their type case. Figure 9.1 diagrams this process.

"Ludlow" type composition is actually a combination of hand-set and machine-set hot type. Rather than use actual type characters as is done with foundry type, the typographer composes lines of type ("linecasting") by aligning molds ("matrices") for all desired characters, in the specified point size, and adding quads to fill out the line. The composing stick full of matrices and quads is then inserted into the Ludlow machine, locked into position, and used as the mold to create a solid line of metal type. Point size may range from approximately 12 to 72 points. When all lines are set and leading added between them, a galley proof for proofreading and a corrected galley for paste-up are made or the type is used directly for printing, as with foundry type. The line of metal ("slug") is then melted down, and the

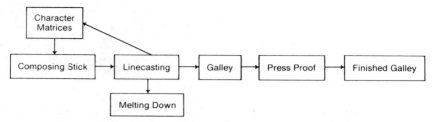

FIGURE 9.2. How hot type is linecasted by hand using character matrices.

matrices and quads are returned to their case for the next project. Figure 9.2 diagrams this process.

Wooden type is also a form of hand-set type, although it is not hot metal. Characters are carved to make a font, much as foundry type, only out of hardwood. Though rarely used today, wood type once enjoyed some popularity because of its ability to produce very large characters ("second-coming type").

The cost of hand-set type prohibits it from being commonly used today. It has some appeal for headline and display type because characters can be set quite large and maintain their quality (unlike enlargements of characters set by mechanical hot type or by cold-type methods). Also, many more printer's marks, borders, and peculiars were developed for hand-set composition than are now readily available on cold-type equipment or in transfer or camera-ready art. If you need only a few words of display type and your typographer has foundry type or Ludlow equipment, consider using it to set that type. The time required may be less than that needed to prepare and use other more modern methods.

Hand-set type has the potential of producing clean characters in large sizes, especially using the Ludlow method that molds new characters with each use. Also, it is "type-high" like machine-set type and can go directly onto a platen or flatbed cylinder letterpress ready for inking, without an intervening press plate. Paper galleys from hand-set type do not, however, enlarge well photographically since rough edges show at more than 10% enlargement. Tight kerning is impossible because of the metal surrounding each character. Because of the amount of time involved, none of the methods covered is practical for high-volume body type. Unless one of them happens to be all that is available or your only source for a particular typeface or special character, you may never be concerned with hand-set hot type for a print project.

MACHINE-SET HOT TYPE

Though it has similarities to hand-set composition, "Monotype" is classified as a machine-set method of producing hot type. A typographer types the words

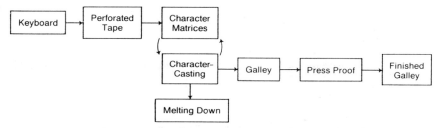

FIGURE 9.3. How hot type is set using character-casting equipment.

to be set on a keyboard, and each character is translated by the machine as a pattern of holes on a paper tape. This perforated tape is "read" by a Monotype caster to select from storage the required matrices. It then casts each character and arranges them into lines of type. The typographer adds quads between words and leading between lines for proper spacing, and the galley is finished. When a project is completed, the metal is melted down, ready for recasting. This equipment sets type from approximately 4½ to 14 points, up to 36 picas wide. Figure 9.3 diagrams this process.

Instead of casting each character, "Linotype" and "Intertype" machines cast a line of type as a unit ("linecasting"). A typographer types the words to be set on a keyboard. The machine selects the required matrices from storage and arranges them into a line of predetermined width (using as many quads as necessary to justify the line), casts it into a metal slug, trims away excess metal, and adds it to the galley. Once cast, matrices are automatically returned to storage, ready to be reused by the machine to produce another slug. All slugs are melted down when the job is finished. This equipment sets type from approximately 5 to 18 points and is generally limited to a 30-pica column unless slugs are butted to create longer lines. Attachments allow these machines to accept perforated Teletype tape in place of keyboard commands. Figure 9.4 diagrams this process.

Linecasting is not used today to any great extent for commercial typesetting, although it offers several familiar, well-designed typefaces. Many city newspapers and book and magazine typographers, however, have retained linecasting equipment as the most efficient way to generate type for direct impression on a letterpress.

Monotype has an advantage over Linotype or Intertype for a job requiring easy editing and correction. Since it casts characters individually, a change can be made in a financial table or business form, for example, without resetting entire lines. Linotype or Intertype allows quick substitution, addition, or deletion of a line of type on a page, which used to be its big advantage over phototypesetting that requires more extensive resetting and paste-up to work in changes. Directories, price lists, and other data that retained a consistent format with minor changes from year to year could easily be updated with hot type. The development of editing capability for cold type

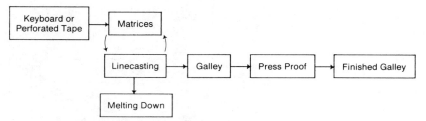
FIGURE 9.4. How hot type is set using linecasting equipment.

and especially of electronic composition, however, has diminished this advantage.

For an extensive project such as those just mentioned, you may want to do a cost and time comparison. Mixing typefaces, type styles, and sizes, leadered tabular matter, and other complexities will add to the cost of a machine-set job; so provide a sample of the copy to be set for a realistic estimate or bid.

COLD-TYPE COMPOSITION

The bulk of type set today for commercial purposes is cold type. Though both cold type and hot type can result in camera-ready galleys, composition processes are different. While hot type relies on hot metal, cold type does not. Manual and impact cold type put ink directly on paper; phototypesetting and electronic composition use a photographic process to transmit characters directly onto photosensitive paper or film. Each of these cold-type methods will be covered in this section with particular emphasis on phototypesetting and electronic composition.

MANUAL COLD TYPE

Manual (also called mechanical) methods of setting cold type include transfer type and calligraphy. They are executed by hand when a project requires a small amount of display type. (See Figure 6.26 for examples of calligraphy and Chapter 6 for general information on its use. Transfer type is covered in detail and illustrated in Chapter 7.)

IMPACT COLD TYPE

The standard office typewriter is one form of impact ("strike-on" or "direct impression") typesetting. It and other machines functioning on the same principle transfer ink directly onto paper to produce a camera-ready galley. This method is commonly used for low-budget projects to be printed by a photocopier, an offset duplicator, or a quick printer. It produces primarily body type, though display type generally no larger than 12 points can be created using capital letters, underlining, or a primary element with large and small capitals. A typewriter with interchangeable elements offers a choice of typefaces and styles. While an executive model is limited to the element that is built in, it offers letterspacing proportioned to the width of individual characters ("proportional spacing"), rather than allotting equal space for all, and it can justify columns. A word processor or word-processing computer can also be used to produce justified camera-ready body type in different typefaces and styles. With any impact equipment, copy may be typed on a standard keyboard and the resulting galleys proofed, corrected, and pasted up to be printed in final form. (Guidelines for preparing camera-ready galleys by these impact methods are presented in Chapter 10.)

Equipment designed especially for impact typesetting is based on the typewriter principle but provides more control and better-quality type. The oldest is the electric Varityper. It offers a multitude of body typefaces, styles, and weights in sizes ranging up to 13 points and can set rules. The Varityper can accommodate two fonts or elements at once so that roman and italic, for example, can be easily interspersed on the same galley. It can vary leading between lines, and column width is limited only by the size of paper used in the wide carriage. Letterspacing may be consistent or proportional, depending on the Varityper model being used. Columns may be justified, although lines must be typed twice to work in appropriate wordspacing.

The Singer Friden Justowriter varies letterspacing and produces a justified column. Copy is first typed on a keyboard to create a perforated paper tape. The proofed and corrected tape is then "read" by a reproducing unit which automatically adds sufficient space between words to produce a justified galley.

IBM's Selectric Composer is another desk-top machine for creating type galleys and, like the others, uses a carbon ribbon only to create characters crisper than those produced by a cloth ribbon. Though many typefaces and styles are available in interchangeable fonts up to 12 points, it does not necessarily reproduce them all in comparable quality. The Composer offers proportional spacing; but the result may still be too much or too little space, depending on typeface, style, and size. Only one font can be used at once; switching from roman to italic, for instance, must be done manually. The Composer can automatically vary leading and justify lines with one manual typing and a second typing automatically done by the machine. A unit with a tape drive uses a magnetic tape much like the Justowriter's paper tape as an intermediate step for proofing and correcting before final copy is set. An electronic Selectric Composer automatically justifies, centers, and executes other formats with one typing.

Each of these impact units has the advantage of a comparatively low purchase cost over more sophisticated, commercial composition equipment.

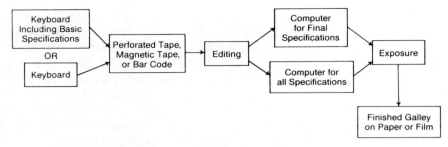

FIGURE 9.5. How cold type is set using phototypesetting equipment.

Some quality is sacrificed, however, and using them may require more skill and patience than the cost savings is worth, especially older Varitypers that demand a regular rhythm of keystroking for proper performance. They are also not practical for setting a lengthy project—unless you have lots of time to do the job.

PHOTOTYPESET COLD TYPE

Because of its speed, cost, and capability, phototypesetting is the most popular method today for type composition, surpassing hot type and all other cold-type processes. All phototypesetting systems use a photographic process to create an image on photosensitive paper or film. The approach to composition for body type varies from that used for display type, and both are different from photocomposing used to produce a finished layout. Specific capabilities and limitations are presented here, including what changes can be easily made on galleys.

Body Type

The elements common to all phototypesetting units used to create body type (generally from $5\frac{1}{2}$ to 12 points) are systems for input, storage, editing, computer control, and output. Their combination is what allows fast, relatively low-cost composition with a great deal of flexibility in face, style, size, weight, and spacing of body type. Figure 9.5 diagrams this process.

Input. To be converted into type, words, numbers, and symbols must first be entered into the typesetting machine. A keyboard similar to that of a standard office typewriter is used for input. At the touch of a key, an electrical impulse starts the process of creating that character on a finished galley. On more automated equipment, the operator may set the machine for typeface, style, size, weight, spacing, column width, and justification, or all specifications may be incorporated later by the computer, depending on

COLD-TYPE COMPOSITION

equipment design. With a more manual machine, the operator sets all these specifications, and makes decisions about wordspacing and hyphenation ("end-of-line decisions") at the time of keyboarding. On most phototypesetting equipment, all copy is entered on the keyboard before proceeding to the next step.

Storage. Information entered on the keyboard is stored for later processing by using paper tape, magnetic tape, floppy disk, or "optical bar recognition" (OBR) which converts each character typed into a bar code. One of these intermediaries can then be "read" at high speed by the equipment as the next step.

Editing. To produce as accurate a galley as possible, phototypesetting machines have some system for editing what is entered before type is actually set. One approach is a visual (or video) display screen (similar to a television screen or computer terminal) that allows the operator to see and correct what has been entered on the keyboard. Another approach produces a typewritten sheet ("hard copy") of what was entered (minus typesetting specifications) so that it can be proofed and a corrected tape made. A separate visual display terminal (or "editing and correcting terminal") is still another method; the tape is "read" onto a screen so an operator can make a corrected tape. This modular approach allows efficient use of the photounit, as does having an off-line keyboard so that output equipment isn't tied up with input activity.

Computer Control. This element of a phototypesetting machine manipulates information entered at the keyboard so that the resulting galley of type will be set to specifications. Words are properly hyphenated and lines justified (unless end-of-line decisions were made manually at the time of keyboarding) and any other remaining specifications incorporated.

Output. To complete the process of producing a galley of type, a photounit uses one of two basic methods: negative projection or cathode-ray tube (CRT) projection. With negative projection, a film negative of the character struck on the keyboard is exposed in a split second to high-intensity light and projected onto photosensitive paper or film, much like a photo negative is exposed in the darkroom to create a print. The type font is a square, drum, strip, or disk image master containing some or all of the characters in a given typeface, style, and weight, depending on how the equipment is designed.

CRT projection is more advanced and faster. This method uses a digital computer to convert a type character into a pattern of lines or dots on a CRT according to an image master. Each line or dot is then projected onto photosensitive paper or film to compose the full character. CRT projection has the advantage of speed and high-quality images.

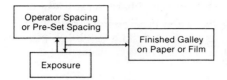

FIGURE 9.6. How headlines are set using phototypesetting equipment.

Display Type

Type that is between 12 and 74 points and/or that must be set in a style commonly reserved for headlines can be set manually or automatically by photodisplay equipment. Though much slower than phototypesetting for body type, display units can tailor letterspacing and wordspacing and select typeface and style from a wide assortment of display fonts. Regardless of manufacturer, all such machines operate by passing high-intensity light through negatives to form characters on photosensitive paper or film. By using different lenses, some can distort the image, screen it, add shadows, or execute other special effects. Figure 9.6 diagrams this process.

Manual. These devices are used in a darkroom, and the operator sees each character and word as they will appear on the finished galley. The negative of each character in the desired typeface and style is positioned according to the operator's judgment about letterspacing, and then exposed to light. The image is either developed then or after additional chemical processing.

Automatic. This equipment is used in regular room light. The operator can't see the actual composition (unless the machine is equipped with a special lens) but must direct the machine to leave a predetermined amount of space between characters and words. The finished galley is either developed inside the machine, or the photosensitive paper or film is transferred to a light-tight container that is taken to the darkroom for final processing.

Photocomposing

This version of phototypesetting allows preparation of complete layouts of ads or magazine pages, for example, without the need for paste-up. It is called photo*composing* and is not to be confused with use in this book of "composition" as a synonym for "typesetting." Characters and specifications are entered on a keyboard to create a paper tape or entered by scanning to appear on a display terminal. The machine operator then specifies the position of each element according to the layout and can change type size, content, and line lengths until satisfied. Previewer capability allows the

COLD-TYPE COMPOSITION

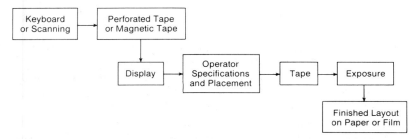

FIGURE 9.7. How a full layout is set using photocomposing equipment.

operator to see actual typeset copy. The resulting magnetic tape is then "read" by a photounit to create a one-piece composition on photosensitive paper or film. Figure 9.7 diagrams this process.

Capabilities

In addition to speed and cost, phototypesetting offers capabilities not available with hot type or the other cold-type methods covered earlier. Because phototypesetting creates images directly onto paper or film with no intervening inking, characters are sharp with crisp edges and corners.

Phototypesetting allows precise, proportional letterspacing so that each character can be set where you want it. The selection of fonts is broad, and typefaces and styles can be quickly changed in the middle of a job. Slanting, condensing, outlining, or extending type is readily accomplished with special lenses. Composition can be right-reading or wrong-reading, and the capability of producing type on film or a complete layout eliminates the need for (and cost of) mechanical artwork. Units offering a wide range of sizes may be used for both body and display type.

The frustration of phototypesetting is not in limited capability but in the wide assortment of equipment now being used with new generations continually being introduced. Type from one manufacturer won't necessarily match that from another, nor possibly even that from another model made by the same company. Type sizes may be offered in set increments only (such as 6, 8, and 10 points but not 7 or 9 points). A unit that can accommodate "S paper" only or a typographer who buys only that type of photosensitive paper produces galleys that will fade and turn brown, sometimes in just a few days. Typographers who invest in more expensive "RC paper" produce archive-quality galleys that are stable.

Typographers aren't the same, either. Some have a thorough knowledge of typography and their equipment, while others are little more than typists. (See later information on selecting a typographer.)

Changing Phototypeset Galleys

Because of different approaches to phototypesetting, certain changes after galleys are set may be inexpensive and others quite costly. The deciding factor is whether or not the copy must be re-entered on the keyboard.

If copy was set on a unit that allowed a tape to be made and typeface, style, size, weight, leading, and line length to be specified later by the computer, you can easily change any or all of these specifications after entry. For example, if you need to reduce column width from 15 to 13 picas, the tape can be rerun with the computer control reset to produce a 13-pica column. Leading can be decreased or increased and other specifications changed, all using the original tape.

If copy was set and specified on the keyboard to prepare a tape, most corrections or changes will have to be made by re-entering copy with different specifications. Ragged copy for which the operator made all end-of-line decisions is especially time consuming to change. Leading can be altered by the computer-control unit, as can point size, but only if column width is changed proportionally. Copy going directly from the keyboard to photosensitive paper or film allows no opportunity for changes. Know what your typographer can easily (and inexpensively) do, especially if you can foresee the need for such changes after composition is under way. Also specify whether or not you want a tape saved for possible use later; without that tape, your job will have to be completely reset, regardless of system design.

ELECTRONIC COLD TYPE

A direct outgrowth of phototypesetting technology, electronic cold type is sometimes referred to as third-generation typesetting (or fourth, depending on who's counting). The distinction between the first generation (hot type) and the second (phototypesetting and other cold type) is clear-cut. Third-generation developments aren't as distinct, however, because they have some elements in common with second-generation systems.

The reference to electronic composition as "computer typesetting" is accurate but somewhat misleading. Phototypesetting makes use of a computer for editing, formatting, and producing finished galleys of what has been entered on a keyboard. The difference with electronic typesetting is that it makes use of a computer for input as well, bypassing the keyboard except for corrections. Electronic typesetting also differs in speed; it is much faster for body type than other typesetting methods because of the greatly reduced need for operator participation at the input stage.

Because of rapid developments in this area, the information presented here is necessarily general. However, it should provide a basis for understanding the technology involved and what is evolving. Figure 9.8 diagrams this process.

COLD-TYPE COMPOSITION

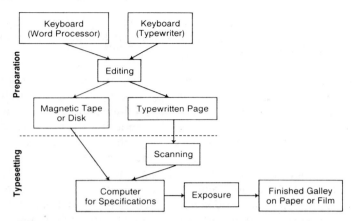

FIGURE 9.8. How cold type is set using electronic-composition equipment.

Input

Copy for electronic typesetting undergoes extensive processing before it is sent to the typographer, allowing highly efficient use of a commercial typographer's time and your budget. Copy is fully edited and proofed before the typographer receives it, either by conventional typing and retyping or by using a word processor or computer with word-processing capability. Copy that is long and repetitive in a simple format is the easiest to input. Complexities (such as Greek and math or a mixture of point sizes) may complicate input, as will a complex format. Input may be from a typewritten sheet for scanning, a magnetic tape or disk, or impulses over a telephone line.

Scanning. A sheet of typewritten copy may be scanned ("read"), eliminating the need to enter information on the keyboard. You can type copy using an element with a bar code for each character which can be scanned by optical-bar-recognition (OBR) equipment for direct input into typesetting equipment. Or you can type it using an optical-character-recognition (OCR) element which can be scanned. The latter choice is more popular because the typed sheet looks like any other product of a typewriter, word processor, or computer printer, except that an OCR element makes simplified, distinct characters that can be accurately scanned. Correcting words or characters is also easier since you don't have to be concerned about correcting the bar code as well.

An operator inserts the typed sheet into the scanner unit and watches the characters appear on a monitor. If the scanner can't recognize a character, it signals the operator to use the keyboard for clarification. Specifications may be typed in code on the first sheet of copy or the operator may enter them ahead of time on the keyboard. This latter more "user-friendly" approach allows you to submit your specifications without learning the coding language and leave translation to your typographer.

Magnetic Tape or Disk. An electronic typesetter may be equipped to accept input from a computer tape or floppy disk, bypassing paper entirely. Depending on supplier capability, you could produce final copy for magazine articles, for example, on a word processor, code it with typesetting commands, and take that disk to a typographer with a compatible typesetting unit or have it converted into a magnetic tape ("media conversion") to be compatible. Another method is to have a magnetic tape made directly (such as of a price list stored on your company's main computer) and use that tape for typesetting. Specification codes need to be incorporated before the tape is fed into the typesetting unit, either by you or by the typographer.

Teletransmission. Another input method relies on a standard telephone and a special attachment (a "modem") to your word processor or computer that allows coded information to be transmitted to a typographer across town or across the country. The modem converts characters to electronic impulses that are carried over a telephone line and reconstituted on the other end by a photounit into type galleys. Using this method of input depends on supplier capability and whether or not a modem can be attached to your word processor or computer.

Computer Control

Regardless of input method, information must be manipulated by a computer so that all specifications are incorporated. This step is similar to that used in phototypesetting. The computer unit may have some storage capacity so that the entire manuscript is entered before any of it is sent on to the photounit for typesetting, or it may operate in "real time" so that a steady flow occurs from input through to type galleys.

Output

Once information has been matched with typesetting specifications, it goes to the photounit, just as in phototypesetting. This unit may be housed in the same cabinet as the computer or be separate. A high-volume shop may feed information from two or more sources into a central computer which, in turn, supplies the photounit. Such a scheme compensates for different rates of operation and uses equipment efficiently.

Storage

Material that is to be retained for possible update or reuse as is can be stored on magnetic tape or disk. Changes can be made by the operator at the keyboard or you can make them yourself if the storage method is compatible with your word processor or computer.

Capabilities

Though speed of body composition and ease of update are strong assets of electronic typesetting, technology now goes beyond body type to systems that compose all elements of advertisements or full-page layouts, eliminating the need for doing paste-up. Units are available for scanning of line or continuous-tone art in black and white or color as well as copy and even electronic color correction and airbrushing, all for output on photosensitive paper or film. In general, they function on a digital principle, using a laser to convert the original image or direct magnetic impulses for composition by a photounit.

Because of system differences, check with your typographer about input options and specification capabilities (such as range of point sizes). If you have the equivalent of more than four typeset pages of text, electronic composition can save time and an estimated 30% or more of the cost over other methods. (An experienced typographer should be able to be more specific about how much of a savings you can expect.) In addition to volume jobs, it is ideal for price lists, rate charts, and catalogs that need to be updated (e.g., to reflect a 5% price increase) because the update can be made by computer.

To use electronic typesetting effectively, you must be able to provide clean, correct material for input. If you (or others approving copy) can't resist making last-minute changes, you'll be better off having the typographer re-enter the manuscript on the keyboard, unless those changes are very minor. Without word-processing equipment, the cost savings of scanning may be outweighed by the expense of having pages manually retyped and carefully corrected for scanning. (See Chapter 10 for information on typing copy for scanning.)

As with any other computer-based technology, its merits are in the knowledge and skill of its human users. Make sure your typographer knows how to use an electronic system to its fullest potential and how to work with you to provide appropriate input. Get acquainted with the process and typographer on a simple project and understand the particular equipment being used.

Finally, don't be dazzled by capabilities you will never use. You will pay for them in higher typesetting fees than would be charged by a shop that meets—but doesn't exceed—your needs.

SELECTING A TYPOGRAPHER

The typesetting method you use is only partially determined by what the project is. How well you select a typographer can add efficiency to any method or frustrate the entire production process. This section introduces various kinds of commercial typographers and suggests criteria for evaluating their services. Also presented is a summary of trade customs of the industry.

KINDS OF TYPOGRAPHERS

Perhaps the most common kind of typographer is the general shop. Virtually any project will be set by a general typographer, from a small ad to a many-page booklet or type not destined for a printing press (such as for audiovisual presentations). Ability to set a particular project depends on equipment, workload, and competence. A general typographer tries to have a broad enough selection of typefaces to meet most needs of its customers. A job that is especially large or that requires an unusual typeface may be referred to a specialty typographer. A trade shop sets type for print projects, and within this category may be specialists in annual reports or other specific printed pieces. A specialized trade shop will have equipment geared for that specialty (such as a system designed primarily for tabular material). The term originated to identify shops serving the printing trade exclusively, although unless a trade shop has a specialty, it is not today appreciably different from a general shop.

An advertising shop specializes in work for advertising agencies and other customers needing a wide array of display typefaces. It also has a selection of body type but concentrates on classic and contemporary faces in larger sizes for headlines. It is geared to meet very precise specifications of its customers, and its services may be priced higher than general or trade shops.

Book/magazine and catalog/directory typographers focus on large projects and offer a limited selection of utilitarian typefaces. Their services are suited to publishers and corporate customers with a high volume of repetitive material to be set.

Look for a shop that can handle the bulk of the material you will have. Once you and a typographer are acquainted and working well together, a long-term association can pay off with better service than you could expect from using a series of shops. If your needs are mostly for advertising type, for example, go to an advertising shop and when you have occasional need for brochure or booklet type, stay with that shop, as long as you can be assured of satisfactory results.

EVALUATING A TYPOGRAPHER

Finding the right kind of shop is the first step toward selecting a typographer. The second is evaluating that shop—the manager and staff—to determine competence, reliability, type selection, cost, and convenience. Even if you live in a one-shop town, look carefully at these criteria because going out of

SELECTING A TYPOGRAPHER

town for typographic services may be more efficient in the long run than attempting to work with a local shop that frustrates production standards and schedules.

Competence

A typographer should have the know-how and equipment to execute your project according to specifications, within estimate. The advent of phototypesetting and electronic composition brought many people into the typography business who weren't steeped in the science of hot-metal composition, nor well trained in these contemporary methods. Though a shop employing only glorified typists may offer an attractive price, don't become a victim of its limited competence. Projects will be successfully completed every time if you patronize a shop employing a manager and staff with a thorough understanding of typography and the particular equipment with which they work. They should also understand the basics of paste-up and other stages of print production so that the type they supply is in accordance with your subsequent production requirements.

Evaluate samples of their type, both galleys and finished printed pieces. Visit the shop and ask to see galleys set that day. Even though the shop may offer proofreading before galleys actually go out, rough galleys should be almost error free. The fewer the mistakes, the faster proofreading will go, the faster corrections can be made, and the less likely that another round of corrections will be required. Samples of finished printed pieces will indicate scope of projects done by that shop and some of its customers. Ask which ones you should contact for references.

Equipment should be adequate for your needs. With phototypesetting, determine when specifications are entered so you will know the time and cost ramifications of changing specifications when you see galleys. For electronic typesetting, determine input capabilities: magnetic tape, floppy disk, teletransmission, and/or typewritten pages for scanning.

You are also entitled to competency in pricing, whether by estimate or bid. Discuss how a price is arrived at, then check references to see if current customers are satisfied with the shop in this regard as well as with promptness and completeness of billing.

Reliability

A reliable typographer meets agreed-to deadlines and schedules workload accordingly. What you don't need is to attempt to do business with a shop that takes in more work than can be completed on time. Reliability also includes having a backup subcontractor in case an emergency causes an overload of work or equipment breaks down. Ask who the subcontractors are and make sure that the typographer accepts full responsibility for their work.

Type Selection

Compare a typographer's sample book with the kind of projects you expect to have. Several typefaces should be offered in various styles, weights, and a range of sizes to satisfy your needs for body and display type. If you must match type used in past projects, study samples closely for an exact or close-enough duplicate. If you can't make a match and are offering several printed pages' worth of type on a regular basis, perhaps the volume of work is sufficient to entice the typographer to purchase that font. Don't be dazzled by an array of exotic typefaces. If you will never use more than a half-dozen serviceable faces, don't help pay the higher overhead costs of a shop that offers a much wider selection.

Cost

Typographic services are priced by the hour or by the job. You may find phototypesetting and hot type charged by the hour, while electronic typesetting is charged by the job. Hourly rates are based on actual operator time; but because electronic typographers sell speed of service, they may charge according to convenience, not time.

Ask about rush and overtime rates and how they are defined. A shop running two shifts seldom needs to push a job into overtime, although its standard rate may be higher than that of a one-shift typographer to compensate for extended service. Also ask about a minimum charge, a fee that will be charged no matter how small the job. It may be based on one-half or one-quarter of a shop's hourly rate or be independently determined.

Convenience

Ease of pick-up and delivery is another factor. If you have a choice of one shop that charges for such service versus a comparable shop that doesn't or that is just down the street, the choice is clear from a budget and schedule standpoint. Geographic isolation, dissatisfaction with local shops, or the need for a specialty typographer may send you out of town. Discuss how long-distance phone calls, pick-up, and delivery would be handled so that you can schedule and budget adequately.

A typographer who offers related services may be more convenient for you than one that does composition only. If you also need paste-up or design services, look for one with a mechanical artist or designer on staff. A typographer with a darkroom and camera equipment to make stats may also be convenient. A full-service printer will most likely have at least some typesetting capability. This combination of services is not only convenient (if performance is satisfactory and you plan to print the piece in that shop) but also may cost less than service from a stand-alone shop with higher overhead costs.

TRADE CUSTOMS

Ownership of elements developed by a typographer to execute your job is retained by the typographer. You own the original manuscript, tape, or disk and the finished galleys. The typographer owns all metal, film, computer programs, or other materials that were used during composition to complete the project.

Alteration charges may be assessed for changes other than correction of typographical errors. If you decide to rewrite a sentence, widen a column, or otherwise alter the original manuscript or specifications after galleys are set, you will pay for the time needed to make those alterations.

Tapes and metal galleys can be stored by your typographer (possibly for a fee) if you expect to update or reuse them later. Unless you ask, however, such materials will be destroyed after the job is completed.

Place your order in writing (as covered in the following chapter). Should you need to postpone or cancel the job, expect to be charged for work completed to date. On a protracted project, a typographer may ask for payment in stages, for example, if weeks are to elapse while galleys are being proofed.

A complete statement of trade customs of the typographic industry is presented in Appendix A. Consult your typographer for other practices followed by that shop.

CHAPTER TEN

Getting Type Set

SELECTING A TYPEFACE   260
   Type Anatomy and Measurement, 260
      Special Type Characters, 261
      Measuring Type, 263
   Classifying Type, 264
      How Type Is Classified, 264
      Classifications, 264
   Selection Considerations, 268
      Appropriateness, 269
      Legibility, 269
      Readability, 270

SUBMITTING COPY   272
   Typing the Manuscript, 272
      Copy for Phototypesetting or Hot Type, 272
      Copy for Electronic Typesetting, 275
      Camera-Ready Galleys, 276
   Estimating Copy Length, 277
      Estimating Before Final Typing, 277
      Estimating After Final Typing, 280
   Specifications, 285
      Basic Specifications, 287
      Additional Specifications, 287
      Procedures, 287
      Providing a Layout, 288

PROOFING GALLEYS 288

    How Proofs Are Made, 289
        Hot-Metal Composition, 289
        Photocomposition and Electronic Composition, 289
    Ordering Proofs, 290
        Kinds of Galleys, 290
        Number of Galleys, 291
    Proofreading, 291
        What To Check, 291
        Proofreader's Marks, 294
        Checking Corrections, 298

Typesetting is the first production step in which raw materials are refined into a semi-final form. Up until this point in the process, words could be revised and design or art altered with relatively minor disturbance to your schedule or budget. Once type is set, however, such changes can impede successful completion of the piece on time and/or within cost projections.

This chapter presents details on the actual process of getting type set, regardless of your choice of composition method or typographer. It begins with the selection of a typeface, offers guidelines for submitting copy in acceptable form, and provides information on proofing galleys.

SELECTING A TYPEFACE

Deciding which typeface to use for a particular project depends on several factors, including what your typographer has available. Some knowledge of how type is constructed, measured, and classified is helpful as a starting point in the selection process.

TYPE ANATOMY AND MEASUREMENT

Knowing the basic parts of a type character, the content of a font, and type measurement will assist you in talking with your typographer about type selection and about any concerns you may have later about type galleys. Figure 10.1 identifies parts of a hot-metal character, and Figure 10.2 shows the various parts of a printed character, whether set by a hot-type or cold-type method. Like many other aspects of print production, these names aren't all universally used. The most common designations have been chosen here from among the synonyms. Perhaps the ones you will most often

SELECTING A TYPEFACE

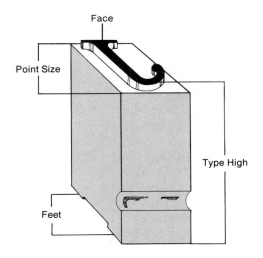

FIGURE 10.1. Basic parts and measurements of a hot-metal character.

use when talking with a typographer are "x-height" (height from baseline to the top of lowercase letters, minus any ascender), "ascender" (an extension above x-height), "descender" (an extension below x-height), and "type-high" (0.918 in. or 2.33 cm from feet to face of a hot-metal character).

Special Type Characters

A "font" (all characters in one typeface and style) usually consists of capital and lowercase letters, numerals, and punctuation marks. It may, however, include special characters which you might use occasionally (sometimes called "sorts" or "peculiars"). Figure 10.3 presents examples of these special characters.

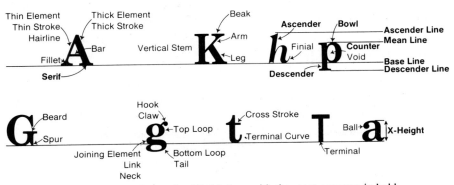

FIGURE 10.2. Parts of printed letters with the most common in bold.

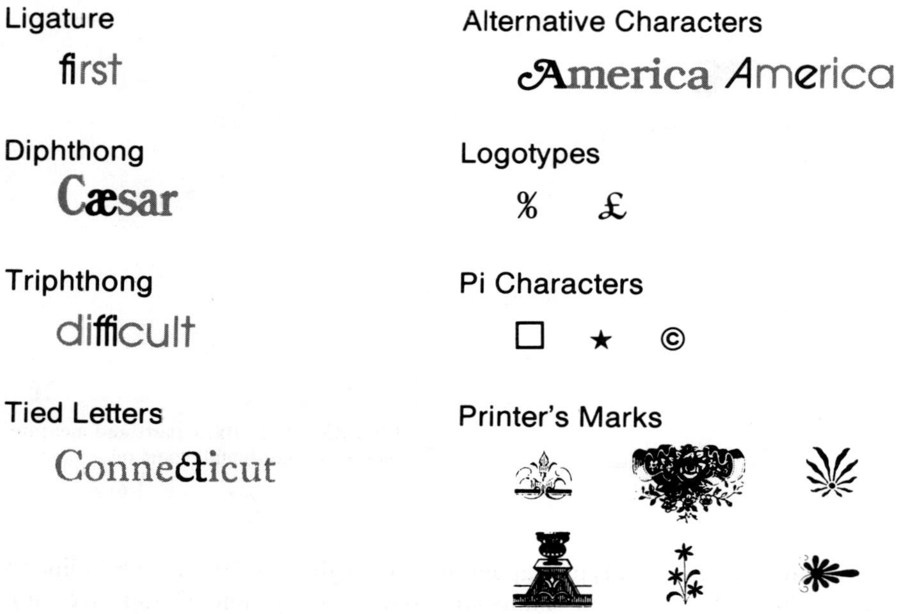

Ligature	**Alternative Characters**
first	*America* America
Diphthong	**Logotypes**
Cæsar	% £
Triphthong	**Pi Characters**
difficult	□ ★ ©
Tied Letters	**Printer's Marks**
Connecticut	

FIGURE 10.3. Examples of peculiar characters available in hot and cold type. (Printer's marks courtesy of West Coast Paper Company, Seattle.)

A "ligature" (or "quaint character") is two or more letters that are connected. Ligatures were developed in hot-metal composition to link "kerns" (portions overhanging the body of closely spaced characters) to adjacent characters so that they wouldn't be accidentally chipped off. A "diphthong" is a ligature composed of vowels, and a "triphthong" consists of three characters. "Tied letters" are linked by an additional stroke similar to handwriting; their use today is commonly limited to display type for a special effect.

"Alternative characters" are commonly seen in newer typefaces. They are distorted versions or ligatures available for special design purposes (usually display) and must be specifically requested in specifications.

"Logotypes" are symbols or trademarks used frequently enough by customers to justify including them on a type font to save the time and expense of constructing them for individual projects. "Pi characters" are unusual symbols most likely not on a regular font. They range from a simple copyright symbol to stars, arrows, and boxes in various sizes and weights.

"Printer's marks" ("ornaments," "dingbats," "fleurons," "printer's flowers," "cabbages," or "vegetables") are small but detailed designs available in hot metal, on a special font, or in transfer or camera-ready art. (The examples shown are from West Coast Paper Company's type museum and illustrate the many printer's marks that may still be available to you in hot metal.)

SELECTING A TYPEFACE

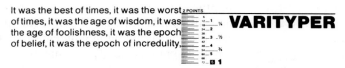

FIGURE 10.4. Using a line gauge, type is measured in points baseline to baseline. (© AM International, Inc. Line gauge courtesy of Varityper Division, AM International, Inc., East Hanover, NJ.)

Measuring Type

The height of a hot-metal character from feet to face is 0.918 in. (2.33 cm). This standardization assures that, regardless of typeface, manufacturer, or method of composition, all characters in a galley of hot type will be even across the surface.

All set type is now measured vertically in points—approximately $\frac{1}{72}$ of an inch (0.0351 cm). Type characters, leading between lines of type, and rules are measured in points. As shown in Figure 10.4, point size is accurately determined by the vertical distance from the baseline of one line of type to the baseline of the next when the lines are set solid (with no extra leading between). To do this baseline-to-baseline measurement, use a "line gauge" (or "linespacing gauge"), a ruler calibrated in point units that can be purchased from an art-supply dealer or possibly provided by your typographer.

If you don't have a line gauge and want to know the point size of existing type of unknown leading, compare it with specimens in the type-sample book provided by your typographer. Some authorities recommend measuring point size from the top of ascender to bottom of descender. This approach cannot be reliably used, however, given variations in typefaces and styles. Ask your typographer for the "E-scale" used by that shop. It shows capital E's in a range of sizes on a clear vinyl base so that the type you're measuring will be visible through it. An E-scale provides a common gauge regardless of typeface or style for the work you do with that shop.

Leading adds points between lines for improved readability and is measured in the same way as type—vertically in points. For example, 10 points of type on 12 points of space (written 10/12) amounts to two extra points of leading. In hot-metal composition, strips of lead are inserted by hand or machine between slugs; in cold-type, lines are automatically set further apart by including leading in the specifications entered during composition.

Alternatives to measuring in points which you might encounter are "nonpareil," a French measure which is the equivalent of 6 points, and "agate," a newspaper measure the equivalent of $5\frac{1}{2}$ points. Columns of advertising in newspapers and magazines may be sold today by the agate line, and any fine print (regardless of actual size) may be referred to as "agate copy." More commonly, you will convert points to picas, inches, or centimeters; one pica consists of 12 points and is roughly the equivalent of $\frac{1}{6}$ in. or 0.4216 cm.

Pica is also used as a horizontal measure of a line of type, as are inches and centimeters. "Em" is another horizontal measure, amounting to the square of the type size (such as 10 points wide by 10 points high); "en" is half an em. These terms are used to specify paragraph indentation and other horizontal spacing. A typographer might base a cost estimate on how many thousand ems are to be set for a job. They may also be used to designate space between words or letters; a 3-em space, for example, has three spaces to the em so that each space is one-third the width of an em.

Figure 10.5 is a partial page from a phototypographer's sample book illustrating the characters and point sizes available in this particular typeface and style. (See later information in this chapter on how to specify special characters and type measurements.)

CLASSIFYING TYPE

Type classifications are used by designers and typographers to group faces with similar qualities as a means of describing in common terms the thousands of typefaces available. Depending on the source consulted, the number of classifications ranges from four to a dozen or more. The list presented here is one of the more simplified versions. Refer to specialized books on typography, should you need more detailed information.

How Type Is Classified

Whether produced by hot or cold composition, type is classified according to its form, proportion, weight, and size. "Form" is a type character's overall shape. Included in form are whether or not the character has "serifs" (finishing strokes at the end of letters) and whether it is "roman" (straight up and down) or "italic" (slanted). Proportion takes into account both shape and letterspacing. As shown in Figure 10.6, proportion can be "normal" (regular), "condensed" (compressed), or "extended" (expanded). Form and proportion may collectively be referred to as type "style."

Weight ranges from extra light to extra bold in some typefaces, as illustrated in Figure 10.7. Size is a factor used in classification systems that separate body type from display. The system presented here doesn't make that distinction but relies on form as the primary determinant.

Two terms used as subclassifications are "series" and "family." A series is all sizes available in a particular typeface, which could range from 5½ to 74 points, for example. A family is the entire range of styles available in one typeface, including different forms, proportions, and weights.

Classifications

The following classifications encompass both body and display sizes, roman and italic forms, and all weights and proportions. Figure 10.8 presents examples of each. Names of typefaces may vary by equipment manufacturer,

VariTyper Phototypesetting Systems | Font No.
Colonial | **7-151**

abcdefghijklmnopqrstuvwxyz
ABCDEFGHIJKLMNOPQRSTUVWXYZ&
1234567890$¢£¹²³⁴⁵⁶⁷⁸⁹⁰$ ⅛ ⅜ ½ ⅝ ¾ ⅞
().,-:;!?/"—-*%#'"

picas 3   6   9   12   15   18   21   24

6 POINT
Typography is architecture, and the typographer is the archi
The building bricks he uses are the type faces, and the morta
is the spacing he selects for his composition. His blueprints a

7 POINT
Typography is architecture, and the typographer is
the architect. The building bricks he uses are the typ
faces, and the mortar is the spacing he selects for hi

8 POINT
Typography is architecture, and the typograph
is the architect. The building bricks he uses ar
the type faces, and the mortar is the spacing h

9 POINT
Typography is architecture, and the typo
is the architect. The building bricks he us
are the type faces, and the mortar is the s

10 POINT
Typography is architecture, and the typograph
is the architect. The building bricks he uses ar
the type faces, and the mortar is the spacing h

11 POINT
Typography is architecture, and the typog
is the architect. The building bricks he use

12 POINT
Typography is architecture, and the ty
is the architect. The building bricks he

14 POINT
Typography is architecture, and the typo
is the architect. The building bricks he use

18 POINT
Typography is architecture, and

FIGURE 10.5. An example of sizes from a typographer's sample book. (© AM International, Inc. Courtesy of Varityper Division, AM International, Inc., East Hanover, NJ.)

COPPERPLATE GOTHIC LIGHT
COPPERPLATE GOTHIC LIGHT CONDENSED

Eurostile
Eurostile Extended

FIGURE 10.6. Examples of normal, condensed, and extended proportion in type.

Megaron Extra Light
Megaron Light
Megaron Medium
Megaron Bold
Megaron Extra Bold

FIGURE 10.7. Examples of type weights from extra light to extra bold.

further confusing classifications. Your best defense is to go by the names your typographer uses and know the synonyms of those faces you use most often. The lowercase "g" is a distinctive character to compare if you are trying to identify a particular typeface.

Old English. Also called Text, Blackletter, Black Text, or Spire Gothic, this group of serifed typefaces resembles early hand-lettering and sets an ecclesiastical tone. Capital letters are ornate and difficult to read, though they can be effective design elements when used as large initial caps to start sections or paragraphs of text. Examples other than those shown are Goudy Text, Old English, and Textura.

Old Style. These typefaces have heavy serifs and little difference between thick and thin strokes within a character. Old Face and Antique are other names for them. Though designed centuries ago, they are still commonly used today by newspapers and magazines for lengthy body type because they are highly readable. Examples other than those shown are Goudy Old Style, Cloister, and Melior.

Transitional. Combining characteristics of Old Style and Modern, this group of serifed typefaces is noticeably more oval than round, with greater contrast between thick and thin strokes and an overall vertical shape. They, too, are commonly used for lengthy text. Examples other than those shown are Caledonia, Century Schoolbook, and Fournier.

SELECTING A TYPEFACE

Old English
Engravers Text
Wedding Text

Old Style
Bembo
Garamond
Times Roman

Transitional
Baskerville
Caslon
De Vinne

Modern
Bodoni
Walbaum

Sans Serif
Chelmsford
Futura Medium
Olive

Square Serif
COPPERPLATE GOTHIC LT.
Lubalin Graph Medium
Stymie Medium

Script
Helanna Script
Kaylin Script
Nuptial Script

Novelty
BROADWAY IN-LINE
Brush
STENCIL

FIGURE 10.8. Type classification and examples of each. (© AM International, Inc. Courtesy of Varityper Division, AM International, Inc., East Hanover, NJ.)

Modern. Strokes are distinctly thick and thin and letters even more oval and vertical. The serif is thin and usually straight. Though a readable body type, these faces make striking headlines. Examples other than those shown are Normande, Firmin Didot, and Craw Modern.

Sans Serif Gothic. Also called English Grotesque or Block, these faces are an outgrowth of Old English, minus the serifs. Strokes are either of one thickness or only slightly thick and thin. Because of their design neutrality, they have many uses, although the simplest versions aren't considered very readable for lengthy text. A job to be run on a gravure press, however, may require their design simplicity. Examples other than those shown are Avant Garde, Franklin Gothic, and Helvetica.

Square Serif Gothic. This classification is distinguished by the boxy shape of its heavy serifs and is also called Slab Serif or Egyptian. It is another group used primarily for display type. A variation is Copperplate

Gothic. It has tiny square serifs and has no lowercase letters. In large and small caps, it is popular for business stationery and cards. Examples other than those shown are Egyptian, Craw Clarendon, and Cheltenham.

Script. Unlike italics which are slanted versions of roman styles, Script (or Cursive) typefaces resemble handwriting and may be ornate or simple. Set in capitals and lowercase, they are popular for invitations and (in heavier weights) for display uses. Examples other than those shown are Brophy Script, Lydian Cursive, and Commercial Script.

Novelty. Also called Decorative, Special, or Occasional, this broad classification includes typefaces too ornate and difficult to read for anything but short headlines or artistic touches to a page. Many are elaborate variations on the more common faces already described and are more likely available in transfer lettering than from your typographer. A shop with a wide selection of novelty faces charges in accordance with the higher overhead of maintaining it, often by the letter. Examples other than those shown are Rustic, Neon, and Playbill (and those shown in Figure 6.25).

SELECTION CONSIDERATIONS

Each print project that you manage requires selecting a typeface (or more than one) that is appropriate as well as legible and readable, the latter two factors being especially important for body type. These considerations are covered in detail here because of their importance to the outcome of the production process. The faces and styles available to you will determine the scope of possibilities. If you must choose from among six body types, keep that limitation in mind as you consider these factors and be prepared to make compromises that wouldn't be necessary if your choices were more numerous.

Also consider them in light of the fact that you could rely on your typographer's judgment about type selection. The greater your knowledge of and experience with the selection process, however, the more you will want to participate or be fully in charge. You can take a project to your typographer, talk about what its for, and ask for advice on a typeface—or simply ask that the typographer make a selection and give you finished galleys. If you aren't satisfied with the choice, *the typographer is not liable*. You must pay for resetting in a typeface you do like.

When in doubt about your selection or that suggested by your typographer, have a representative sample of your copy set. Evaluate it for design, style, size, and weight, as well as column width, justification, linespacing, and other legibility and readability factors. Sample type is a good idea on any project that amounts to several galleys of type or that involves unusual specifications. For a large project, your typographer may give you a sample on speculation—at no charge in anticipation of doing the work for you if you like what you see.

Appropriateness

Like other elements that go into print production, type may be used to meet one or more objectives. The key to selecting appropriate type is to match type objectives with those of the printed piece overall.

Type may be used to attract attention, for example, or to create a mood, to stimulate the reader to some action, or to encourage reading. The type you select for headlines or other display purposes might primarily fulfill the first three objectives and body type the last one. Type for a brochure intended to sell furniture should reinforce that objective by attracting, stimulating, and creating a feeling of solid quality. Type for a lengthy engineering report intended to inform should encourage reading and also establish a tone of authority. Though the same typeface may be appropriate for text and headlines in a larger size, not all typefaces and styles are interchangeable in this way. An intricate design may be quite effective to attract attention in a headline but not simple enough to encourage reading of page after page of text.

The mood of a typeface should be appropriate to the subject and audience. Type should support a responsive design and be sensitive to your audience. The same lively typeface appropriate for an announcement of the company picnic would not be appropriate for an announcement of poor quarterly earnings.

Also keep in mind the paper and press on which your project will be printed. For instance, a highly textured paper to be run on a press incapable of producing high quality mandates a typeface that will come through the process intact. Select one that has medium to thick strokes that won't be lost on the textured paper and one with wide bowls that won't clog with ink. Your choice is not restricted when using a smooth or finely textured paper on a utilitarian press or any paper on a quality press.

Legibility

The ease with which one character can be distinguished from another is type legibility. Familiarity is a large factor in legibility; regardless of how intricate or distorted a character may be, if we see it frequently enough, it is legible. A personal signature is a ready example; individual letters may be ill-formed but the signature fully legible to a familiar eye.

Intricacy of design is a legibility factor when a typeface or style isn't familiar. Look again at the examples in Figure 10.8. Old English typefaces with their swirls and extra strokes are not as readily deciphered as the sans serif group. Style and size differences are other factors. A perpendicular, roman letter is easier to identify at a glance than an italic, as are medium to bold weights compared with the extremes of ultra light or extra bold. Letters set in 6-point type can't be distinguished as quickly as those set in 10-point type, and extremely large sizes are not legible when viewed up close.

Legibility is jeopardized or destroyed when characters are distorted, either in design or by special composition. Psychedelic lettering was artistic in its day but illegible without concentrating on the images. Backslanting, varying the baseline so letters jump around, staggering type horizontally, interlocking or overlapping type, and screening it back with a heavy pattern are other distortions that can make characters illegible.

Readability

The ease with which words and sentences can be comprehended through to completion of the message is type readability. This factor is especially important to consider when selecting body type since display type isn't required to be nearly as readable because of its brevity and size. You can also take readability risks when your message is short or when your readers are so interested in the message that they will persevere.

Readability is established by typeface, line length, and spacing between lines, words, and characters. (It is also influenced by design considerations covered in Chapter 6: paper finish and color, ink color, and quality of printing.) Study a type-sample book and/or a sample galley with these factors in mind.

Readable Typefaces. Extreme sizes of type are difficult to read as characters, words, and sentences. Of two typefaces the same size, however, the more readable one will have a higher x-height—more to the body of characters than to ascenders or descenders. A full, regular typeface is easier to read than one with condensed or extended proportions, as is one that isn't ultra light or extra bold.

A combination of capitals and lowercase is considered more readable than all capitals or all lowercase, whether the words are in text or headlines. A medium face is easier to read in quantity than a bold or light face. Though sans serif characters are more legible when a limited amount of type is to be comprehended (such as on a headline or directory listing), roman type is more readable for a quantity of copy because serifs help break letters apart for greater legibility.

A typeface that doesn't call attention to itself is more readable than one that is distracting by design or distortion (as covered under legibility). This quality of being invisible or neutral applies more to body type than display and also relates to mixing typefaces and/or styles within a printed piece. A crazy quilt of faces and styles might draw attention but doesn't assist the communication process (see Figure 6.14). Select different faces and styles that harmonize and use them consistently throughout a piece, rather than switching from one combination to another.

Readable Line Length. Optimum line length (or column width) for readable body type or display is one and one-half times the length of the lowercase alphabet in a given typeface and style. Type set up to 25% narrower or

50% wider is still within an acceptable range of readability. For example, if the optimum line length is 16 picas, the material could still be read if the column were 12 or 24 picas wide, though probably not as readily.

Short lines (generally 10 words or less) are considered more readable if they are justified. Longer lines can be unjustified on the right; a line of any length that isn't flush left is a challenge to reader perseverance. Also difficult are extreme "runarounds" in which type is staggered to accommodate illustrations, photographs, or a design whim. The eye becomes accustomed to a given column width and is jarred out of rhythm by shifts in width. A "widow" is another distraction—a line left over at the top of a column that is less than half the column width. (Technically, any such line at the end of a paragraph is a widow, although only a widow at the top of a column is now considered disruptive to readability and a design blunder.)

Keep line length in mind when setting financial tables or other tabular material. Your reader's eye won't jump the gap between columns unless those columns are close enough together to avoid confusing the eye or unless you use screened bars or leaders as guides. A leader can be a row of periods or hyphens. Leave a space between these characters, rather than set a solid row of periods or hyphens, so that a leader doesn't become a distracting element itself.

Readable Spacing. Lines of type should be spaced sufficiently far apart so that each line can be read distinctly. If ascenders and descenders touch or if the eye must bridge an excess of space, readability is jeopardized. A general guideline for linespacing is to add a half-point to a full point of leading between lines of type up to 8 points in size and a full point to 2 points from 8 through 12 points. In larger sizes, experiment to see what is readable; lines should be close enough so that the eye readily moves from one line to another.

Words should be spaced to avoid wide gaps between them that break up eyeflow. If you see "rivers" of white space in a type galley (a vertical or diagonal path of white through the galley), words are spaced too far apart.

Letterspacing should be close enough that characters in a word are seen as a unit but not touching (unless two or more are designed as a ligature). Though hot-metal composition may leave too much space between some letter combinations, cold-type methods generally produce acceptable letterspacing in body type (as does a skilled typographer who knows how to make adjustments). Because letterspacing is more noticeable in display type, sketch out the letters or revise a trial galley to indicate how you want it "kerned" or spaced.

Break up a lengthy sentence or paragraph into smaller ones to improve readability. (City newspapers use very short sentences and paragraphs for this reason.) Columns of type that have breathing room on a page are more readable than those that are close together and close to top, bottom, and side edges. Lengthy columns of type can be divided with subheads, art, or other devices suggested in Chapter 7 to make the material more palatable. Instead of a regular paragraph indentation of two or three spaces, set paragraphs flush left and leave an extra line between them to air out a column. An alternative is a "hanging indentation," the opposite of regular paragraph indentation. The first line is set full measure and all others in the paragraph are indented on the left.

Also keep in mind how a piece will be bound. A column of type too close to the fold or spine is difficult to read because it will roll into it vertically.

SUBMITTING COPY

The process of preparing final, approved copy for typesetting starts with typing the manuscript in a form appropriate for the typesetting method to be used and acceptable to your typographer. Knowing approximately how long the manuscript will be when set will help you select a typeface and size in accordance with the design and refine design as necessary to accommodate the amount of type you will have. Once all specifications are determined, they must be thoroughly conveyed to your typographer in understandable form.

TYPING THE MANUSCRIPT

For a project requiring a quality, finished look and compact text, copy should be typeset. Typewritten galleys may be satisfactory for a simple printed piece not having these requirements. Procedures are presented here for typing the manuscript for phototypesetting or hot type and for electronic scanning, along with information on preparing your own camera-ready galleys using a typewriter, word processor, or computer printer.

Copy for Phototypesetting or Hot Type

For print projects involving only a few pages of typed copy or several pages with many editorial changes on them, a form of photocomposition or hot-metal typesetting will probably be your choice. By any of these methods, an operator takes your manuscript and retypes it. Follow these guidelines to use your typographer's time efficiently, help assure accurate galleys, and avoid being charged extra for "penalty copy" that is difficult for the operator to read.

Paper and Spacing. Using a black ribbon (preferably carbon) and a clean typewriter, type copy double spaced on one side of a standard sheet of white paper, leaving at least 1½ in. (3.81 cm) of margin all around. Handwritten copy is acceptable only if legible and limited to a few words. Try to end each page at the close of a paragraph.

SUBMITTING COPY

To be set justified
You are cordially invited to an open house at our new offices on Thursday, March 12, 9-11 a.m.

You are cordially invited to an open house at our new offices on Thursday, March 12, 9-11 a.m.

To be set line-for-line
You are cordially invited
to an open house
at our new offices
on Thursday, March 12,
9-11 a.m.

You are cordially invited
to an open house
at our new offices
on Thursday, March 12,
9-11 a.m.

To be set line-for-line, centered
You are cordially invited
to an open house
at our new offices
on Thursday, March 12,
9-11 a.m.

You are cordially invited
to an open house
at our new offices
on Thursday, March 12,
9-11 a.m.

FIGURE 10.9. Copy with spacing specifications as submitted for typesetting (left) and the resulting galleys (right).

If copy is to be justified, it can be typed the full width of a manuscript line. If it is to be set "line-for-line," however, it must be typed with exactly the words you want on each line or copy broken by slashes to indicate where new lines are to start. Figure 10.9 shows how type for an invitation would look if marked for justification and how it should be typed if certain words are specified for each line.

Clean, Organized Copy. The fewer insertions, deletions, transpositions, and other changes that are made in copy after it is typed, the easier it will be for your typographer to read and retype. The faster a job is set, the less it will cost. Don't expect your typographer to be an editor or proofreader. Check spelling, punctuation, grammar, and numerals, and make any other editorial changes before final typing; then proof the manuscript to catch typographical errors.

Make any last-minute changes in colored pen or pencil. Print or write clearly between lines of copy or in the margin, not on the back, on an attached sheet, or anyplace else that would confuse or slow the operator. A lengthy insertion should be typed, the copy cut apart where it is to appear, and the insertion taped or glued into place. Your typewritten page won't look especially neat; but the operator won't have to flip through pages to find what you want inserted and then go back to the main copy.

Small pieces of copy should be taped or glued to a full-sized sheet of paper so they won't be lost or overlooked. If you have pages of tabular or other material that is distinct from the main copy, add them at the end of the manuscript or clip them together as a separate unit. Typesetting can then be done more efficiently unit by unit.

Send your typographer original copy or a clean photocopy or carbon. Don't submit handwritten material or a blurred copy unless you are prepared to pay for the extra time that will be required for the operator to read it. Always retain a copy of the manuscript in case the original is lost or your typographer needs to discuss some point with you by phone.

On a large job, your typographer may subcontract part of the work to another supplier or divide it between operators in that shop. Save the time and expense of having the typographer photocopy the manuscript by providing a copy as readable as the original.

Marking Copy. To help the typographer keep track, number each page of your manuscript consecutively in the upper left-hand corner. The notation "p. 2 of 6" indicates the current page as well as the total number of pages to be set. This dual numbering system lets the typographer know immediately the size of the job and if any pages of the manuscript are missing. Should you need to add a page after copy is typed, change all page numbers or label the insert "p. 2a."

At the bottom of each page of copy, write "(more)" to indicate that more copy follows. In addition to numbering the last page, signal the end of the manuscript with "###" or the old newspaper designation "-30-." Also in the upper left-hand corner, identify ("slug") each page (such as "Phoenix Brochure") so that pages of different projects don't get mixed. Full numbering and identification are especially important if your lengthy manuscript is set by more than one operator, with galleys to be assembled later in proper sequence.

Your typographer will probably prefer that you clip all pages together in order, rather than staple them. The loose manuscript can then be stacked in a page holder and each sheet quickly turned down when that page is set. If sheets are stapled, the operator must stop to flip each one and reposition the copy.

Special Markings. Words to be emphasized or otherwise treated differently should be underlined on the typewriter or by hand, as described in detail later in this chapter. Indicate on the first page of the manuscript what a typed underline means: "emphasis italic" or "emphasis boldface," for example. If copy contains a mixture, do the underlining by hand (see Figure 10.15); it will take you extra time but will prevent operator confusion and error.

Don't ask your typographer to take the time to determine if a word split at the end of a line is to be hyphenated or if it was simply split between syllables when typed. Try to avoid hyphens at the end of manuscript lines; but when a word is split, type a hyphen if the parts are to be joined when set (e.g., "note-book" to be set "notebook") and an equal sign if the parts are to be hyphenated (e.g., "no=hitter" to be set "no-hitter").

Copy for Electronic Typesetting

Copy to be scanned and typeset by an electronic, computerized system is prepared somewhat differently from copy that will be retyped on a keyboard by an operator using a photocomposition or hot-metal method. Here are guidelines to follow if the copy is to be scanned. Verify them with your typographer before preparing the final manuscript and ask about additional requirements of the particular equipment to be used.

Paper and Spacing. Type on one side of white bond paper (preferably xerographic paper because it is especially opaque for high contrast) that is at least 20 lb (30 g/m^2) so that it doesn't curl as it goes through the scanner. Leave at least one space between typewritten lines; a scanner can't accurately read copy that is too closely spaced vertically. Use a new, black, carbon-film ribbon. Make certain lines are even and horizontal; a line with one or more words higher or lower than the rest or typed at an angle may throw off a scanner. Type line-for-line copy as described earlier for photocomposition or hot-type manuscripts. Margins should be consistent page to page; the scanner can't read closer than a half-inch (1.27 cm) to the edge of the paper.

The Typing Element. Check with your typographer to confirm which elements for your typewriter, word processor, or computer printer the scanner can read. Many styles are too fine or intricate for scanning, and copy typed on 12 pitch may be too tight to be read correctly. Select a simple, block pica style, an elite typed on 10 pitch, or a special OCR element. Characters should not touch, should be free of smudges or globs of ink caused by a dirty element, and should be complete, with no nicks. When in doubt, ask your typographer to test a sample of your typed copy.

Clean, Original Copy. Any editorial changes on the final typed copy for scanning will require operator intervention to make. The operator will have to follow along on the manuscript and monitor to the point of a change, and then manually make the change on the keyboard to correct the tape or disk before it goes to the composing unit. (Such intervention is also necessary when the manuscript is set directly from a computer tape or disk.) Therefore, the cleaner the copy going to the typographer or transferred onto tape or disk, the more efficient computer typesetting will be.

Make sure that the manuscript is immaculate—thoroughly edited and proofread (as covered for photocomposition or hot-type manuscripts). All changes must be typewritten and incorporated into the main copy, not inserted between lines or in the margins. Unless you can provide a photocopy free of specks and shading, give your typographer the original manuscript for scanning.

Marking Copy. Designate the job name, page numbers, end of copy, and other signals to the typographer in red or blue pencil or a felt-tip marker. The scanner will read only (and all) typed copy; so markings must be by hand instead of being typed, as may be done for other methods of typesetting. Attach a separate page showing all type specifications and notes to the typographer. Don't mix written specifications with copy. Your typographer may instruct you to type certain signals to the typesetting unit, however (such as to triple space before and after a word you want set in boldface).

Camera-Ready Galleys

To prepare camera-ready galleys on a typewriter or computer printer, start with a clean typing element with no nicked or clogged characters and use smooth, dull, white xerographic or bond paper. Mimeo or other rough paper will "pick"—paper fibers will pull loose upon impact to create a fuzzy appearance. Lightly mark any guiding lines (such as column width or center) in nonrepro blue or mark them in dark ink or pencil on a separate sheet so that the lines show through when the sheet is placed underneath the galley you are typing.

Ribbons and Printers. Use a black carbon-film ribbon; a cloth ribbon will splay or spread characters and not produce fully legible galleys. Store the ribbon in a cool place to prevent it from drying out and leaving carbon dust on your galleys and inside your typewriter or printer. If you're using a computer printer, make sure it is letter quality. Trying to read galleys produced on a dot-matrix printer is worse than suffering through the product of a cloth ribbon. Set your typewriter or printer to give a heavy impression, without being so heavy that the paper is cut upon impact.

Typefaces and Sizes. Avoid thin, intricate typefaces. A standard block face (such as Courier or Prestige Elite) will give you clean, crisp galleys. If you need to have both roman and italic styles, select faces of comparable weight or double-print an italic that appears lighter than its roman equivalent. Select a typing element to which you have access later, should you need to add to or change galleys or match the typeface on a subsequent project. Either a pica or elite element is readable. To keep characters from touching, set an adjustable typewriter or printer to 10 pitch for a pica element and to 12 pitch for an elite or pica. Neither size should be typed as tightly as 15 pitch. Use proportional spacing if your equipment has that feature; the resulting galleys will have a more professional look than is possible if each character is allotted the same amount of space, especially if the column is justified.

Emphasis. To add emphasis to body type or to create headlines on a typewriter or printer, underline words or use all capital letters. For extra

SUBMITTING COPY

emphasis, create boldface by double- or triple-printing or (if possible) by "shadow printing," in which the second impression is slightly out of registration from the first to make heavier characters.

Justification. The easiest format for typewritten galleys is flush left and ragged right, unless your word processor, computer, or executive typewriter will justify automatically. You can justify galleys manually by typing the copy once as close to the full column width as possible, marking the number of spaces to be added to each line, and inserting those extra spaces as you retype. Another way is to use wide paper, type a line on the left half, and then add necessary spaces as you retype on the right half, line by line. This procedure creates two columns—the one on the right a finished galley and the one on the left a throwaway.

Corrections. Erasures or a correcting ribbon or strip that masks a mistake may leave smudges or marks that the camera will pick up when a negative or printing plate is made. Instead, use a thin coat of opaque correcting fluid that fully covers over a mistake or a correcting tape that lifts off the carbon film. Another alternative is to type corrected words or phrases on white gummed labels or strips and stick them over mistakes or on white paper and glue them over. When using either of these methods, type two or three words beside the mistake and stick them as a unit over the affected area to assure alignment. To be even more precise, also type words from the line above or below the mistake.

Reducing Typewritten Galleys. Text set on a typewriter or printer will more closely resemble typeset galleys if it is clarified by a slight reduction. An effective reduction of pica type is 85% of the original, or expressed another way, 15% less (reduce only 5 or 10% for elite). Type galleys 15% wider and longer than the desired final size, complete mechanical artwork accordingly (see Chapter 11), and ask your engraver or printer to reduce the image by 15% when making the negative or printing plate. The result will be crisper type, in the desired finished size.

ESTIMATING COPY LENGTH

The easiest and most accurate method of estimating how much text you will have when copy is typeset starts before final typing of the manuscript so that margins can be set according to the character count of the typeface, style, size, and weight you have selected. Length can also be estimated after final typing of copy has taken place, though not as accurately.

Estimating Before Final Typing

"Character count" refers to how many average characters will fit into one pica of space. The fastest way to determine character count is to consult a

Copyfitting Data
Friz Quadrata Medium

Point Size	Characters Per Pica	Lowercase Alphabet Length in Points
5	5.2	60.3
6	4.4	78.3
7	3.7	91.3
8	3.3	104.3
9	2.9	117.5
10	2.6	120.5
11	2.4	143.5
12	2.2	156.6
13	2.0	169.6
14	1.9	182.7
15	1.7	195.8
16	1.6	208.8
17	1.5	221.8
18	1.5	235.9

FIGURE 10.10. An example of a character-count table from a typographer's sample book.

table supplied by your typographer for all the typefaces, styles, sizes, and weights offered by that particular shop.

Character-Count Table. Figure 10.10 is a simplified character-count table for Friz Quadrata medium roman. Use it to estimate copy length as follows, assuming, for example, that type size is 10 points and column width is 20 picas:

1. Look up 10-point type and the corresponding characters per pica: 2.6.
2. Multiply 2.6 by 20 (the column width) to get characters per 20-pica line: 52.0.
3. Set your typewriter margins to produce lines of 52 characters each and type the manuscript.
4. Count the number of full and partial lines for a close estimate of how many lines you will have when copy is typeset in a 20-pica column of 10-point Friz Quadrata medium roman.

Accounting for Leading. To translate number of lines into how deep the column of type will be, you must first decide how much space to leave between lines of type. Figure 10.11 converts points to vertical picas, and

Points	Picas	Inches	Centimeters
1	0.0833	0.0138	0.0351
2	0.1667	0.0277	0.0703
3	0.2500	0.0415	0.1054
4	0.3333	0.0553	0.1405
5	0.4167	0.0692	0.1757
6	0.5000	0.0830	0.2108
7	0.5833	0.0968	0.2460
8	0.6667	0.1107	0.2811
9	0.7500	0.1245	0.3162
10	0.8333	0.1383	0.3514
11	0.9167	0.1522	0.3865
12	1.0000	0.1660	0.4216
13	1.0833	0.1798	0.4568
14	1.1667	0.1937	0.4919
15	1.2500	0.2075	0.5271
16	1.3333	0.2213	0.5622
17	1.4167	0.2352	0.5973
18	1.5000	0.2490	0.6325
19	1.5833	0.2628	0.6676
20	1.6667	0.2767	0.7027
21	1.7500	0.2905	0.7379
22	1.8333	0.3043	0.7730
23	1.9167	0.3182	0.8081
24	2.0000	0.3320	0.8433
25	2.0833	0.3458	0.8784
26	2.1667	0.3597	0.9136
27	2.2500	0.3735	0.9487
28	2.3333	0.3873	0.9838
29	2.4167	0.4012	1.0190
30	2.5000	0.4150	1.0541
31	2.5833	0.4288	1.0892
32	2.6667	0.4427	1.1244
33	2.7500	0.4565	1.1595
34	2.8333	0.4703	1.1946
35	2.9167	0.4842	1.2298
36	3.0000	0.4980	1.2649

FIGURE 10.11. Conversion of vertical points to picas, inches, and centimeters.

then to inches and centimeters. For example, assume that type will be set 10/11—10 points of type on 11 points of space—and be 35 lines deep:

1. Refer to the table to find that 11 points is equal to 0.9167 picas (0.1522 in. or 0.3865 cm).
2. Multiply 0.9167 picas by 35 lines to find that type depth will be 32.08 picas (5.33 in. or 13.53 cm).

For a major or frequent project, type copy on pages ruled according to the number of characters in a line of your type and either photocopied or printed in quantity. With this guide, all you need to do is count the number of typed lines and multiply by points per line for an estimate of depth.

Alphabet Length. When you don't have a table of character counts for a particular typeface, style, size, and weight, you can calculate approximate count if you know the length in points of its lowercase alphabet, a to z. Figure 10.12 converts alphabet length into approximate characters per pica.

In the example just presented, the lowercase alphabet of 10-point Friz Quadrata medium roman extends to 130.5 points, according to the character-count table (Figure 10.10). If you don't have such a table, ask your typographer for the lowercase alphabet length. Find that alphabet length on the master table (Figure 10.12), or the next longest length to it (129). This typeface, style, size, and weight then averages 2.6 characters per pica.

Using Sample Type. If you have neither kind of table, you can estimate copy length by using a sample of type already set in the face, style, size, and weight you want:

1. Measure a line of existing type and mark where it would end if set in the column width you want.
2. Retype the line on your typewriter up to that mark. Do three or four lines to get an average count.
3. Set your right typewriter margin to correspond.
4. Type your manuscript and count the number of lines you have.

Estimating After Final Typing

If you are handed copy already typed in its final form, you can still estimate its length (though not as accurately) using character or word count:

1. Count the number of characters in a few lines to find the average characters per line. (An alternative is to find the average line length in inches and multiply that number by 10 characters per inch if copy was typed at 10 pitch or by 12 if typed at 12 pitch.)

Lowercase Alphabet Length In Points	Picas 1	10	12	14	16	18	20	22	24	26	28	30	32	34	36	38	40	42
61	5.00	50	60	70	80	90	100	110	120	130	140	150	160	170	180	190	200	210
62	4.92	49	59	69	79	89	98	108	118	128	138	148	157	167	177	187	197	207
63	4.87	49	58	68	78	88	97	107	117	127	136	146	156	166	175	185	195	205
64	4.80	48	58	67	77	86	96	106	115	125	134	144	154	163	173	182	192	202
65	4.75	48	57	67	76	86	95	105	114	124	133	143	152	162	171	181	190	200
66	4.70	47	56	66	75	85	94	103	113	122	132	141	150	160	169	179	188	197
67	4.65	47	56	65	74	84	93	102	112	121	130	140	149	158	167	177	186	195
68	4.60	46	55	64	74	83	92	101	110	120	129	138	147	156	166	175	184	193
69	4.55	46	55	64	73	82	91	100	109	118	127	137	146	155	164	173	182	191
70	4.50	45	54	63	72	81	90	99	108	117	126	135	144	153	162	171	180	189
71	4.45	45	53	62	71	80	89	98	107	116	125	134	142	151	160	169	178	187
72	4.40	44	53	62	70	79	88	97	106	114	123	132	141	150	158	167	176	185
73	4.35	44	52	61	70	78	87	96	104	113	122	131	139	148	157	165	174	183
74	4.30	43	52	60	69	77	86	95	103	112	120	129	138	146	155	163	172	181
75	4.25	43	51	60	68	77	85	94	102	111	119	128	136	145	153	162	170	179
76	4.20	42	50	59	67	76	84	92	101	109	118	126	134	143	151	160	168	176
77	4.15	42	50	58	66	75	83	91	100	108	116	125	133	141	149	158	166	174
78	4.10	41	49	57	66	74	82	90	98	107	115	123	131	139	148	156	164	172
79	4.05	41	49	57	65	73	81	89	97	105	113	122	130	138	146	154	162	170
80	4.00	40	48	56	64	72	80	88	96	104	112	120	128	136	144	152	160	168
81	3.95	40	47	55	63	71	79	87	95	103	111	119	126	134	142	150	158	166
82	3.90	39	47	55	62	70	78	86	94	101	109	117	125	133	140	148	156	164
83	3.85	39	46	54	62	69	77	85	92	100	108	116	123	131	139	146	154	162
84	3.80	38	46	53	61	68	76	84	91	99	106	114	122	129	137	144	152	160
		10	12	14	16	18	20	22	24	26	28	30	32	34	36	38	40	42

FIGURE 10.12. Approximate characters per pica. Use this table to determine approximate characters per pica based on the length of the lowercase alphabet in a given typeface, style, weight, and size. Find lowercase-alphabet length on the left (letters a through z). Column 1 shows characters per pica; other columns give approximate character count for different column widths measured in picas.

Lowercase Alphabet Length In Points	Picas 1	10	12	14	16	18	20	22	24	26	28	30	32	34	36	38	40	42
86	3.75	38	45	53	60	68	75	83	90	98	105	113	120	128	135	143	150	158
87	3.70	37	44	52	59	67	74	81	89	96	104	111	118	126	133	141	148	155
88	3.65	37	44	51	58	66	73	80	88	95	102	110	117	124	131	139	146	153
90	3.60	36	43	50	58	65	72	79	86	94	101	108	115	122	130	137	144	151
91	3.55	36	43	50	57	64	71	78	85	92	99	107	114	121	128	135	142	149
93	3.50	35	42	49	56	63	70	77	84	91	98	105	112	119	126	133	140	147
94	3.45	35	41	48	55	62	69	76	83	90	97	104	110	117	124	131	138	145
96	3.40	34	41	48	54	61	68	75	82	88	95	102	109	116	122	129	136	143
98	3.35	34	40	47	54	60	67	74	80	87	94	101	107	114	121	127	134	141
100	3.30	33	40	46	53	59	66	73	79	86	92	99	106	112	119	125	132	139
102	3.25	33	39	46	52	59	65	72	78	85	91	98	104	111	117	124	130	137
104	3.20	32	38	45	51	58	64	70	77	83	90	96	102	109	115	122	128	134
106	3.15	32	38	44	50	57	63	69	76	82	88	95	101	107	113	120	126	132
108	3.10	31	37	43	50	56	62	68	74	81	87	93	99	105	112	118	124	130
110	3.05	31	37	43	49	55	61	67	73	79	85	92	98	104	110	116	122	128
112	3.00	30	36	42	48	54	60	66	72	78	84	90	96	102	108	114	120	126
114	2.95	30	35	41	47	53	59	65	71	77	83	89	94	100	106	112	118	124
116	2.90	29	35	41	46	52	58	64	70	75	81	87	93	99	104	110	116	122
118	2.85	29	34	40	46	51	57	63	68	74	80	86	91	97	103	108	114	120
120	2.80	28	34	39	45	50	56	62	67	73	78	84	90	95	101	106	112	118
122	2.75	28	33	39	44	50	55	61	66	72	77	83	88	94	99	105	110	116
124	2.70	27	32	38	43	49	54	59	65	70	76	81	86	92	97	103	108	113
127	2.65	27	32	37	42	48	53	58	64	69	74	80	85	90	95	101	106	111
129	2.60	26	31	36	42	47	52	57	62	68	73	78	83	88	94	99	104	109
132	2.55	26	31	36	41	46	51	56	61	66	71	77	82	87	92	97	102	107
135	2.50	25	30	35	40	45	50	55	60	65	70	75	80	85	90	95	100	105
	1	10	12	14	16	18	20	22	24	26	28	30	32	34	36	38	40	42

FIGURE 10.12. (Continued)

Lowercase Alphabet Length In Points	Picas 1	10	12	14	16	18	20	22	24	26	28	30	32	34	36	38	40	42
138	2.45	25	29	34	39	44	49	54	59	64	69	74	78	83	88	93	98	103
142	2.40	24	29	34	38	43	48	53	58	62	67	72	77	82	86	91	96	101
146	2.35	24	28	33	38	42	47	52	56	61	66	71	75	80	85	89	94	99
150	2.30	23	28	32	37	41	46	51	55	60	64	69	74	78	83	87	92	97
154	2.25	23	27	32	36	41	45	50	54	59	63	68	72	77	81	86	90	95
158	2.20	22	26	31	35	40	44	48	53	57	62	66	70	75	79	84	88	92
162	2.15	22	26	30	34	39	43	47	52	56	60	65	69	73	77	82	86	90
166	2.10	21	25	29	33	38	42	46	50	55	59	63	67	71	76	80	84	88
170	2.05	21	25	29	33	37	41	45	49	53	57	62	66	70	74	78	82	86
175	2.00	20	24	28	32	36	40	44	48	52	56	60	64	68	72	76	80	84
180	1.95	20	23	27	31	35	39	43	47	51	55	59	62	66	70	74	78	82
185	1.90	19	23	27	30	34	38	42	46	49	53	57	61	65	68	72	76	80
190	1.85	19	22	26	30	33	37	41	44	48	52	56	59	63	67	70	74	78
195	1.80	18	22	25	29	32	36	40	43	47	50	54	58	61	65	68	72	76
200	1.75	18	21	25	28	32	35	39	42	46	49	53	56	60	63	67	70	74
206	1.70	17	20	24	27	31	34	37	41	44	48	51	54	58	61	65	68	71
212	1.65	17	20	23	26	30	33	36	40	43	46	50	53	56	59	63	66	69
218	1.60	16	19	22	26	29	32	35	38	42	45	48	51	54	58	61	64	67
225	1.55	16	19	22	25	28	31	34	37	40	43	47	50	53	56	59	62	65
233	1.50	15	18	21	24	27	30	33	36	39	42	45	48	51	54	57	60	63
241	1.45	15	17	20	23	26	29	32	35	38	41	44	46	49	52	55	58	61
250	1.40	14	17	20	22	25	28	31	34	36	39	42	45	48	50	53	56	59
260	1.35	14	16	19	22	24	27	30	32	35	38	41	43	46	49	51	54	57
270	1.30	13	16	18	21	23	26	29	31	34	36	39	42	44	47	49	52	55
280	1.25	13	15	18	20	23	25	28	30	33	35	38	40	43	45	48	50	53
295	1.20	12	14	17	19	22	24	26	29	31	34	36	38	41	43	46	48	50
	1	10	12	14	16	18	20	22	24	26	28	30	32	34	36	38	40	42

FIGURE 10.12. (Continued)

Lowercase Alphabet Length In Points	Picas 1	10	12	14	16	18	20	22	24	26	28	30	32	34	36	38	40	42
310	1.15	12	14	16	18	21	23	25	28	30	32	35	37	39	41	44	46	48
325	1.10	11	13	15	18	20	22	24	26	29	31	33	35	37	40	42	44	46
340	1.05	11	13	15	17	19	21	23	25	27	29	31	34	36	38	40	42	44
360	1.00	10	12	14	16	18	20	22	24	26	28	30	32	34	36	38	40	42
380	0.95	10	11	13	15	17	19	21	23	25	27	29	30	32	34	36	38	40
400	0.90	9	11	13	14	16	18	20	22	23	25	27	29	31	32	34	36	38
425	0.85	9	10	12	14	15	17	19	20	22	24	26	27	29	31	32	34	36
450	0.80	8	10	11	13	14	16	18	19	21	22	24	26	27	29	30	32	34
475	0.75	8	9	11	12	14	15	17	18	20	21	23	24	26	27	29	30	32
500	0.70	7	8	10	11	13	14	15	17	18	20	21	22	24	25	27	28	29
	1	10	12	14	16	18	20	22	24	26	28	30	32	34	36	38	40	42

FIGURE 10.12. (Continued)

2. Multiply the average characters per line by the number of lines in the manuscript, counting full lines and equivalents. This figure is the total number of characters in the manuscript.
3. Use a character-count table or existing typeset copy (as just mentioned) to determine how many characters of your selected typeface, style, size, and weight will fit on each line when set.
4. Divide the total number of characters in the manuscript by this figure to get the number of lines of type.

For example, assume that the average characters per line of typewritten manuscript is 30 and the number of full or equivalent lines is 100. The total number of characters in the manuscript is then 3000 (30 characters × 100 lines). If copy is to be set in 10-point Friz Quadrata medium roman in a 20-pica column, the character count per line will be 52 (2.6 characters per pica × 20 picas). Divide 3000 by 52 for a total of 57.69 or 58 lines of typeset copy.

A less accurate approach to estimating length of copy already typed is to calculate word count. One double-spaced page contains 200 to 300 words, or an average of 250, depending on margins, headings, and type size:

1. Multiply 250 by the number of pages in the manuscript for a total count.
2. Calculate the average number of words per line in the selected typeface, style, size, and weight.
3. Divide the total count for the manuscript by this figure for an estimate of the number of typeset lines.

SPECIFICATIONS

The specifications you send along with the manuscript to your typographer must be accurate and comprehensive if the resulting galleys are to be what you want. To assure that both these requirements are met, get into the habit of marking copy as well as including a summary of specifications (as suggested in Figure 10.13) for anything other than a routine, familiar project. It prompts you to provide basic specifications (Section A), any additional ones necessary for a project that is demanding in detail (Section B), and procedural information (Section C). A rough layout is a required part of the specifications package (depending on complexity) when you expect full composition to bypass some or all paste-up work.

On a small or routine job going to a familiar typographer, including a separate list of specifications may not be necessary. However, on a large or complex project—and one that is to be scanned—the separate list is added assurance that your specifications will be relayed clearly and completely.

The advent of photocomposition brought convenience and lower cost but also its own vocabulary. Though your typographer may use older, more

Type Specification Information

For _Sample Project_

A
- Body Type _12/14 Souvenir Medium_
- Headlines _12 pt. bold_
- Leading Between Lines _14 points_ Leading Between Paragraphs _28 points_
- Column Width _20 picas maximum_
- Indentation _1 cm_
- Justified _____ Rag Right _✓_ Rag Left _____ Centered _____

B
- Cutlines _8 pt. Souvenir Italic - see layout for column widths_
- Subheads _____
- Continued Lines _____ Continued Headlines _____
- Bylines _8 pt Souvenir Italic_
- Footnotes _____
- Introductory Material _____
- Page Numbers _7 pt Megaron Medium_
- Initial Caps _____
- Large/Small Caps _____
- Volume/Number/Date _____
- Tabulation _____
- Emphasis _Italic_
- Kerning _on headlines_
- Rules _____
- Leaders _____
- Litho in U.S.A./Canada _____ Publication Code _____
- Copyright/Trademark _____

C
- All Copy Attached _✓_ More to Come _____
- Number Galley Copies _3_ Number Originals _1_
- Waxed _wax both originals and one set of galleys_
- Need First Proofs by _Wednesday, 8/16_ Need Corrected Galleys by _Friday_
- By _John L. Smith_ Phone _555-1234_

FIGURE 10.13. An example of a form to organize typesetting specifications.

SUBMITTING COPY

familiar words, don't be surprised to be asked to use the following alternatives in your specifications:

Newer Term	Older Term
Quad left	Flush left
Quad right	Flush right
Quad centered, center quadding	Centered
Interline spacing	Leading
Intercharacter spacing	Letterspacing
Photoproof	Proof
Photoprint, photorepro	Repro Proof

Basic Specifications

As noted in Section A of Figure 10.13, typeface, style, size, and weight must be designated for all copy to be set, including body, headline, subhead, cutline, and any other type. Column width should be specified in picas, and leading between lines and paragraphs expressed in points. How columns are to appear is specified by marking justification: justified, rag right, rag left, or centered. Mark the copy with these specifications, as shown in Figure 10.14. Italic words are underscored with a straight line or marked with "ital"; boldface words are underscored with a wavy line or marked with "bf" or "bold." To distinguish specifications from copy, circle specification markings. (See the following section on proofreading for details on markings.)

Additional Specifications

Section B of Figure 10.13 itemizes additional specifications that may be required on more detailed typesetting jobs. Though a project requiring all of them would be rare, this section does help assure that no fine point is overlooked, down to how to set your company's code for a particular printed piece or the size and position of a copyright or trademark symbol. Mark these specifications on copy as well.

Procedures

Your typographer needs to know the procedural information suggested in Section C. Indicate if all copy is attached or if more is to come and when you need to see the first set of galleys and the final, corrected set. Specify the number and kind of galleys you need for proofing and dummying and if you want any waxed (covered in detail later). This information should not be duplicated on the manuscript. It could be relayed to your typographer by phone; however, putting it in writing leaves no room for misunderstandings. An alternative to showing these requirements on a specification form is to include them on a purchase order.

12/14 Souvenir medium, 20 picas maximum, flush left, rag right.

A Tale of Two Cities

☐ It was the best of times, it was the worst of times, it was the age of wisdom, it was the age of foolishness, it was the epoch of belief, it was the epoch of incredulity, it was the season of Light, it was the season of Darkness.

☐ It was the spring of hope, it was the winter of despair, we had everything before us, we had nothing before us.

FIGURE 10.14. An example of simple specifications for copy as it would be submitted for typesetting showing type size, face, style, and weight, leading, column width, justification, headline treatment, indentation, and spacing between paragraphs.

Providing a Layout

Your typographer will need to see how you want elements of an ad or a magazine page positioned, for example, if you request photocomposing to save paste-up time and expense. Copy should be typed and specifications noted as on any other type job; a sketch or layout is an additional requirement, not a substitution.

Also include a layout if you want type to run around an illustration or other art that juts into a column or if you want it set in an unusual shape. A rough sketch or example of what you need is probably adequate if your typographer has experience with such special work and if you discuss it element by element. Otherwise, you or your designer will need to do a more precise layout. Expect to be asked to react to a preliminary galley so that adjustments can be made before the final version is satisfactorily executed.

PROOFING GALLEYS

Proofing of type occurs at two stages of production: when galleys are checked ("galley proofs") and when camera-ready pages ("page proofs") are checked. As far as the type itself is concerned, the two proofings have the same objective of checking that type is accurate in content and specifications. This section presents information on how proofs are made, the various kinds of proofs you can request, and how to do proofreading. (See Chapter 11 for guidelines for proofing mechanical artwork.)

HOW PROOFS ARE MADE

Galley proofs are simply copies of galleys of type. How they are made depends on the composition method used to create the type and on the condition of original galleys.

Hot-Metal Composition

When enough lines of type are set to fill a metal tray ("galley"), the type is locked in place and inked, a sheet of paper laid over it, and pressure applied to transfer the type image onto the paper. A small proof press is used for this purpose. Each galley proof may be faintly numbered at intervals from top to bottom so that the original type can be located quickly to correspond with numbered galley segments on the mechanical dummy. Some proofing and correcting may be done by the typographer before proofs go to the customer, although probably at an extra charge.

Photocomposition and Electronic Composition

Because the process of transferring an image onto photosensitive paper or film is the same for each of these typesetting methods, the making of proofs is the same. A typographer may send you the original to proof so you can check it closely, without being distracted by the quality of a photocopy. If you are to proof from the original (or request to), handle it carefully because if it is correct as set, it will be used on the mechanical. If not, the whole galley or segments of it will be corrected and rerun to create a fresh original, a costly process given the expense of photosensitive paper but perhaps not as costly as the extra paste-up time needed to piece in corrections.

The simplest copy of a galley for proofing is a photocopy made on xerographic paper by an ordinary office copier. Though it may slightly enlarge or reduce characters and deposit specks of toner on the paper, a photocopier is adequate for most projects. Should you need a clearer copy, request that proofs be made on a Diazo, Ozalid, or Bruning machine, depending on what your typographer has available.

Proofs of galleys or full compositions set directly onto film are made just as proofs of negatives that are ready for the press. Blueprint or Dylux paper is used to create a blueline and brown paper to create a brownline or Vandyke, again depending on supplier equipment. These methods produce a proof similar in quality to a photocopy. If you need one that is exact in every detail, either have the original galley set on photosensitive paper or use the film and photosensitive paper to shoot a stat, an exact paper duplicate.

Some typesetting systems produce a typewritten copy of what was entered on the keyboard for proofing before the type is set. Though this approach makes efficient use of supplies, it doesn't allow you to see how the actual galley will look. It is little different from the operator's catching errors on the monitor. Your typographer may proof and correct galleys before you see them, although for an extra charge. A shop that customarily proofs and corrects galleys may base its price on doing so, whether or not you use the service.

ORDERING PROOFS

Several kinds of proofs can be made, depending on your needs for proofreading and dummying. Though hot-type composition may generate different kinds of proofs in addition to those mentioned here, these are the ones now commonly used. Specify how many of which kinds you want when copy is submitted for typesetting.

Kinds of Galleys

If you ask for a galley proof or "checker" to proofread, you will receive a copy of the galley on newsprint if made from hot type or one of the kinds of photocopies just mentioned if made from cold type. A "pink proof" is a galley proof on colored paper (which may or may not be pink) for purposes of dummying. It shows the same type and is merely a way to distinguish a galley for dummying from a galley for proofreading.

A "rough proof" is one that has not been proofed and corrected by the typographer. If you don't want to pay for that service, ask for this kind of galley proof. "Glassine" or tissue proofs are made on semi-transparent paper so that type can be placed over the layout to check fit. Graphic designers often request this kind of proof if they are working with an exacting layout.

A "reproduction proof" ("repro") is made on coated paper (if from hot type) or on photosensitive paper (if from cold type) to create an exact duplicate of the original galley. A repro can be used for proofing and, if perfect, used immediately for pasting up the mechanical. This kind of proof is requested when production time is short and when a backup galley needs to be readily at hand in case of an error during paste-up.

A "waxed galley" has a thin coating of hot wax applied to the back for ease in moving segments of the galley around during dummying and paste-up. Once positioning is fixed, the galley is burnished and the wax holds it in place. Any original or copy of a galley can be waxed (though some designers and mechanical artists prefer to use instead rubber cement or a spray adhesive).

Number of Galleys

Each person who will be proofing galleys should have a set to read, thereby allowing simultaneous proofing. An alternative when time allows is to route the galleys, although receiving an already-marked set may cause the next proofreader to slacken off. Request an additional set of galleys on which to consolidate all corrections and changes for the typographer. Another set should be designated for dummying, and original galleys reserved for mechanical artwork. For a project requiring especially detailed paste-up, you may want a spare set of originals in case of damage during production.

On a newsletter project that you will be proofing and dummying yourself and that a freelance mechanical artist will paste up, for example, request three sets of galleys—one to read, one to send back with corrections and changes, and one to use in dummying. Request two copies of corrected galleys—one to read and one to send back if further changes are needed—and the original for the mechanical artist to paste up.

PROOFREADING

The process of proofreading may be a one-person or two-person operation (or you can use a tape recorder as a substitute for another person). You can read and mark proofs by yourself, hire a freelance proofreader for an extensive project, or you and another person can take turns reading original copy out loud and marking corrections on the proofs. Either method has the same objective: to make sure that type is accurate. One-person proofing is less time consuming than two-person and, with experience, can be very accurate. For the inexperienced, however, or for a complex or technical job, having two proofreaders involved is worth the time invested.

What To Check

Type that is accurately set is correct in content and in specifications. These elements can be checked simultaneously on a routine project but are better checked separately on a project that is new or demanding in both areas.

Checking content consists of determining if the entire manuscript has been set—each word, line, paragraph, number, headline, cutline, and any other material sent in for composition—and if everything is set accurately. Words should be properly spelled and hyphenated, punctuation complete, and numbers in order.

Checking specifications consists of determining if all elements are set as you want them—typeface, style, size, weight, spacing, indentations, column width, and all the other requirements you noted on the manuscript and/or specification form. This process may require measuring and checking against sample type.

"Author's alterations" are changes made in the original manuscript after

Marks	Meaning	Marked in Manuscript	Set in Type
Size and Style of Type			
lc /	Lowercase	lc It was the Best of times	It was the best of times
caps ≡	Capital	caps it was the best of times	It was the best of times
clc or cap & lc or u & lc	Initial caps and lowercase	clc It was the best of times	It Was the Best of Times
ital ___	Italic	ital It was the best of times	It was the *best* of times
bf ~~~	Boldface	bf It was the best of times	It was the **best** of times
bf ital ≈	Boldface Italic	bf ital It was the best of times	It was the ***best*** of times
Position			
] or ⌋	Move to right	It was the best] of times	It was the best of times
[or ⌊	Move to left	It was the best [of times	It was the best of times
] [	Center	] It Was the Best [ ] of Times [	It Was the Best of Times
⌇	Ragged margin	It was the best of times, ⌇ it was the worst of times, ⌇ it was the age of wisdom ⌇	It was the best of times, it was the worst of times, it was the age of wisdom
tr ∧ or ⌒ or ⌒⌐	Transposition	tr It was teh best of times	It was the best of times
⌒ or ↶ over	Run over into next line	It was the best of ↩ times	It was the best of times
⌒ or ⌐ back	Run back into preceding line	It was the best of ↪ times	It was the best of times
Spacing			
◡	Close up entirely	It wa◡s the best of times	It was the best of times
#	Extra space	# Itwas the best of times	It was the best of times
⌢ or ⌢#	Close up leaving some space	It⌢was the best of times	It was the best of times
Insertion and Deletion			
⌒	Delete	It was the ~~very~~ best	It was the best

FIGURE 10.15. Common proofreader's marks organized by function. The mark and any variations are on the left, an example of its use in the middle, and the typeset result on the right.

Marks	Meaning	Marked in Manuscript	Set in Type
͡ or (⌒)	Delete and close up	It was the best of times͡ it was the worst of times	It was the best of times, it was the worst of times
stet ___ stet	Let stand as was	stet It was the ~~very~~ best stet It was the ~~very~~ best	It was the very best It was the very best
∧	Insert copy in margin	the It was∧best of times	It was the best of times

Paragraphing

¶ or ₽ or ⌐	Begin a paragraph	It was the best of times.¶It was the worst of times.	It was the best of times. It was the worst of times.
⊃ or run in	Run together as one paragraph	It was the best of times.⊃ ⊂It was the worst of times.	It was the best of times. It was the worst of times.
⌐ or ⌐ or flush ¶	No paragraph indention	⌐It was the best of times. It was the	It was the best of times. It was the

Punctuation

⊙ or ⊗	Period	It was the best of times⊙	It was the best of times.
⌃ or ,/	Comma	It was the best of times⌃	It was the best of times,
(:) or :/ (;) or ;/	Colon, semicolon	It was the best of times(:)	It was the best of times:
⌄/pos or ⌄/ ⌄ ⌄ or ⌄ ⌄	Apostrophe or single quotation mark	pos ⌄It Was the Best of Times⌄	'It Was the Best of Times'
⌄⌄/ quotes or ⌄⌄/	Quotation marks	quotes ⌄⌄It was the best of times⌄⌄	"It was the best of times"
?/ or query	Question mark	?/ It was the best of times⌃	It was the best of times?
!/	Exclamation point	!/ It was the best of times⌃	It was the best of times!
-/ or =/	Hyphen	-/ It was the age of wis⌃ dom	It was the age of wis- dom
( ) parens or (/)	Parentheses	parens (It was the best of times)	(It was the best of times)

Diacritical Marks, Signs, Symbols

&	Ampersand	It was the best of times & ~~and~~ it was the worst of times.	It was the best of times & it was the worst of times.

FIGURE 10.15. (Continued)

Marks	Meaning	Marked in Manuscript	Set in Type
....	Leader	best..............worst wisdom......foolishness belief..........incredulity	best............worst wisdom....foolishness belief.......incredulity
... or *** or ⊙□⊙□⊙	Ellipsis	It was the best... It was the best⊙□⊙□⊙	It was the best... It was the best...
⚹	Asterisk	It was the best of times⚹	It was the best of times*

Miscellaneous

a̸ o̸	Lowercase a and o when handwritten	a̸/ It was the best of times o̸/ It was the best of times	It was the best of times It was the best of times
(more)	Copy continued on next page	It was the best of (more)	It was the best of times
correction	Correct marked letter or word	e/ It was the best of times	It was the best of times
⟲	Do reverse of what's been done	(One) It was the best. (Two) It was the worst.	1. It was the best. 2. It was the worst.
### or -30 or end	End of copy	It was the best of times. ###	It was the best of times.
OK w/c OK a/c	OK with corrections, as corrected	OK w/c It was the best of times	It was the best of times

FIGURE 10.15. (Continued)

type has already been set. A typographer is entitled to charge extra to make such changes, though not to correct typographic errors and other mistakes of the operator. Keep author's alterations to a minimum for this reason, although if you see awkward wording or an editorial mistake of any kind, make the change on the galley. The longer you delay in the process of print production, the more costly changes become.

Proofreader's Marks

Figure 10.15 presents the various proofreader's marks that are recognized by typographers. These marks are used to edit copy as well as to proof type. They are numerous, and (in some instances) more than one mark means the same thing. Ask your typographer for a list of proofreader's marks commonly used by that shop to avoid confusion. Shown is the mark and any alternate(s), proofread type using the mark, and the final, corrected type, all illustrating the most common marks. Figure 10.16 illustrates those proofreading symbols for which you may have occasional use.

Marks	Meaning	Marked in Manuscript	Set in Type
Size and Style of Type			
wf ○ or wf ═	Wrong font	wf It was the best of ti(m)es wf It was the best of ti̲mes	It was the best of times It was the best of times
sm cap ═	Small capital	sm caps IT WAS THE BEST OF	IT WAS THE BEST OF
cap & sc ═	Initial capital and small capital	cap & sc It was the best of	IT WAS THE BEST OF
lf ○	Lightface type	lf It ⓦas the best of times	It was the best of times
rom ○	Roman type	rom It ⓦas the best of times	It was the best of times
∨ or ∧	Superior letter or figure	∨1 It was the best of times∨	It was the best of times¹
∧2	Inferior letter or figure	∧2 H₂O	H₂O
Position			
⌊___⌋	Lower to baseline	⌊was⌋ It was the best of times	It was the best of times
⌈‾‾‾⌉	Raise to baseline	⌈was⌉ It was the best of times	It was the best of times
∥ ═	Straighten baseline or align horizontally	∥ It was the best of ti̅m̅e̅s̅	It was the best of times
∥ fl, ∥ fr, ∥ ∥	Align vertically or justify	∥ It was the best of times, it was the worst of times	It was the best of times, it was the worst of times
3 2 4 1	Rearrange in order numbered	2̲ 1̲ 4̲ 3̲ Was it of times the best	It was the best of times
⌐ reset ¬ up	Reset to eliminate widow	reset It was the best of up times, it was the worst of times, it was the age of up wisdom.	It was the best of times, it was the worst of times, it was the age of wisdom.
Spacing			
solid	Set lines solid	solid It was the best of times, it was the worst of times, it was the age of wisdom	It was the best of times, it was the worst of times, it was the age of wisdom
# > or lead > or ld >	Additional space between lines	# >It was the best of times, >it was the worst of times, # >it was the age of wisdom	It was the best of times, it was the worst of times, it was the age of wisdom

FIGURE 10.16. Less frequently used proofreader's marks.

Marks	Meaning	Marked in Manuscript	Set in Type
ld	Take out space between lines	ld> It was the best of times, ld> it was the worst of times, it was the age of wisdom	It was the best of times, it was the worst of times, it was the age of wisdom
∨ or ⌣	Less space between words	It ∨ was ∨ the ∨ best ∨ of	It was the best of
∧# or /	More space between words	It#was#the#best#of#times	It was the best of times
thin# / or hair#/	Thin or hairline space between letters	thin# It w|as the best of times	It was the best of times
l,/s	Space out between letters to measure	14 to 15 picas It was the best of times	It was the best of times
∧∨ or eq #	Equalize space between words	It ∨ was ∨ the∧best∨of∨times	It was the best of times
en quad	En quad space or indention	en quad 1. ∨ It was the best. 2. It was the worst.	1. It was the best. 2. It was the worst.
☐ or em quad	Em quad space or indention	☐ It was the best of times, it was the	It was the best of times, it was the
☐☐☐	Insert em quads in number shown	It was the best of ☐ times, it was the worst ☐☐ of times,	It was the best of times, it was the worst of times,

Insertion and Deletion

| see copy ∧ | Refer to copy for missing material | see copy It was∧of times | It was the best of times |
| Ⓐ | Copy to be inserted | It was the best of times, Ⓐ
it was the age of wisdom
Ⓐ *it was the worst of times,* | It was the best of times,
it was the worst of times,
it was the age of wisdom |

Paragraphing

| 2 ¶ | Indent number of em quads shown | 2 ¶ It was the best of
times, it was the worst
of times, it was the | It was the best of
times, it was the worst
of times, it was the |
| 2 hang in or 2 hang in or ⌐2⌐ | Hanging indention number of em quads shown | It was the best of times,
2 it\|was the worst of times,
it\|was the age of wisdom | It was the best of times,
it was the worst of
times, it was the age |

FIGURE 10.16. (Continued)

Marks	Meaning	Marked in Manuscript	Set in Type

Punctuation

Marks	Meaning	Marked in Manuscript	Set in Type
⊥/en or ⊤/n or ⊢en⊣	Short dash or en dash	⊥/en It was the best of times ∧ it was the worst of times	It was the best of times - it was the worst of times
⊥/em or ⊤/m or ⊢⊣	Medium dash or em dash	⊥/em It was the best of times ∧ it was the worst of times	It was the best of times — it was the worst of times
2/em or 2/m or ⊢2/em⊣	Long dash or 2-em dash	2/em It was the best of times ∧ it was the worst of times	It was the best of times —— it was the worst of times
[/]	Brackets	[/] It was the best of times, [it was] the worst of times.	It was the best of times, [it was] the worst of times
},{}	Brace	brace { It was the best of times, it was the worst of times.	{ It was the best of times, it was the worst of times.

Diacritical Marks, Signs, Symbols

Marks	Meaning	Marked in Manuscript	Set in Type
¨	Umlaut or diaeresis	¨ Köln	Köln
´	Acute accent	´ Día	Día
`	Grave accent	` à la carte	à la carte
^	Circumflex accent	^ Rôti	Rôti
₅	Cedilla or French c	₅ Garçon	Garçon
~	Tilde (Spanish) or til (Portuguese)	~ Mañana	Mañana
—(bar)	Bar mark or fraction	— 1/2 bar	½
/(shill)	Virgule, separatrix, solidus, stop mark or shill mark	/ ½ shill	1/2
‾, ⌣	Macron, breve	mā•cron	mā•cron
	Footnote symbols:	† It was the best of times, † it was the worst of times.	It was the best of times,† it was the worst of times.
	Star or asterisk * Dagger † Double dagger ‡ Section § Parallels ‖		

FIGURE 10.16. (Continued)

Marks	Meaning	Marked in Manuscript	Set in Type
Miscellaneous			
✕	Broken type	It was t̶h̶e̶ best of times	It was the best of times
⟲	Thin type or right-side up type	⟲ It we̶s̶ the best of times	It was the best of times
⊥ or ↓ or ⊥̣	Push down lead that prints	⊥ It was the be█t of times	It was the best of times
(G?) (F?)	Question of grammar/fact	(G?) It ⟨were⟩ the best of times	It was the best of times
(Qy au?), (?), (Qu?), (Qy ed) or (Qu ed)	Question to author/editor	(Qy au?) It was the best of times	It was the best of times
fî or a͡e	Use the diphthong or ligature of these two letters	f͡i first	first
see l/o	See layout	It was the best of times, □ it was the worst of times, □ it was the age of wisdom	It was the best of times, it was the worst of times, it was the age of wisdom
cq	Correct as is	(cq) Twas the best of times	Twas the best of times

FIGURE 10.16. (Continued)

Checking Corrections

The typographer makes corrections and changes you have marked and sends you corrected (or revised) type in the form of only the lines affected or of the entire galley with corrections incorporated. You can specify which you prefer; the first saves some typesetting expense and allows paste-up to start while you wait for corrections, although the second saves paste-up time because small strips of corrected type don't have to be worked into the original. When only affected lines are reset, the operator determines how many must be reset to incorporate each correction but should provide a minimum of two lines for accurate alignment on the mechanical.

Read all material that has been reset, not just the word or line that was corrected. Other errors might have been made in the process. Use proofreader's marks to indicate additional corrections. If they are extensive, ask for another set of corrected galleys to confirm that changes have all been made.

CHAPTER ELEVEN

Preparing Mechanical Artwork

MAKING A MECHANICAL DUMMY   301
    When To Prepare a Mechanical Dummy, 302
    Ruling Dummy Sheets, 303
    Dummying Type, 303
        Fitting Text, 304
        Writing Headlines, 306
        Writing Cutlines and Legends, 310
    Dummying Art, 311
    Finishing the Dummy, 312

PASTING UP MECHANICAL ARTWORK   313
    Engraver and Printer Requirements, 314
        Second Color, 314
        Art Position, 314
        Separations, 314
        Line and Halftone Combination, 315
        Gripper Margin, 315
        Sequence, 315
        Schedule, 315
    Offset Paste-Up Techniques, 315
        Tools and Supplies, 316

Adhesives, 316
Ruling Paste-Up Sheets, 317
Positioning and Handling Type, 318
Line Art, 320
Screens, Tints, and Other Line Work, 322
Halftones, 324
Preparing Art, 328
Color Paste-Up, 332
Corrections, 335
Proofing, 335
Cleaning and Protecting Mechanical Artwork, 336
Paste-Up for Other Presses, 336
Letterpress, 336
Gravure, 337
Screen Printing, 337
Evaluating Mechanical Artwork, 337

"Mechanical artwork" consists of type and all other elements in place "camera-ready"—ready for conversion into negatives for press plates or directly into plates. Also called simply the "mechanical," "artwork," or "board," it includes the basic text for a printed piece as well as provisions for halftones, screens, and multiple colors. "Paste-up" is another term often heard as a synonym, although it doesn't take these frequently needed provisions into account. The term may, however, be appropriately used to describe the process of executing artwork.

The mechanical is prepared according to a "mechanical dummy" or rough mechanical that specifies to the person pasting up the job how each page or panel is to be made up. On a project to be printed directly from metal type, the dummy guides the arrangement of that type and other elements ("page makeup"). Because such projects are not now as common in commercial print production as those to be reproduced by offset, this chapter concentrates on executing mechanical artwork destined for that or other photographically based printing methods. It starts with how to make a dummy as a guide to the paste-up process. Information is pertinent if you prepare the dummy and/or mechanical yourself or if you supervise their completion by a graphic designer or mechanical artist. It may also be selectively applied if photocomposing equipment is to be used to create complete artwork on photosensitive paper or on film (as covered in Chapter 9).

Making a mechanical is one of the most tedious procedures in print production, requiring close attention to spacing, placement, and alignment. However, the resulting artwork is one of the strongest influences on quality of the finished piece because of the following benefits:

1. It keeps type and line-art elements intact as one unit. They are less likely than separate items to be lost or damaged when sent to an engraver or printer for conversion into press negatives or directly into plates.

2. Proper positioning of images on the printed page is assured. When preparing mechanical artwork, elements can be rearranged until they are exactly where you want them to appear.

3. Mechanical artwork saves money by allowing proofing of in-place elements at a stage in production when making changes is relatively easy and inexpensive. Once a project moves into engraving, changes are much more costly. If undetected until the job is finished, mistakes are very time consuming and expensive to correct.

4. A mechanical shows you what a piece will look like when printed, before it's on the press. A detailed comp can't show as much as artwork presenting exact type, sizes, and positioning. This advantage is especially important if people not experienced in visualizing how a printed piece will look must approve its layout. A photocopy of the artwork, trimmed and folded per specifications, can make their okay an educated approval. (So can a proof of the press negative, of course, though it comes at a later stage of production when changes are more costly.)

5. A mechanical belongs to you to use again as is or to "cannibalize" (borrow from) for parts for a later job. Technically, press negatives and plates prepared by a printer in order to print the job belong to the printer; but you own the raw materials, including the artwork.

MAKING A MECHANICAL DUMMY

A mechanical dummy shows how finished mechanical artwork is to be prepared. It incorporates all type and art elements, presents exact measurements, calls out where different colors are to appear, and provides other details necessary for accurate execution of mechanical artwork for a particular printed piece.

It is not the only type of dummy used in print production, however. A "printing dummy" ("blank-page dummy" or "bulking dummy") provided by a graphic designer or printer shows you actual paper, size, sequence, fold, binding, and bulk. An "imposition dummy" provided by a graphic designer, printer, or bindery shows you the sequence in which a piece is to be folded or assembled. When thoroughly prepared, a mechanical dummy provides all this information (except actual paper) and can be passed from supplier to supplier as production progresses.

A mechanical dummy is itself an outgrowth of earlier planning about overall layout and specific sizes and placement. It may be based on thumbnails, rough sketches, a longstanding format for a regular publication, or a more polished "comprehensive dummy" ("comp") used primarily as a design tool. The latter piece closely refines layout and indicates with either pencil lines or "Greeking" (nonsense type or Latin words) where body and display type are to be positioned and how art is to be sized and placed.

A comp prepared by you or a graphic designer can be used to guide paste-up if it is concise and fully detailed. An alternative if further refining is needed is to use it as the basis for the mechanical dummy after text type is set. A copy of galleys can then be placed over the areas indicated for text on the comp to confirm fit and division of text into columns. After display and cutline type is set, a copy of it may also be positioned on the comp to confirm placement, weight, and size.

WHEN TO PREPARE A MECHANICAL DUMMY

A mechanical dummy can save you time, money, and frustration if it is adequately prepared, whether adapted from a comp, partially prepared along with a comp to check an especially intricate design, developed from more vague concepts about layout, or extracted from an established format for a regular publication. The larger and more complex a print project is, the more valuable it becomes so that all elements are accommodated and placed and sized according to the layout. A dummy is not, however, always required. Selective use can save time if you would make it yourself and money if you would ask a graphic designer to make it.

A small job (such as a simple announcement) that has minimal type and art can be accurately pasted up without a detailed dummy, especially if you had the type set in a block spaced as you want text and display elements to appear in the printed piece. Whoever does the mechanical (or makes up hot-type forms for direct printing on a letterpress) can easily work from a thumbnail or rough layout on such a project, as long as you have clearly indicated finished size and margin allowances.

A repetitive job (such as a directory) that has page after page laid out in the same format doesn't need a dummy either. The person doing paste-up or making up hot-metal pages can simply be instructed to fill each page to a given depth until all type is incorporated and to repeat any art (such as column rules) in the same location on each page.

You don't need to do a dummy yourself if you have asked your mechanical artist or printer to decide on page layout. This procedure leaves many decisions to a supplier but is an option if you lack design skills and funds to buy the assistance of a graphic designer. On a newsletter project, for example, you can decide which articles are important enough for the front page and roughly on which pages others should go, collect accompanying illustrations and/or photographs, and hand the project over to your mechanical artist or printer to determine a layout. Ask to approve a detailed comp or dummy, however, to avoid unpleasant surprises.

RULING DUMMY SHEETS

Dummy on paper that has been accurately ruled to indicate all dimensions, edges of pages or panels, margins, column width, and gutters. Check with your printer about the amount of room to be left clear for the press gripper (generally a half-inch or 1.27 cm). The image you want printed ("live matter") should start below the gripper margin. Also ask your engraver or printer if measurements for the dummy and resulting artwork should be in picas, inches, or centimeters. If materials used in preparation of press negatives (as covered in Chapter 12) are ruled in inches or centimeters, artwork measurements should correspond. Column and art measures may be in picas because type is set that way; but the space between them, gutters, and margins should be calculated in inches or centimeters.

To save considerable time on an extensive or regular project, rule a master and have a supply of dummy sheets printed (or photocopied if they are small). When accurately printed in nonrepro blue, sheets at least 1 in. (or 3 cm) larger all around than the finished size of the piece can also be used for paste-up. (You may want to print some on less-expensive, medium-weight paper for dummying, rather than printing them all on heavy paper or board suitable for paste-up.) If the project is sizable enough, your printer may contribute preprinted sheets to make the print estimate more attractive. Don't try to save money by going to a low-quality printer for dummy/paste-up sheets for a job to be printed by another shop. The same printer should do this aspect of the project as well so that you are certain of providing a mechanical that matches your printer's press and the binding equipment to be used.

An accurately ruled dummy will lead to an accurately ruled mechanical. Use a T-square as a guide or make close measurements with a ruler to ensure that the image area is squared. Because a dummy is only a mock-up, rules can be drawn in any color of pen or sharp pencil, as you prefer.

DUMMYING TYPE

Though you need to have art elements in hand and at least a rough idea of their size and placement, start dummying by positioning the type; then make revisions as needed to accommodate final decisions about art (unless its placement is more important than that of the type). A dummy is prepared after text type is set. Only if you know exactly where display type and cutlines will appear on the page should they be set ahead of dummying; otherwise, you may face additional charges to reset them to final specifications and a delay in the production schedule. Set non-text type after position, width, depth, and desired size, style, and weight have been determined in the process of doing the dummy.

Reserve one set of galley proofs for dummying, either fully corrected proofs or rough proofs. When using the latter, make certain that all corrections affecting number of lines are marked and that paragraphs are indicated so that you can proceed with confidence to do the dummy. Trim proofs to eliminate excess paper around type and have them waxed to hold them in place as you work and allow repositioning as necessary. Alternatives to wax are thin rubber cement or a light coating of spray adhesive.

With scissors and tape in hand, you are now ready to dummy text type. Refer to any existing comp, thumbnails, or rough layout to determine placement. In the case of a regular newsletter or newspaper with an established format, decide which articles should go on the front page, which on the back (as the second-most prominent position), and which on the inside. Following the layout guidelines presented in Chapter 6, construct each page to work in text type and spaces for display type, art, and cutlines.

Generally, an article should appear on a page in its entirety to avoid confusing and/or losing the reader by continuing (or "jumping") it elsewhere in the issue. If you must continue it, however, continue no more than one article on that page, guide the reader with a "(continued on page ___)" line, a small jump headline (a shortened version of the main headline), and a "(continued from page ___)." Jump an article to a later page; jumping backward is awkward and doubly confusing. An article continued logically from one column or one page to the next should not have "continued on" or "continued from" lines.

If you don't yet have some of the text set, type missing copy so that you can closely estimate how much space it will require, using the character count for your particular typeface, style, size, weight, and your column width (as explained in Chapter 10). Indicate with a "slug line" (two or three identifying words) and penciled arrows drawn from beginning to end of the space the area that type is to occupy when set.

Standardize spacing within the piece and issue to issue if the project is a periodical or one in a series. To do so, determine what spacing is most effective between headline and text, between subheads and text, between articles, and between text, cutlines, and art. Measure vertical space in points from the baseline of the line of type above to the baseline of the next line of type below so that spacing isn't influenced by ascenders and descenders (see Figure 10.4). Measure horizontal space from the edge of type to the edge of art. To assure consistent spacing time after time, note what the spacing is, revise it as necessary after you see the first issue in print, and then compile a master guide for dummying and paste-up by marking each kind of spacing on a copy of the piece for your files.

Finally, note on the type where corrections are to be inserted if you have not dummied with fully corrected galleys. Circle the areas and indicate that corrections are still to come ("to cum" or "TK").

Fitting Text

With a format that calls for even margins at the top and bottom of each page or panel, dummying toward that end can be frustrating, especially for the beginner. Type is often too long to fit the allotted space or woefully too short. The following tips will help you fit text type in such situations.

Lengthening. To lengthen copy that you can determine is too short before the manuscript is typeset, consider setting paragraphs fully flush to the margins (not indented or ragged) with extra space between them. The column of type will then be longer. This option will work only for isolated projects in which consistency isn't a factor, not for one of a series of brochures or one page of a multipage report. If an established format calls for indented paragraphs with no space between them, don't mix or switch styles just to solve an immediate space problem. You might, however, add another half or full point of leading between lines or a few extra points between paragraphs to lengthen text; only your closest readers might notice. Don't mix leading, however, within the same page or spread. Another possibility is to allow more space between letters; type that is tightly kerned will set longer if letterspacing is relaxed. Several small paragraphs will set longer than one long one and be more readable.

To lengthen copy that is already set, try cutting text apart between paragraphs and adding a little more space. Make sure the amount is consistent throughout the column or page by doing baseline-to-baseline measurements. The lengthening tips just mentioned as possibilities before type is set may also be used afterwards, but only if you are willing to pay extra for resetting.

To increase the space taken up by an article in a periodical or by a section in a brochure or report, occupy more room with the headline. Use a two-line headline instead of one, write a "kicker" (short introductory headline) and a main headline, or extend the headline over more columns. Don't give an article more prominence than it's worth, however, by overdoing its headline. Add a subhead at the beginning of an extensive article or intersperse two or more subheads within the article. Occupy one column with a headline and run the article in the remaining columns to the right of it; this approach is most effective if the unit is at the bottom of the page and separated from articles above it by a rule or extra space.

Use art within the article; the space you'll need to leave above and below it will probably be more than would be taken up if the art were placed to the side, at the beginning, or at the end. Box or rule in an article for the same reason; the margin that you'll need to leave between the type and the box or rule will help fill space. Illustrations and photographs can be used larger to make up for short text. You must be alert, however, to enlarging them beyond their value.

On an employee or member publication, consider filling a large gap in text with a public-service ad for United Way or another charitable cause that provides such camera-ready "filler." What you use should be timely (e.g. during the organization's annual fund drive) and the art compatible with the quality of your publication.

Shortening. Making type occupy less space is more difficult. Earlier tips on paragraph indentation and leading can be reversed when copy is set if you know it will be long. Point size and letterspacing can be reduced; but don't present a mixture within the same page or spread. Although you risk extra charges for author's alterations after type is set, paragraphs can be combined or words selectively deleted to decrease the number of lines of type. When deletions are made close to the end of a paragraph, less resetting is needed. If copy was written in inverted pyramid style with the most important information at the beginning and the least important at the end, whole paragraphs may be cut from the bottom up without serious damage to content. Always reread it after making deletions, however, so you are sure copy still communicates.

Work in long text by using less room for headlines—one line instead of two, one or two columns across instead of three. Delete kickers and subheads to save more space. Make art smaller or take it out entirely. Place it at the beginning or end of an article instead of in the middle and, if the format allows, put it in the margin, reserving all column space for the text.

When space is extremely tight, set aside articles that can wait until the next issue ("holdovers"). If the message won't be out of date later, you can gain valuable space by keeping them in reserve. Reread holdovers, however, before using them next month to see if wording needs to be changed to make the information current.

Writing Headlines

Once the text is in place and space reserved for art, you are ready to write headlines. Whether you do the dummy yourself or have a designer do it, you should write the headlines or have the person who wrote the text do so. Though length, weight, and size of headlines are graphic decisions, content is an editorial responsibility.

A headline is needed for each article in a newspaper, newsletter, or other periodical. The title alone may suffice for a flyer or folder; but any brochure, report, or more extensive piece with multiple sections should have a headline (or "heading") for each, except for any opening paragraphs that are obviously introductory.

Content. A headline is intended to call attention to accompanying text and summarize its main point. It may be in straightforward news style ("Board Promotes Five") or take a lighter approach ("Five Climb the Ladder"). Either way, the reader should be able to tell at a glance the subject of the text that follows. A headline may be an abbreviated sentence with subject and verb or a bulletin (or label) consisting of a word or two. The abbreviated-sentence form is historic and generally considered to carry more information than the bulletin form. The simplicity of the latter, however, has an appeal (especially with designers) for its greater ability to attract the eye.

If you choose to use a bulletin headline in a periodical, follow it with an abbreviated-sentence subhead in smaller type that fulfills the objective of previewing text. An alternative is to use the bulletin as a kicker, followed by the main headline. Figure 11.1 presents examples.

Whenever possible, headlines should be written in the present tense to lend immediacy to text. Words that are understood (is, are, the, a, an) are eliminated, a comma substituted for conjunctions, and single quotation marks used instead of double ones when a headline includes a direct quote. These practices condense a headline to basic information and use allotted space for maximum communication. For example, "Painters and Potters Are To Exhibit at the Fair" should be condensed to "Painters, Potters To Exhibit at Fair." In a bulletin format, the bulletin or label might be "Fair" with "Painters, Potters To Exhibit" presented as a subhead.

Capitalization. A mixture of capital and lowercase letters in headlines is considered more readable than all capitals, all lowercase, or capitalizing only the first letter of the first word (see Chapter 10). Use a consistent style throughout a printed piece. Any word with four or more letters, any proper noun, and any word that starts a line should be capitalized. "To" as an infinitive should always be capitalized but be lowercase when used as a preposition.

Counting. A well-written headline fills or nearly fills the space allotted for it and doesn't extend into the margin (unless by design). A headline having two or more lines should be as close to even in length as possible. The break should come between phrases whenever feasible, and a prepositional phrase should not be split between lines.

The length of a headline is calculated using "unit count" (or "character unit")—the approximate width of each character in relation to others, regardless of typeface, style, weight, or size. An accurate unit count will tell you whether or not a headline will fit its space when typeset.

Though unit count varies slightly by typeface and style, the relationship of one letter to others is the same. Unit counts for capital and lowercase letters, numerals, and punctuation marks are presented in Figure 11.2. Small capitals may be counted the same as lowercase letters, and a space counted as one.

Weight, proportion, and size also influence how much room a headline will occupy, though not its unit count. A bold 18-point headline that counts 20 units will set longer than a light 18-point headline with the same count. An extended style will set longer than one that is condensed. A 36-point line will set longer than a 16-point line. Within each of these headlines, however, the relationship of characters to one another remains the same.

Refer to the page from a type-sample book (Figure 10.5) to see the origin

A **Painters, Potters To Exhibit at Fair**

ned libiding gen epular religuard on cu umdnat. Improb pary minuiti potius inflam dodecendense videantur, Invitat igitur vera aequitated fidem. Neque hominy infant au

Et harumd dereud facilis est er expedit im soluta nobis eligend optio comque nihil d maxim placeat facer possim omnis es inis dolor repellend. Temporem autem

B **FAIR**
Painters, Potters To Exhibit

expedit distinct. Nam liber a tempor cum impedit anim id quod maxim placeat facer dolor repellend. Temporem autem quinsu saepe eveniet ut er repudiand sint et mol

C Lorem ipsum dolor sit amet, consectet nonnumy eiusmod tempor incidunt ut lab

Artists ad minimim venia oris nisi ut aliquip
ire dolor in repreh
Exhibit t, vel illum dolore e o odio dignissim delenit aigue duos dolor et molestais excep provident, simil tempor sunt in culpa qui

D **Painting, Pottery**
Artists at County Fair

esse molestaie son consequat, accusam et justo odio dignissin molestais exceptur sint occaec officia deserunt mollit anim id

l illum dolore eu fugiat nulla ji blandit praesent lupatum del upidat non provident, simil te aborum et dolor fugai. Et haru

E Painters, Potters — See Them, More at Fair

Lorem ipsum dolor sit amet, consectetu tempor incidunt ut labore et dolore magn veniami quis nostrud exercitation ullam commodo consequat. Duis autem vel eum

lipscing elit, sed diam nonnumy eiusmod liquam erat volupat. Ut enim ad minimim or suscipit laboris nisi ut aliquip ex ea ire dolor in reprehenderit in voluptate velit

F **Artists
At Fair**

Temporem autem quinsud et aur office de saepe eveniet ut er repudiand sint et mol earud rerum hic tenetury sapiente delect asperiore repellat. Hanc ego cum tene ser

Artists at Fair Endium caritat praesert
quaerer en imigent cupidat a natura proficis
autend unanc sunt isti. Lorem ipsum dolor
g elit, sed diam nonnumy eiusmod tempor

Artists at Fair Ut enim ad
l exercitation ullamcorpor suscipit laboris
consequat. Duis autem vel eum irure dolor
lit esse molestaie son consequat, vel illum

G Artists at Fair

Lorem ipsum dolor sit amet, consecte nonnumy eiusmod tempor incidunt ut lat erat volupat. Ut enim ad minimim veni ullamcorpor suscipit laboris nisi ut aliquij

FIGURE 11.1. Headline forms: regular abbreviated sentence (A), bulletin with subhead (B), cut-in (C), kicker and main headline (D), centered (E), side variations (F), and margin (G).

Unit Width Chart

Lowercase		Capitals		Numerals/Punctuation Marks	
Character	Count	Character	Count	Character	Count
a	1	A	1½	1 through 9	1
b	1	B	1½	Fractions	1½
c	1	C	1½	Subscript, superscript numerals	1
d	1	D	1½	$, ¢, £	1
e	1	E	1	&, %	1½
f	½	F	1	) (	½
g	1	G	1½	. , : ; ! / en-	½
h	1	H	1½	?, #, *, em-	1
i	½	I	½	' '	½
j	½	J	½	2 em-	2
k	1	K	1½		
l	½	L	1	Small capitals same as lowercase	
m	1½	M	1½	Space	1
n	1	N	1½		
o	1	O	1½		
p	1	P	1		
q	1	Q	1½		
r	1	R	1½		
s	1	S	1		
t	½	T	1½		
u	1	U	1½		
v	1	V	1½		
w	1½	W	2		
x	1	X	1½		
y	1	Y	1½		
z	1	Z	1		

FIGURE 11.2. Unit counts per character.

of unit count. Units per character indicate how wide each character in a given typeface and style will set. Character units roughly translate into the following values for counting headlines:

Character Units	Headline Count
7 or less	½
8–11	1
12–16	1½
17 or more	2

Extended or condensed styles or any typeface or style tightly kerned has different values although, again, the relationship of characters to one another remains the same.

Figure 11.3 shows how counts are applied to a headline set in 18-point Souvenir medium roman. In this particular typeface, style, and weight, one pica can accommodate 1.5 units. Thus a headline with 20 units will set to

S a l e s   U p   f o r   Q u a r t e r
1½+1+½+1+1+1+1½+1+1+½+1+1+1+1½+1+1+1+½+1+1=20

FIGURE 11.3. Determining the length of a headline using unit count.

13.33 picas (20 divided by 1.5). If the space available is 14 picas, this headline will fit. If only 12 picas are available, it will be too long.

A headline schedule can be a quick reference when dummying a regular job. Cut representative samples out of a previous issue, tape them to a sheet of paper or card, and note the total count of each version. Use this schedule to avoid having to calculate each time the count for spaces to be filled. If you are working on an extensive job yet to be printed and thus have no samples, make a list specifying this information. To be safe, have your typographer set samples to confirm your counts; the slight charge could prevent costly errors later.

Slugging. When a headline is written and the count checked to make certain it fits, either write the headline in full on the dummy where it belongs or write the first couple of words. This "slug" is a guide during paste-up to assure that the typeset headline will be put in the proper place.

Writing Cutlines and Legends

The words that explain what is happening in a photograph or identify who is in it make up the cutline. (Technically, a cutline appears below or beside art while a caption appears above it, although "cutline" is now commonly used to identify either.) A legend explains what is presented in a graph, chart, diagram, map, or other technical illustration and may include a key to what different colors or patterns mean.

A photograph or illustration requires a cutline or legend when any doubt could exist about what it shows. Individuals in a group photograph need to be identified, while a sketch of one person who is obviously the subject of a surrounding story does not (although purists would argue that a cutline is required even then). A legend is definitely needed to explain different colors or patterns used in an illustration, even if all other information is self-explanatory.

Start a cutline or legend with the point you want to emphasize most. It may be the person's name ("Harry Smith sifts through the rubble of his burned-out warehouse."), an important fact that the art underscores ("Arson has claimed another victim, this time Harry Smith's warehouse."), or the mood of the art ("Picking up the pieces is going to be tough for Harry Smith.").

A cutline or legend should be brief (possibly name only), not a repeat of

MAKING A MECHANICAL DUMMY

details in the text. It should augment the message of the art, not overpower it. Whenever possible, it should be written in the present tense so that the art appears current. Even an historic photograph is seen with fresh eyes and may have a present-tense cutline; use past tense, however, if present-tense wording seems awkward.

Break up a long cutline of a group of people by identifying the back row in one short paragraph, for instance, and then identifying the front row in another. Identification within photographs is generally understood to be from left to right. To identify people or objects not in neat rows, consider going clockwise or insetting a small, simple line drawing with each figure outlined and numbered, and then repeat those numbers in an adjacent cutline. This inset approach may also be effective to identify a grid of photographs by repeating the grid in miniature with the spaces numbered and providing an adjacent numbered cutline (see Figure 11.4 for an example).

Place a cutline or legend below, above, or to the side of its art and as close to it as possible. A legend that explains a graph and also its different colors may have the explanation below, for instance, and the color code within the graph as long as data aren't obscured. The easiest way to help the reader identify which cutline belongs to which photograph is proximity; the words below the photograph relate to it, not to the photograph across the page. Directions can be indicated within the cutline ("Above," "Far right," "Left," etc.). In a well-designed page or spread, arrows pointing from cutlines to their corresponding photographs aren't necessary and give the spread an amateurish look.

For readability, try to avoid a lengthy cutline that explains numerous photographs in one block. Split them into two or more smaller blocks if the layout allows. Follow the guideline for column width presented in Chapter 6 so that a cutline or legend stays within a readable width for its point size.

DUMMYING ART

Before deciding on final placement of type, you will need to determine at least an approximate size of any illustrations and/or photographs that must be accommodated on the dummy. You may refine size now or wait until the mechanical is finished. The latter is a wiser choice if size might be altered by the more exact placement that occurs during paste-up, as opposed to during the dummy stage. While doing the dummy, you should at least eye art to make sure that it will work how and where intended.

You may use photocopies or stats of art to check position and size. They will also help you decide how art should be "cropped"—what portion of the image will be reproduced in the printed piece. When you are sure that the art you have selected will fit, move ahead to paste-up and follow instructions presented later in this chapter on how to crop, scale, and mark art.

To show on the dummy where halftones or line art is to go on the finished piece, rule a box around the area, connect the corners with diagonal lines to

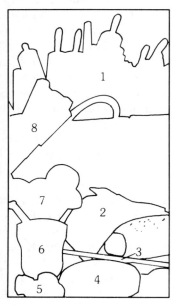

1. Rolling Pins 2. Pastry Bags 3. Carving Knife 4. Butter Stamp 5. Petit-four Cases 6. Mortar and Pestal 7. Wooden Utensils 8. Spatulas

FIGURE 11.4. An example of an outline system for identifying parts of a complex photograph.

make an X, and slug it (such as "Baseball Art Here"). If you have several related illustrations or photographs, number or letter them and write their codes in the corresponding boxes. Or photocopy them to approximate size and glue the photocopies in place. For line art to be added during engraving, you might photocopy to-size stats or roughly sketch the image in the space.

FINISHING THE DUMMY

Each page, panel, or spread should be laid out on the dummy completely and in actual size, with no element or measurement overlooked. All text, display, and cutline type and all illustrations and/or photographs should be accommodated; but don't forget smaller elements.

Though dummy pages may be done in numerical or reading order from beginning to end, the mechanical will probably be done in the sequence necessary for printing and binding pages in proper order. Therefore, the mechanical artist must rearrange dummy pages to be in imposition sequence and will need page identifications to do so correctly. If page numbers are to be printed, number pages with the dummy copy of those typeset numbers. If the piece is brief and sequence is understood so that page numbers aren't necessary, note on the dummy what that sequence is to be (such as "Panel A," "Panel B").

Include dummy type for running heads or feet—type at the top or bottom of pages in a lengthy publication that repeats a chapter or section title or the month of publication in the case of a periodical. Also include legends for charts, graphs, or maps and sketch in any art (such as arrows on a diagram) you need the mechanical artist to draw.

Note areas to be screened or tinted and the ink values to be used in percent of full value (e.g., "10% black"). Indicate whether type is to be surprinted on or reversed out of screened or tinted areas or art. In the latter case, be specific about placement of type within art; sketching it over a to-size photocopy or photostat of the art is helpful.

Mark all dimensions, even if dummy sheets were carefully ruled. Provide exact measurements for outside margins, center gutter, space between columns, and column width, in addition to any other measurements required by the design. Note the finished size and where any folds will fall so that accurate trim and fold lines can be added to the mechanical. Indicate any perforation lines and whether they are to be printed and die-cut or die-cut only, and provide a rough or finished outline of any other area to be die-cut.

Call out elements that are to be pasted up on an overlay (see later information on multicolor paste-up). If not specified, they will be part of the basic artwork, to be separated during engraving. Finally, note if anything is missing and when the mechanical artist may expect to have it to finish the project.

PASTING UP MECHANICAL ARTWORK

Mechanical artwork is a camera-ready composite of all type and line-art elements that are to be printed. It includes text, display, and cutline type, line illustrations, and rules exactly in position, ready for the camera to make a press negative or a press plate. It may also include "windows" for insertion of halftone negatives during engraving and "overlays" of elements to appear screened or in a second color.

This section opens with a summary of questions that need to be answered before paste-up begins to make sure that engraver and printer requirements are met. Step-by-step paste-up techniques will help you execute artwork yourself or supervise a mechanical artist or graphic designer doing the work. Information presented here is geared to preparation of artwork for an offset press; exceptions for other kinds of presses are covered near the end of the chapter, followed by a review of the qualities of a well-prepared mechanical.

ENGRAVER AND PRINTER REQUIREMENTS

Before doing the mechanical for any project that isn't routine or that will be given to a new supplier for engraving and/or printing, find out what is preferred and required. Asking first may save unnecessary work on your part, extra work by the engraver and/or printer, and a missed deadline.

Second Color

Are elements to be printed in a second color to be on an overlay or part of the basic artwork? As detailed later, colors may be preseparated by using an overlay or left to the engraver to separate by making multiple negatives of the basic artwork and masking out areas according to color. While the use of an overlay saves some time later, it may not produce as close a register as the second method. A quality printer will prefer to have separation done on the negative whenever the layout allows to assure accurate registration on the press.

Art Position

How should the position of separate art be indicated? The position of line art to be reduced or enlarged to fit a space or of photographs to be converted into halftone negatives must be called out in some way on the mechanical. Your engraver may prefer a rough sketch or a photocopy of line art of approximate size lightly glued in the area. Photographs may have windows or ruled boxes only.

Separations

Shall four-color separations of art be prepared or does the printer prefer to do them or supervise their production? A printer who has had difficulties with the quality of customer-supplied separations may prefer (or even require) that you relinquish supervision of this phase of production. If so, the printer will select an engraver to do the separations, provide specifications, and share with you the result for your approval (and probably add a handling charge to the bill).

Line and Halftone Combination

How should a combination of line and halftone art be prepared? The line art could be sized and mounted on a board (or on the artwork itself) and a window for the halftone done on an acetate overlay and registered into position. Your engraver may recommend a different procedure.

Gripper Margin

How much space should be allowed for the press gripper and on which edge? Depending on the press to be used, a gripper margin may be a half-inch (1.27 cm) wide, probably at the top (and possibly also at the bottom) of the sheet. The area must be free of any image that you want printed.

Sequence

Should pages be pasted up in imposition or numerical sequence and how many per paste-up sheet? Even though a dummy is made in numerical or reading sequence so that facing pages can be visualized together, the mechanical should probably be done in imposition sequence (or "pasted sheeter signature") so that, when the signature is folded and bound, pages will be in order. A thick publication, however, may be assembled by the engraver into imposition sequence. With such a project, margins will vary to account for the greater distance outside (as opposed to inside) pages must wrap around the spine or how they roll at the gutter into the binding. Your engraver may prefer to do such precise placement for "creep allowance" (or "shingling") starting with individual pages, rather than try to work with artwork in which every page is measured the same. Size of the process camera to be used to make press negatives, size of the press, printer preference, and ability to align elements on a large mechanical dictate how many "up" (how many pages or panels per sheet) the mechanical should be.

Schedule

When is the job scheduled to be on the press? A missed press date because of slow paste-up means a missed deadline for the finished piece. Especially on a major project requiring several hours of engraving and press time, ask your engraver and printer before paste-up begins when it must be ready in order to stay on schedule.

OFFSET PASTE-UP TECHNIQUES

While only some of the following techniques for executing artwork for offset would be used for a simple brochure, most of them would be used for a multipage magazine or other more complex project. They are presented here in sequence so that you can review those that apply for orderly preparation of a particular mechanical yourself or supervision of a graphic designer or mechanical artist. Information on paste-up for other types of presses is presented later in the chapter.

Tools and Supplies

Though artwork for a very simple, low-budget project can be made with a sheet of paper and a keen eye, more ingredients are required if the result is to be of higher quality. One requirement is a firm, squared surface in a well-lighted area, either a drafting or light table or a portable board larger than the piece to be pasted up. A T-square is used to align elements horizontally and a right-angle triangle butted up against it to align them vertically. A separate triangle should be used as a cutting edge so that the triangle used for aligning elements is reserved intact for that purpose. Guidelines are drawn with a nonrepro blue pen or sharp pencil. An art knife is ideal for lifting pieces of type and for trimming and dividing galleys, though you may prefer scissors for some cutting needs. A cutting surface separate from the paste-up board protects the latter surface from being marred; gluing pieces on uninked waste paper also protects the paste-up surface.

A metal or quality plastic ruler is used for measuring margins, gutters, columns, and windows, and a line gauge or ruler is used for measuring smaller dimensions (such as baseline to baseline between headline and text). A bottle of opaquing liquid is convenient for eliminating spots, as is a glue pick-up for cleaning away excess rubber cement. Solvent or thinner is used to clean glue, wax, or other residue from the finished artwork, and a lint brush is used to sweep away hair and lint from the work area.

You will need a burnishing tool to apply even pressure to secure all elements in place. It can be a burnishing roller, a wide burnishing stick, or a smooth ivory stick. Finally, have on hand a roll of tissue paper or sheets of lightweight, unprinted paper to cover and protect the completed artwork.

Adhesives

Several adhesives are available to affix type and other elements to a mechanical. People who do a lot of paste-up work tend to develop a preference, and new materials on the market may prove more satisfactory than old standbys. An adhesive for artwork should be evenly distributed with no lumps that create an uneven surface. It should adhere elements securely enough so that corners don't curl or small pieces don't fall off. At the same time, it should allow manipulation; you need to be able to move text and headlines around until they are positioned exactly as you want, without having the adhesive dry or wear out before you're finished.

Probably the easiest and least expensive adhesive is hot wax applied to the back of text and display type by the typographer or the person doing paste-up. If properly applied with a machine that rolls on the hot wax or with a hand-held waxer, coverage is even and the soft wax allows virtually unlimited manipulation. It holds firmly when burnished and doesn't discolor type or art over time.

Thin rubber cement (one part cement to four parts thinner) is another low-cost alternative, although it may discolor after several months. It is brushed on the back of type that can then be positioned and repositioned much like waxed galleys. The thinner the solution, the more type can be manipulated but the less securely it will hold when burnished and dried in place.

A light coating of spray adhesive is also effective, though more costly. Like rubber cement, its ability to allow manipulation and to adhere depends on how much you use and how long you allow it to dry before setting to work. If formulated for mounting photographs and other art, it should not discolor.

Finally, clear tape can be used on artwork, although not on a project you consider of high quality. For a quick flyer or small typewritten newsletter to be printed on an office offset duplicator or by a quick printer, small pieces of tape will hold elements in place. Avoid inadvertently adding fingerprints at the same time by using an art knife in one hand to cut off small pieces from a strip of tape held in the other hand. Burnish the tape well so that it won't curl and leave a mark that will show up on the printed piece.

Ruling Paste-up Sheets

A quality mechanical should be done on smooth, dull, white paper or board (bristol or illustration) larger than the image area or preprinted heavy paste-up sheets (as covered earlier in relation to the dummy). Paper or sheets should be sufficiently heavy to avoid wrinkling when elements are adhered to them. An alternative if increments are accurate for the job is a sheet of grid paper printed in nonrepro blue and mounted on a board to add stability. Another is to paste up on a sheet of film shot from precisely ruled artwork. Though an expensive alternative, this approach ensures absolute measurements, especially when several pages must be pasted up to the same specifications.

To rule a paste-up sheet for an individual project or as camera-ready artwork for printing of dummy/paste-up sheets, align the board with a T-square and tape it to your working surface on at least two corners or sides. Using a T-square and aligning triangle, draw a vertical center line and a top trim line in nonrepro blue (if doing an actual paste-up sheet) or in black ink with a ruling pen (if doing camera-ready artwork for printing of sheets). These lines are the reference points for all other measurements: gutter, columns, horizontal and vertical rules, outside margins, and other dimensions. Mark the center fold with a dotted line outside the image area. Indicate where edges are to be trimmed with solid horizontal and vertical lines outside the image area or with full crosses to be masked over or opaqued out

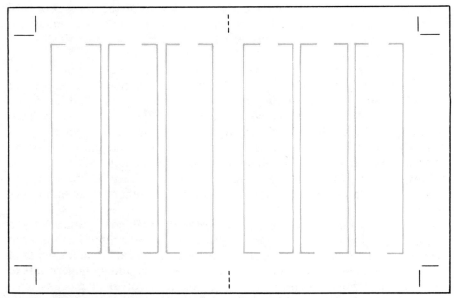

FIGURE 11.5. A paste-up sheet showing trim and fold lines outside of the image area and nonrepro-blue guidelines within.

before press plates are made. See Figure 11.5 for an illustration of these ruling procedures.

Position the image area so that you will have at least 1-in. (or 2.5-cm) margins outside of it to place registration and other marks and to make notations to the engraver or printer. Excessive margins are wasteful of paste-up board and of film when press negatives are shot, and an oversized mechanical may be too large for the camera's copyboard.

Positioning and Handling Type

With the paste-up sheet aligned on your working surface, trim away excess paper from text, display, and cutline type using the art knife and a straight-edge or scissors. Starting at the top of the page to organize and protect your work, loosely position the first tier of type as indicated on the dummy, comp, or rough sketch that is your guide. When all elements for this portion of the page are accounted for and you're sure they will fit, go back to the top and work your way across, carefully positioning each piece. Use the T-square for horizontal alignment, the combination of triangle and T-square for vertical alignment, and the ruler or line gauge for spacing between elements (see Figure 11.6). Protect the lower portion of the paste-up sheet with a clean piece of paper and move it up to protect finished areas as you work down the page.

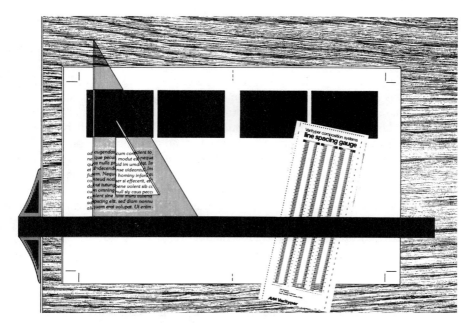

FIGURE 11.6. How a paste-up sheet and equipment are used to execute mechanical artwork. (Gauge © AM International, Inc. Courtesy of Varityper Division, AM International, Inc., East Hanover, NJ.)

Handle type carefully so that it isn't damaged while you are manipulating it. Never roll or crease a galley or strip of display type; store it flat or in a file folder away from light so that it will be suitable when you are ready to do the mechanical. If you are using typewritten copy that is easily smudged, spray it with a fixative. Use an art knife to lift a piece of type and reposition it to keep finger contact to a minimum. Clear your working area of potential hazards (such as a cup of coffee) which can ruin both type and paste-up sheets.

When reusing type from something that is already printed and the ink on the reverse side shows through, mount the type on black paper to neutralize showthrough. The image will then be visually consistent and the type easier to handle because of the extra support. Carefully trim away the backing so that no black is visible or use opaquing liquid on the edge.

"Standing type" is picked up from a previous mechanical and used again. Examples are a logo, a return address, and a mailing-permit number used in each issue of a magazine. Standing type can be reused effectively to save typesetting and/or stat costs until it is discolored or so worn that the image is distorted. An efficient way to ensure that frequently used type will always be available for artwork is to make up a sheet of it. For instance, logos in various sizes can be grouped on a sheet and several stats made, each a quality reproduction. (When using hot type, quality copies can be made on a proofing press using heavy glossy paper.)

If you have picked up standing or other type or art from an old mechanical, note on its tissue-paper overlay or outside the image area that it has been cannibalized. Without such a warning, a mechanical might inadvertently go in for press negatives again with elements missing.

The final step in positioning is to burnish type into place so that it adheres securely. Use a burnishing roller, burnishing stick, or smooth ivory stick with which you can apply firm, even pressure and go over the entire mechanical, working from the center outward. To protect type, use a sheet of paper between it and the burnisher during this process. Once all type has been burnished into place, meticulous mechanical artists go one step further to assure quality reproduction by trimming edges of galleys to bevel them and make them uniform. A beveled edge creates a sloping surface that won't cast a shadow when the press negative is made.

Line Art

Black-and-white drawings, graphs, rules, or decorative borders are examples of line art often used to add interest and information to an otherwise type-only printed piece. Because they are converted into the same press negative as type, they are part of the basic artwork.

Illustrations. All line illustrations are either assembled as elements of the basic artwork or space is allowed so that your engraver can readily insert them before the line press negative is made. A simple drawing, for example, may be reduced to size and pasted onto the artwork, while one requiring a screen or special positioning (such as wrapping it around type) is better left to your engraver to reduce and drop into place.

Unless you or your illustrator is a competent artist and you are absolutely certain of what the art should be, don't draw directly on the mechanical. A mistake in position, size, or content is costly to rectify. The work must be started over with a fresh paste-up sheet, or the art redrawn separately on a piece of paper large and opaque enough to mask the error when glued over it.

To add line illustrations to artwork yourself, start with art that is either the size you want it or larger so that it becomes sharper when reduced. If you must resize it or if you must make a quality reproduction of original art (if you are using a non-copyrighted drawing from an old book, for instance), make a to-size photostat ("stat") of it for your mechanical.

The most common stat for reproduction is a simple photographic copy of line art (although a stat of continuous-tone art is also possible). Using a camera designed for this purpose, your typographer, engraver, or printer can quickly and inexpensively make a positive or negative stat for you on glossy or matte photosensitive paper.

A positive stat may be made directly from the original art or with an intervening film negative. A direct positive is less expensive and quite satisfactory for a project not requiring a negative. An example of one in which having a negative is useful is a sheet of repro-quality logos; when you run out of your supply, the negative can be used to shoot more, with no loss of quality and no expense of reconstructing the sizes that appeared on the original logo sheet. A negative is also necessary if you want to reverse or "flop" (use in the opposite direction) type or art.

Glossy paper is the more commonly used and what you will get unless you specify otherwise. Matte picks up some gray tones and can copy a photograph or other continuous-tone work, although not in reproduction quality. Matte stats are used to indicate placement, size, and content of halftones on artwork; a simple photocopy readily fulfills the same purpose at a much lower cost, although size will not be as exact.

When specifying art for a stat, indicate the desired finished size in picas, inches, or centimeters and the percent of reduction or enlargement that size represents of the original. Specify whether or not you want a negative, and when having a stat of continuous-tone art made, whether you want to convert it to line only on glossy paper or maintain some gray tones on matte paper.

An illustration drawn or statted to size is glued and burnished into place on the mechanical. If it is to be added later, placement must be indicated. One way is to glue a photocopy of actual or approximate size to the mechanical and label it "For Position Only" or "Not for Repro." The engraver will then know position and content, although the actual size must be clearly noted on the art. Another way is to do a "keyline" (or "key line") in which the key or principal lines are sketched on a tissue-paper overlay to indicate placement, content, and approximate size. A keyline is especially useful to call out portions of the art to be printed in a second color or to isolate the part of it that is to be reproduced if the whole image is not to be used.

Rules and Borders. These forms of line art are added directly to the paste-up sheet using transfer art, typeset lines, or a ruling or artist's pen. Unless you are a skilled artist, you will find transfer or typeset rules, borders, or leaders easier and much more predictable. A possible alternative to adding rules at the paste-up stage is to have your engraver scribe them on the negative (see Chapter 12 for details).

Transfer rules, borders, and other lines are available from an art-supply dealer in lift-off or tape form in many sizes and designs. The lift-off variety is the more uniform and less susceptible to error. Transfer rules, borders, and other lines are added after text, display, and cutline type are in place on the mechanical and after their exact position has been determined.

Using a nonrepro blue line as a guide, carefully lay down the transfer rule or border and check alignment with a T-square. You may have to reposition it several times before it is straight and ready to burnish into place. A rule or border longer than is needed gives you room to grip it on each end while you work. To provide a stable strip with which to work, lift off two or three adjacent rules, align one where you want it, and cut away the others. When using lines set by your typographer, have them waxed or add your favorite adhesive and treat them as you would the transfer kind. Keep the work area dust-free to avoid inadvertently adding hair or lint to artwork as you add rules.

To construct a box, rule in all four sides, then cut away the excess at the corners with an art knife. A well-finished corner is mitered, and the sides pulled together to fill any gap or the slit carefully inked in.

To trim evenly a series of lines (such as on a form having a block of blank lines) or rules at the top and bottom of a page, position and burnish them all; then use your T-square and cutting triangle to trim them as a group exactly the same in one motion. If several lines to appear in a block can be laid out with the same spacing as on the transfer sheet or typeset with precise line-spacing, cut out and position the entire block, rather than lay in each line individually. Adding lines to artwork is tedious enough, and this simple shortcut saves considerable time when the layout allows.

Lines that are to be drawn by hand should be done before any other element is added to the mechanical. Early ruling also assures a clean surface, without the possibility of skipping caused by unseen spots of wax from galleys or oil from fingerprints. Ruling pens and artist pens (such as a Rapidograph) are notorious for leaving blobs of ink or running out just as the work is nearly finished. In skilled hands, however, they produce work more precise than transfer lines, and without the accompanying thin layer of plastic. Inking a box or rule in the middle of a page requires close measuring so that the work can be done at the start of paste-up. To some extent, opaquing liquid can be used to cover mistakes, although a line inked over an opaqued area may not be as intensely black as in other areas. An alternative when using a paste-up board with multiple layers of white is to cut away the top layer to eliminate a mistake. Rules that are to bleed off the trimmed page should extend $\frac{1}{8}$–$\frac{1}{4}$ in. (0.32–0.64 cm) beyond the trim line on the mechanical.

Screens, Tints, and Other Line Work

Though still line techniques, screens and tints are not provided for on the basic artwork but rather on an overlay of clear, heavy acetate. Hinged at the top to the paste-up sheet with masking tape, an acetate overlay lays over the basic artwork so that it can be shot separately to make a second line negative. A sheet of lightweight paper may be substituted for the acetate; but it is harder to see through for alignment. Another choice is heavy Mylar or polyester that is less susceptible than acetate to changes in size caused by heat or humidity. For simplicity's sake, acetate will be referred to here to distinguish a clear overlay from a protective tissue-paper overlay.

A screened area is formed by black ink of less than full coverage. A tinted

PASTING UP MECHANICAL ARTWORK

area is formed by colored ink of less than full coverage. Both are done with film screens. For example, an entire page may be screened at 10% black to make it appear gray overall, or part of a drawing may be tinted to add a light coloring. Multiple screens on overlays can create different shades of black or color individually or in combination.

Although a screened or tinted area can be produced by using transfer art available in several designs or intensities, a higher quality is achieved by letting your printer supply the screen at the time press plates are made. A printer uses precise, finely constructed screen unmatched by transfer art to produce crisp, clear results.

To define an area to be screened or tinted, use orange or red transparent plastic "masking film" (or "stripping film") available in peel-off or coated acetate sheets (such as Zip-a-Tone, Rubylith, or Amberlith). Peel off a piece of the masking film sufficient to cover the area, burnish it in place on the acetate overlay (or use the acetate-sheet variety), and then cut away the excess, using a sharp art knife or single-edged razor blade and a cutting triangle (or French curve or template if the shape is irregular). The masking film is photographed as black and reversed by the camera to create a clear area on the negative surrounded by black. An appropriate tint screen is then inserted into the clear area when a press plate is made (see Chapter 12).

Type to be printed over a halftone is also positioned on an acetate overlay. If it is to be surprinted, the original black-on-white type is used. If it is to be reversed, the original type must be converted to white-on-black, either during paste-up by putting a reversed stat on the overlay or at the time press negatives are made by shooting the type on an overlay as a positive. Type to be surprinted over or reversed out of a screened area is glued on the basic artwork and the screened or tinted area designated on an overlay, as just covered.

Whenever an overlay is used, it must be registered with the basic artwork so that elements are properly aligned when press negatives are made. To do so, purchase a sheet or roll of transfer registration marks (commonly a circled cross). Place at least three registration marks on at least two sides of the basic artwork, confirm that the acetate overlay is positioned exactly as you want it, and place corresponding registration marks on it. The crosses should align on each pair of marks, as shown in Figure 11.7. An alternative is to consult your engraver and printer about using "pin registration," a standardized system in which holes punched in original artwork are aligned with pins to make and strip press negatives and used again when plates are made to assure accurate registration at each step.

Label the overlay by wrapping a piece of masking tape around an edge and writing on it what you want done in ink or heavy pencil, for example, "10% black" or "Reverse." Reinforce what you want on a specification sheet or purchase order. All elements on an overlay will be treated as specified; you must do an overlay for each different set of specifications. An overlay doesn't have to cover the entire mechanical if only a small portion is affected

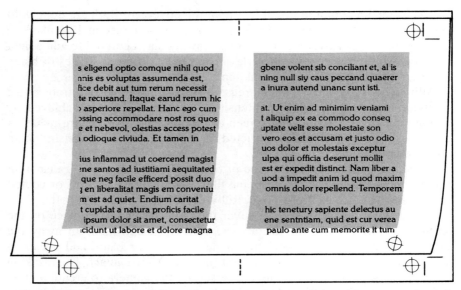

FIGURE 11.7. Pairs of registration marks on the basic artwork and acetate overlay help assure match-up during printing.

(such as one column or one page only). You must have room in the margins, however, for at least three sets of registration marks.

Assuming that center and trim marks were made when the paste-up sheet was drawn, other markings outside the image area may now be needed to assure satisfactory reproduction of a particular project. Fold marks are short dashed lines outside and on both sides of the image area showing where the piece is to be folded in the bindery. A die-cut perforation line is designated with a dotted line in the margins or if a box (such as a coupon), by a dotted box on the tissue-paper overlay. The shape of any other die-cut is marked on the tissue-paper overlay, and an exact drawing of it is provided to be used to make the die or select an appropriate one from standard dies on hand.

Halftones

A black-and-white or color photograph or other continuous-tone art may be converted for use on the basic artwork but is more often handled separately. This section presents information on both processes, although the latter is emphasized as the preferred practice for quality print production.

Prescreened Prints. Converting a halftone to a line negative has the advantage of allowing full assembly of a piece on the mechanical and one-step production of a press negative. This process saves engraver time and corresponding costs, although it is used selectively because an obvious loss of image quality is the tradeoff. Prescreened prints are reserved for jobs to be run on newsprint or other coarse paper incapable of fine reproduction of highlights, shadows, or a broad range of gray tones. They also make halftone reproduction possible when an engraver has no halftone camera or has one with limited reduction capability.

Photomechanical transfer (PMT) paper is used in one such process. Developed by Kodak, it reproduces a halftone to as fine an image as 100-line (40-cm) screen—not up to the standards of quality offset but adequate for newspapers. The photograph is reduced or enlarged during the process, and special paper is used. Also called "diffusion transfer," the PMT method may in addition be employed to create a comprehensive quality proof of artwork to show for approval before negatives are made or to duplicate an entire mechanical, for instance, if a job is to be printed out of town and you want backup artwork in case of loss or damage during shipment.

Another Kodak development for prescreening a halftone for direct paste-up results in a Velox. A regular halftone negative is made to size of the original art using a screen as fine as 120 (48 cm). Instead of waiting to strip the halftone negative into the line negative, a contact print on Velox paper is made of it. The Velox is added to the mechanical, saving engraving time and expense.

By either method, quality is lost, although the dot pattern isn't as pronounced on a Velox as on a PMT. Highlights, shadows, and some gray tones are sacrificed. Whites and blacks can be restored to a degree by extra negative work or retouching; however, the choice of prescreened prints over separate halftone negatives becomes questionable if such special effort is needed. A Velox can be the answer if a poor-quality halftone print is all you have available; retouching after screening may be easier than trying to do it before. When requesting a prescreened print, indicate the type of press on which it will be run so that the amount of contrast can be gauged to press capability.

Halftone Windows. Continuous-tone art that is to be reproduced by the halftone process explained in the following chapter must be provided for on the mechanical. A "window" is the device used for this purpose. It is simply an area corresponding in size and shape to the cropped, scaled halftone. The area is reproduced as a clear space on the negative, ready for the halftone negative to be stripped in. The process of constructing a window and the reason for doing so are like that covered earlier for screens and tints, except that the window is on the basic artwork, not on an overlay.

A window is made in one of three ways (see Figures 11.8, 11.9, and 11.10). One is to use orange or red transparent plastic masking film securely burnished onto the mechanical and cut with a sharp art knife or razor blade to the size and shape desired, with the excess peeled away. The film is interpreted by the camera as black, resulting in a clear hole to be filled by the engraver with a halftone negative. It is transparent so that you can see through it to position it according to nonrepro-blue guidelines. Also, the

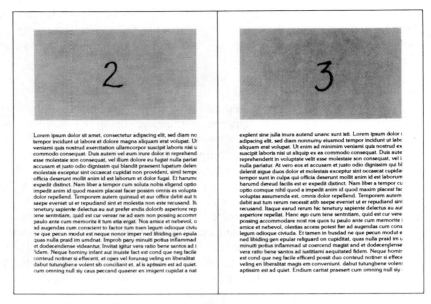

FIGURE 11.8. Masking film used to create windows on mechanical artwork.

number, letter, or other identification for the art to occupy that window can be written on the paste-up board and be seen by the engraver through the transparent film.

Another way to make a window is to rule the space in black ink using no masking film. This approach may be preferred by your engraver if the halftone negative must be very precisely positioned, for instance, if it is to have a narrow, even margin of white around it to stand out on a screened page. This method may also be the better choice if the window is to be an odd shape. A pen is easier to control than an art knife unless you're experienced at cutting curves or sharp angles. If you elect to lay in masking film for an odd window, make sure the image area is covered. Don't be concerned with fully outlining its shape (as long as adequate window is allowed) unless the window is to be dropped out of a screen, in which case precise shape is critical.

A third way to designate a window is with a keyline. The image area can be accurately drawn on the tissue-paper overlay, leaving to the engraver the detailing of the edge of that area.

Keep in mind that the cropped or scaled image must be sufficient to fill the window, and have some left over for taping the halftone negative to the line negative. An image that is too small will leave a gap. For example, if you plan to use a photograph full frame and it measures 30 × 50 picas, it will be too small to fill 31 × 51 picas unless it is enlarged to provide a minimum of one-half pica of clear space for secure taping on at least two sides. An

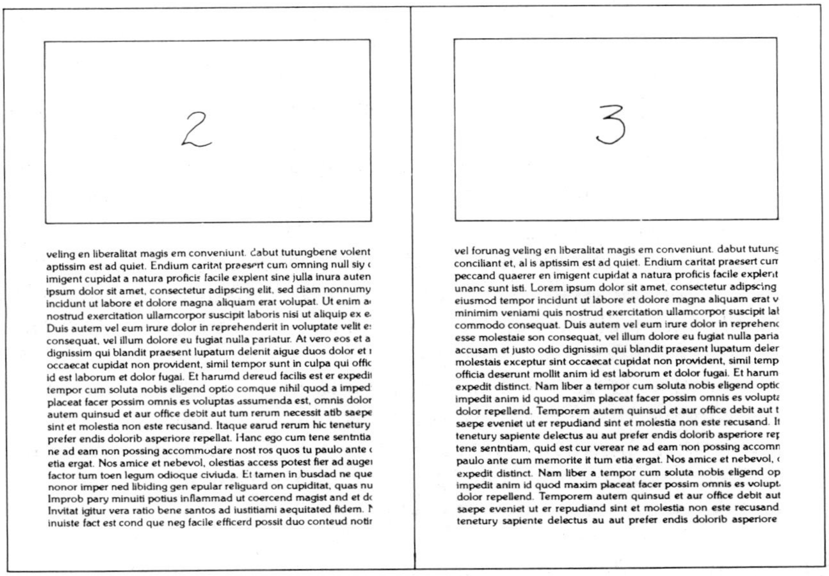

FIGURE 11.9. Precisely ruled boxes used to indicate on mechanical artwork where halftones are to appear.

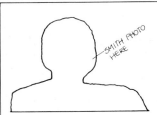

FIGURE 11.10. Keylining on the overlay used to indicate where an irregularly shaped halftone is to appear.

alternative is to pay the additional cost of having a halftone negative made of the full page with art in position to be sandwiched with the line negative. This method allows more layout flexibility and retains the entire negative area in case you want to shift the image or resize the window after seeing a proof of the press negative.

Slug windows to their corresponding halftones using numbers, letters, words, or phrases. Numbers or letters leave no room for confusion about which art goes where, while words or phrases may if a piece is to include several related halftones. Make a list of labels, pages, and sizes for halftones so you can readily clear up any confusion as negatives are being assembled. To be doubly sure halftones are properly placed, lightly tape photocopies of them to their corresponding windows.

A halftone that is to bleed or run off the page on one or more sides requires a window that extends beyond the trim line of the piece. Some of the image is then trimmed away after the piece is printed. The window should extend beyond the trim line $\frac{1}{8}$ to $\frac{1}{4}$ in. (0.32–0.64 cm) to ensure that a sufficiently large area is printed to account for trim.

Silhouettes. Halftone silhouettes drop the background in whole or in part. Silhouetting using masking film should be done during paste-up unless your engraver has agreed to opaque out background on the negative. (See Chapter 7 for an explanation of executing artwork for silhouettes.)

Preparing Art

Following is advice on cropping and "scaling" (sizing) art as well as on marking it in preparation for engraving. Though this detail work may be done at the dummy stage of production, the result is more accurate if you wait until final size and placement are determined by the mechanical.

Cropping Art. An illustration or photograph should be cropped if you don't want to use the full image in the printed piece. For example, you may want to show only two persons out of a group of five or to focus attention on only one part of a drawing. Like copy editing, art editing or cropping is a process of eliminating the extraneous and emphasizing the important. The following guidelines apply to cropping art yourself or to supervising a graphic designer doing so.

A piece of art may be used in its entirety ("full frame"), cropped on all four sides, or left as is on one or more sides. If the entire image communicates information, attracts, or sets a mood, it should be reduced or enlarged to fit the space available and reproduced full frame. Often, however, some part of the image is not necessary or distracting and needs to be eliminated.

To see how an illustration or photograph will look when cropped, block out unwanted portions with four sheets of paper or cut yourself a pair of heavy L-shaped cropping angles, as shown in Figure 11.11. Move them

PASTING UP MECHANICAL ARTWORK

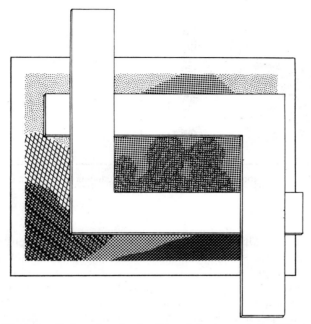

FIGURE 11.11. L-shaped angles for determining how art should be cropped.

around until the image remaining is well-composed for your purpose (see Chapter 7 for guidance on composition) and approximately the same shape as the space reserved for it on the mechanical. If you decide that the ideal crop results in a vertical image but the mechanical calls for horizontal proportions, you will have to alter the way it is cropped or move type around to reshape the space. In anticipation of such maneuvering, evaluate the probable fit of art at the dummy stage when moving elements is much easier and cheaper.

Scaling Art. Measure the space available and adjust your cropping to be in the correct proportion. To do so, use a simple diagonal-line method of scaling or a proportion scale or wheel. Art proportion is based on the principle that when art is reduced or enlarged from side to side, it must be reduced or enlarged a corresponding percentage from top to bottom. You cannot change size in one dimension alone and maintain the same proportion.

The diagonal-line method of proportioning art employs a ruler and a sheet of paper (as illustrated in Figure 11.12). First, measure the image area as you have cropped it and draw a square of that size on the paper (not over the art because you might damage it). Draw a diagonal line from the lower-left to the upper-right corner and extending beyond. Next, measure the width of the space on the artwork that the art is to occupy. Mark this distance on the drawn square, either within the original square if the space is smaller or

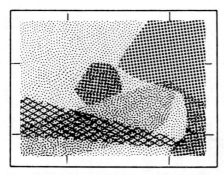 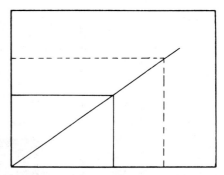

FIGURE 11.12. The diagonal method for enlarging or reducing art, starting from the cropped image area.

beyond if it's larger. Draw a vertical line from that point perpendicular to the width line until it meets the diagonal line. Complete the new square by filling in the top horizontal line until it intersects with the diagonal line. The new pair of lines translates the original proportions into the finished size.

A proportion scale or wheel simplifies this process. Available from an art-supply dealer or possibly from your typographer or printer, a proportion scale consists of two numbered wheels joined in the middle so one can be turned while the other is held constant (Figure 11.13). The inside wheel designates original size and the outside reproduction size.

To use a proportion scale, measure the width of the original art as cropped. Match that number on the inside wheel with the width of the space reserved for the art on the artwork. Hold both wheels stationary and locate the original depth of the art on the inside wheel. The number across from it on the outside wheel will be the finished depth when the art is reduced or enlarged.

The final step in scaling art by either method is to determine the percent of reduction or enlargement necessary, calculated as percent of original size. If you use a proportion scale, percentage is automatically calculated for you, as shown in Figure 11.13. The arrow at the top of the window points to the percent of reduction or enlargement. If you use the diagonal-line measure, you will need to calculate the percentage according to the following equation and example. Use the same equation, whether starting with width or depth:

$$\frac{\text{Finished Width}}{\text{Original Width}} = \%$$

For example,

$$\frac{10 \text{ picas}}{40 \text{ picas}} = 10 \text{ divided by } 40 = 25\%$$

PASTING UP MECHANICAL ARTWORK

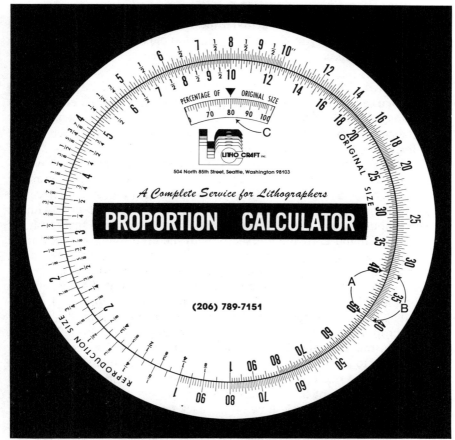

FIGURE 11.13. A proportion wheel for determining enlargement or reduction of art showing original size (A), reproduction size (B), and the percentage of the original the reproduction size will be (C). (Courtesy of LithoCraft, Seattle.)

If you know original and finished sizes of one dimension and original size of the other, calculate the missing finished dimension as follows:

$$\frac{\text{Original Width}}{\text{Original Depth}} = \frac{\text{Finished Width}}{\text{Finished Depth}}$$

For example,

$$\frac{40 \text{ picas}}{60 \text{ picas}} = \frac{10 \text{ picas}}{X}$$

$$40 \times X = 60 \times 10 \text{ or } 40X = 600$$

$$\frac{40X}{40} = \frac{600}{40} \qquad X = 15$$

When expressing dimensions of art, paper, or other print-production elements, always list width first or underline it. One of the few standardized practices in the industry in North America, it avoids confusion about which dimension is which.

Marking Art. When a piece of art is accurately cropped and scaled, it must be marked so that it will be properly reproduced by the engraver. It may be marked at the dummy stage if you are sure size won't change as the job is pasted up or marked later when artwork is completed. See Figure 11.14 for examples of the following procedures.

"Crop marks" are straight lines or arrows placed in the margins of photographs or well out of the image area of illustrations using a dark graphite pencil, a pen, or a wax pencil (though the latter isn't nearly as accurate because of the width of the wax line). If the art is all image area with no margins, mount it on a larger piece of paper and make crop marks on the paper. They should correspond exactly with the crop that you found most pleasing and appropriate and be clearly visible. If, as is usually the case, corners are to be right angles, the image area as indicated by crop marks should be squared. In other words, if width is marked at the bottom of a photograph as 30 picas, it should also be 30 picas at the top, not 31 or $29\frac{1}{2}$. Check that the crop is square by measuring between crop marks, not between a crop mark and the edge of the image. Especially with a full-frame photograph, the edge may be irregular and not a valid reference point.

If the art is to be straightened or reproduced in an unusual shape or if you plan to use only a small portion of it and want to avoid confusion about your intent, place a piece of tissue or other thin paper over it and *lightly* rule or sketch what you want. Don't draw on the back or front of the art.

Finally, crop with pen or pencil, not with scissors or tape. Send the whole image in for a stat or halftone negative, even if you want to reproduce only part of it. Cutting or taping can ruin the art for future use, create a ragged edge when it is reproduced, and leave no margin for stripping during image assembly.

Cropping, Scaling, and Marking Transparencies. When art is in color-transparency form, extra effort is involved in preparing it because the actual transparency may be too small to crop accurately and marks on the mounting too large to be precise. One choice is to determine enlargement or reduction percentage, and then do the cropping when you see proofs of the color separations. Another is to have an inexpensive color or black-and-white print made for scaling and cropping purposes only, reserving the transparency for separating and sizing according to markings on the print.

Color Paste-Up

A project involving two or more ink colors requires special attention during paste-up to ensure that the job will be successfully printed. The procedure to

PASTING UP MECHANICAL ARTWORK 333

FIGURE 11.14. The image area to be reproduced is indicated by crop marks in the margin (left) or on the tissue-paper overlay (right) when any confusion might exist.

be followed in preparing a color mechanical depends on whether or not the colors touch and if the ink used is opaque (a given or simple color) or transparent (a process color).

Loose Register. A project with loose color register has two or more opaque colors that don't print closer than $\frac{1}{16}$ in. (0.16 cm) of one another ("spot color"). An opaque or flat ink is used to produce distinct, individual colors (such as black text surrounded by a green border). This type of color mechanical may be constructed in one of two ways. It may be done as one piece with colors to be separated by the engraver called out on the tissue-paper overlay, or elements to be printed in the second color may be pre-separated at the time of paste-up on an acetate overlay. Figure 11.15 illustrates these two methods for pasting up a job with loose color register.

The latter method may be preferred when the second color is to be a tint over type or art or when a portion of a drawing or other illustration is to be in a second color. Your engraver can advise you about which procedure to use. Despite being called *loose* register, the colors are still aligned; they are simply spaced apart. Upon close examination, however, registration is not as tight as is needed for touching or transparent colors.

Lap Register. A project with lap color register has two or more opaque colors that touch. Negatives must provide for the inks to overlap slightly to account for paper expansion and registration variation during the course of a long run. If the inks don't overlap on the negatives, they may leave an uninked gap on the paper.

Artwork should be done in one piece and keylined on the tissue-paper overlay ("Colors touch; provide for register") so that the engraver will allow

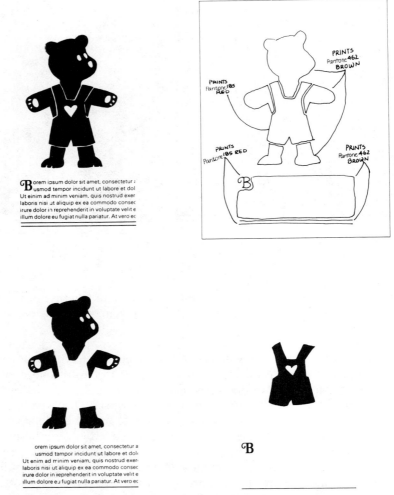

FIGURE 11.15. One-piece mechanical artwork for loose register marked for separation by the engraver (top) or preseparated on an acetate overlay (bottom).

for overlap when negatives are made. An alternative is to provide original keyline art with the line precisely and smoothly drawn between colors. Negatives are then made allowing for overlap the width of the line. Figure 11.16 illustrates these methods of pasting up a job with lap color register.

Another alternative is to cut masking film on an overlay in the desired shape and provide for overlap by the darker color. The amount of overlap is from $\frac{1}{32}$ in. (0.08 cm) for dark inks to $\frac{1}{64}$ in. (0.04 cm) or less for light inks and the highest-quality printing. Again, consult with your engraver about paste-up for lap register.

PASTING UP MECHANICAL ARTWORK

FIGURE 11.16. One-piece mechanical artwork on the left is marked for the engraver to provide for lap register. The pair on the right provides a precise keyline.

Hairline Register. A project with hairline register has two or more transparent colors printed at precise angles to one another to give the visual illusion of a new color. It is required for all full-color process printing. It is also necessary when printing white ink on colored paper to provide an opaque window for subsequent printing of a halftone in the area. Very close register is critical to prevent a noticeable color difference at the edge of the image. The mechanical is constructed as one piece, and separate negatives are made by the engraver based on keyline instructions.

Corrections

The preferred method to lay in type corrections on artwork is to cut out and lift away the incorrect words or lines of type and insert the correction trimmed to fit the hole. The result is a flat surface and no opportunity for shadows to show up on the press negative. While corrections may be placed over the basic artwork, the overlapping elements may create a shadow, requiring tedious opaquing of the negative. Typeset at least two lines of type or the whole paragraph to correct a word or letter so that the correction can be aligned properly. One word or letter is difficult to handle and may not adhere well enough to the mechanical to stay in place.

A marred letter or line already pasted up can be corrected with a fine drawing pen or felt marker *if you are very careful.* Because ink on glossy photographic paper dries slowly, it can be easily smudged when not left undisturbed until dry or sprayed with a fixative.

Proofing

Artwork can be proofed by rereading and remeasuring the artwork itself. Proofing content and alignment with a photocopy, however, has the advantage of presenting an even surface that doesn't distract the eye with trim lines or corrections. Though exact measurements and quality of type should be checked on the mechanical, overall alignment can also be confirmed by looking at the artwork through its protective tissue-paper overlay that screens out distractions. Another way is to lay a clear acetate grid over the mechanical, aligned with top trim and center lines. A look at type and art through the grid will tell you if they are straight. Double-check that all elements are in place by referring to the dummy and to any pieces of type that are left over before you assume they are all scrap.

Cleaning and Protecting Mechanical Artwork

Lightly rub artwork with cotton or tissue dampened in solvent or rubber-cement thinner to clean away excess wax or bits of rubber cement. (Use this process on typeset or statted material only, not on typewritten or hand-lettered type.) For larger rubber-cement deposits, lift them off with a commercial pickup or one made from allowing a dollop of rubber cement to harden. Use opaquing liquid to eliminate any persistent spots or cut them away if you're using a paste-up board with more than one layer of white paper.

A spray fixative can be used as a chemical protection for the surface but is necessary only if typewriting or ink has been used on the mechanical. Typeset text, display, and cutline type and stats of art won't smudge with normal handling. Whether you use a fixative or not, protect the entire face of your artwork with a tissue-paper overlay or a lightweight, unprinted sheet.

While the tissue paper protects, it also provides an ideal sheet for instructions to your engraver or printer. (Slip a sheet between it and the mechanical as you write so that your instructions don't bleed through, however.) Circle elements to be printed in a second color and note all ink colors and screen and tint percentages. Circle any small piece of type (such as a short correction) so that it is handled carefully. On a major or unfamiliar job, reiterate these instructions on a separate specification sheet or purchase order. Some mechanical artists and designers add a second protective overlay of heavy paper before wrapping artwork in kraft paper or inserting it in a large envelope for delivery to the engraver or printer.

PASTE-UP FOR OTHER PRESSES

The preceding information applies to preparing artwork for offset production. Mechanicals for other types of presses are slightly different.

Letterpress

Although most commercial projects run on a letterpress now use artwork as the basis for press negatives and plates, metal type and separate halftone

PASTING UP MECHANICAL ARTWORK

engravings may be used directly. Ask your printer which procedure is to be followed and confirm whether type to be surprinted over, reversed out of, or mortised into art should be overlaid on the art or on the mechanical.

Gravure

All line work should be included on the basic artwork with areas in a second color constructed on an acetate overlay. Because the entire gravure negative carries a halftone screen, all elements on the mechanical must be able to tolerate that screen; thin lines or fine serifs will appear ragged and are, therefore, not advised. Transfer shading films and Ben Day shading screens should not be used because the combination of screens won't be pleasing. A hairline register is difficult to maintain throughout a rotogravure run.

Screen Printing

Prepare a one-piece mechanical for screen printing. Halftones and screens or tints don't reproduce well, and relatively simple art is preferred. If a photographic stencil is to be used, line art should be prepared fully as for offset and colors preseparated. If a hand-cut stencil is to be used, letters and designs need not be filled in by the artist, and a skilled screen cutter can separate colors.

EVALUATING MECHANICAL ARTWORK

How well mechanical artwork is prepared directly affects what comes off the press. The following points summarize how it should be executed; review them before proceeding to engraving and printing of your project. The checklist presented in Figure 11.17 can be used as a guide to evaluation:

1. Content has been carefully proofread and corrected. Errors in writing and typesetting have been rectified, as have mistakes in sequence or placement of elements made during paste-up.

2. Sizes are exact and complete. All measurements are as specified, and spacing between parallel elements (such as between headline and text type) is consistent. Windows for halftones correspond with the halftones as cropped and scaled.

3. All elements align vertically and horizontally, exactly perpendicular to or parallel with the edge of the paper as it is to be trimmed (unless type or art is purposely diagonal). Windows for halftones made of masking film are straight with crisp corners.

4. Windows for halftones that are to bleed into the margin are $\frac{1}{8}$ to $\frac{1}{4}$ in. (0.32–0.64 cm) beyond the trim line.

5. All pieces of art that are part of the mechanical are properly sized.

```
                              Artwork Checklist

        For _____

        _____ Copy Proofed              _____ Illustration Placement Indicated/
        _____ Rules/Boxes in Place              Coded
        _____ Pagination Included       _____ Overlay(s) Registered to Basic
                                                  Artwork
        _____ Standing Type Inserted    _____ All Elements Securely Burnished
        _____ Litho/Publication Code Included
                                          _____ Artwork Cleaned
        _____ Gutter/Margin Spacing Consistent
                                          _____ Art/Artwork Protected by Tissue
        _____ Union Bug Placement Indicated
                                          _____ Trim/Fold/Die-cut Lines Marked
        _____ Headline/Paragraph Spacing
                Consistent                _____ Color Overlaps/Color Breaks
                                                  Marked
        _____ Windows/Rules Trimmed     _____ Screen Values Noted
        _____ Windows Cut Straight/Crisp
                                          _____ Colors Noted
        _____ Bleed Windows Extended    _____ Artwork Meets Special Supplier
                Beyond Trim Line                  Requirements
        _____ Art Accurately Cropped/Scaled/
                Marked                    _____ Other _____
                                          _____ Other _____
        _____ Windows Coded

        Date_____  By_____
```

FIGURE 11.17. An example of a checklist for proofing mechanical artwork.

Separate pieces are correctly cropped and sized and clearly identified, as are the areas on the mechanical they are to occupy.

6. Each layer of a mechanical is registered so that elements will be exactly aligned. At least three sets of registration marks are in place, or alignment has been assured by using pin registration.

7. Artwork is burnished thoroughly so that all elements are securely in place. Corners don't curl, and all bubbles, bumps, and wrinkles are ironed out with a burnishing tool.

8. Artwork is clean. It is free of smudges, dirt, extraneous marks in ink or pencil that will be picked up by the camera, and bits of glue or wax that attract lint or dirt. The surface is fully protected by a tissue or other lightweight paper overlay.

9. Full instructions are noted. Center, trim, fold, and blind perforation marks appear outside the image area. Overlays for screens and multiple colors are labeled with exact screen values and ink colors.

10. Artwork corresponds with what the engraver and printer require so that the resulting press negative and/or plate can be used successfully.

CHAPTER TWELVE

Image Assembly for the Press

LINE NEGATIVES   341
   Line Photography, 341
      Ganging Line Work, 342
      Separating Flat Colors, 342
   Evaluating Line Negatives, 343

HALFTONE NEGATIVES   344
   Halftone Photography, 344
      Halftone Screens, 344
      Ganging Halftones, 348
   Evaluating Halftone Negatives, 348

COLOR NEGATIVES   348
   How Colors Are Reproduced, 349
   How Colors Are Separated, 350
      Direct Color Separation, 352
      Indirect Color Separation, 353
      Electronic Color Scanning, 353
      Dupe and Assembly Separations, 354
   Color Correction, 354
   Evaluating Color Separations, 355

Types of Proofs, 355
What To Check, 356
Factors Influencing Color Reproduction, 357
   Quality of Original Art, 357
   Paper, 358
   Ink, 358
   Engraver Skill, 359
   Type of Press, 359
   Color Perception, 359

STRIPPING AND PROOFING   359

Stripping Techniques, 360
   Basic Assembly, 360
   Single-Flat Assembly, 362
   Multiple-Flat Assembly, 363
   Halftone Assembly, 364
   Pin Registration, 366
   Imposition, 366
Changes on the Negative, 368
Negative Proofing, 369
   Proofing Techniques, 369
   Checking a Proof, 370
Hot-Metal and Gravure Assembly, 372
   Hot-Metal Composition, 372
   Gravure Film Positives, 373

PLATEMAKING   373

Offset Platemaking, 374
   The Basic Process, 374
   Offset Plates, 374
Letterpress Platemaking, 377
   Original Plates, 377
   Duplicate Plates, 377
Gravure Platemaking, 379
Screen Stencils, 381
   Hand-Cut Stencils, 381
   Photographic Stencils, 381

The print-production process is well under way when type and any accompanying art are ready for conversion into a form required for the press. With the popularity today of offset printing, that form is most likely a printing plate made from an intermediate film negative of artwork and art.

This process is referred to as engraving, although only certain types of platemaking are true engraving. More specifically, it is photoengraving since graphic arts photography (as opposed to creative photography) is an integral part of all the automated processes to be covered.

It may be accomplished by your printer or by a specialty supplier called a photoengraver or simply an engraver. The choice depends on skill and equipment available from your printer versus what a specialist offers as well as on your requirement for ownership. Negatives supplied by your printer belong to that printer, according to trade customs (see Appendix A). Buy them outright from an independent engraver or negotiate to buy them from your printer if you might need them again, especially if you don't expect to do subsequent work with the same printer. Store them flat away from light, as your printer will do for a period of time with printer-owned negatives if you request.

With an emphasis on offset production, this chapter presents information on the making of line and halftone negatives, color separations, and proofs; hot-metal and gravure assembly; and preparation of printing plates for the major types of presses. Included are ways to evaluate the elements to assist in the wise purchase of image-assembly services.

LINE NEGATIVES

A "line negative" is a copy on film of black-and-white artwork produced by line photography. The original may be type only or include rules, drawings, or other illustrations. The characteristic that distinguishes a line negative from a halftone negative is the absence of gray tones; it is exclusively black and white. It may not, however, be void of color. For example, a flyer to be printed with black type and blue rules requires two line negatives—one for the black type and a second for the blue rules.

The "line photography" used to create a line negative captures on film the image of the original artwork, much as happens with snapshots except that the camera used for line photography picks up only black and white. (This process is also called "simple black-and-white photography," "single-color photography," "reproduction photography," or "high-contrast photography.")

LINE PHOTOGRAPHY

To make a line negative, an engraver places original artwork on the copyboard of a "process camera" ("copy camera" or "graphic arts camera").

When exposed to light and developed, the high-contrast film in the camera converts the black portion of artwork to transparent (clear) and the white portion to opaque (black). The resulting negative is the reverse of the original. When a printing plate is made from it, transparent areas will pick up and print ink while opaque areas will not (see Figure 12.1). A similar procedure is followed when using a "stat camera," a small self-contained camera for making photostats of line art or type for paste-up.

A process or stat camera reproduces an exact image of flat artwork the same size as the original or a reduced or enlarged version. The extent to which a particular camera can alter size depends on its design. An engraver will have camera capability to match customer volume and variety of work. A horizontal process camera with its copyboard on a track may accommodate film up to (or even larger than) 4 ft or 4 m^2 and occupy considerable floor space for the bellows and track with the camera back enclosed in a darkroom. A vertical process camera or a stat camera entirely in a darkroom takes up less space but accommodates smaller film and has more limited resizing capability. Either process-camera design may use high-contrast sheet film or (for faster production) roll film.

Ability to alter size may range from 50% of the original (or half the size), up to 200% (or twice the size) for a small print shop. Larger operations or engraving specialists should have an expanded range. To assure accurate sizing, always specify finished size *and* exact percent of original (such as "Reduce to 65.5% of original" or "Enlarge to 124% of original").

Another method of line photography is "contact printing" to produce contact positives or negatives. The original positive or negative is placed in direct contact with film and exposed to the same size.

Ganging Line Work

Pieces of artwork and/or art that are to be sized the same and require no special individual attention by the engraver may be "ganged" or shot as a group to save time and money. They are simply grouped on the copyboard, as many as film size will allow, and one line negative shot. When cut apart, the negatives can be used for separate projects. Examples of line work that cannot be ganged are mechanicals with different reduction or enlargement specifications and art combined with very differently colored paper, colored ink, or a combination of both that requires individual attention.

Separating Flat Colors

More than one color of ink may be called for on a particular project, but it is still line work only if no halftones are involved. A spot of green here and there to add interest to black type or a tint of green in the background is an example of flat or mechanical color separated by making two line negatives,

LINE NEGATIVES

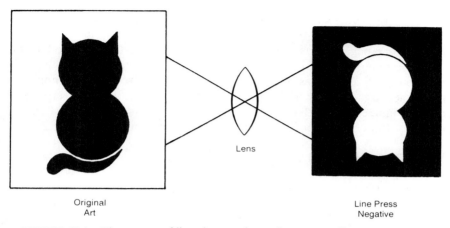

FIGURE 12.1. The process of line photography used to create a line press negative.

one for each color. When converted to press plates, one prints with black ink and the other with green.

Elements to appear in flat colors may be on the basic artwork or pre-separated on an overlay. See detailed information later in this chapter on specific procedures used to assemble flat-color negatives for platemaking.

EVALUATING LINE NEGATIVES

A well-prepared line negative is consistently clear throughout with no distortion of images from side to side or corner to corner. It includes only "live matter"—the images you want to print—not extraneous markings, spots, or scratches. It is the size specified, and colors are broken out as indicated in your instructions.

A skilled engraver should pay attention to amount of lighting, camera exposure time, aperture setting, focus, proper reduction or enlargement, type of film and developer used, developing time, and processing temperature. Lack of attention produces a negative that is lighter or darker than the original, inconsistent in intensity, distorted around the edges, the wrong size, low in contrast, or off-color.

To assure a quality line negative, start with a well-executed mechanical and line art. If you must include newspaper clippings, convert them first to photostats so that they are clear, clean black-and-white images. You could instead ask your engraver to give them individual attention by shooting them separately and stripping them into the line negative. Colored art on medium- to dark-colored paper should also be shot separately and stripped in so that a filter can be used to elicit the most contrast possible.

HALFTONE NEGATIVES

A "halftone negative" is a copy on film of continuous-tone art produced by halftone photography. It differs from a line negative in that it captures shades of gray as well as black and white. It differs from a continuous-tone image negative (used to make a photographic print) in that it is screened to give the image a dot pattern. Halftone photography also encompasses preparation of full-color art for the press (to be covered in the next section).

HALFTONE PHOTOGRAPHY

The process of halftone photography is based on an illusion—the eye's perception of a group of large dots as darker than a group of small dots. Halftone photography converts an image into a pattern of dots spaced equally apart but of different sizes, as illustrated in Figure 12.2. The name derives from the fact that only a portion (or "half") of the full tones of the original is translated into a dot pattern.

This pattern is necessary for the majority of printing methods because most presses can't vary density of ink to make shades of gray. The larger the dots, the more the ink coverage and the darker the area appears. Conversely, the smaller the dots, the less the ink coverage and the lighter the area appears. Thus a tonal range from almost black in shadows to almost white (or pure white for offset) in highlights is created.

Simply stated, the process of converting a black-and-white photograph or other continuous-tone art into a halftone negative starts with putting the art on the copyboard of a process camera. A halftone screen is then placed over the film in the camera, the step that distinguishes halftone photography from line photography using the same process camera. After adjusting for same-size, reduced, or enlarged reproduction, the film is exposed through the halftone screen and developed. The resulting negative is composed of dots in different sizes, one per opening in the halftone screen selected. The size depends on the intensity of light reflected from the art; a dense shadow area reflects less light and creates a larger printed dot than a less-dense highlight area. This process is illustrated in Figure 12.3.

Halftone Screens

The halftone screen placed over the film in the camera is gauged by the number of dots per inch or centimeter counted along a diagonal line. For example, a 100-line screen has 100 dots per inch or 40 dots per centimeter. Approximate inch and centimeter equivalents are presented in Figure 12.4.

Halftone screens are available for commercial work ranging from 50 to 300 lines/in. (20–120 lines/cm), although we most commonly see the result of screens between 120- and 166-line (48–66 cm) in commercial printing. In general, a line screen of 85 (34 cm) or less is used on newsprint and other

HALFTONE NEGATIVES

FIGURE 12.2. The enlarged portion of the eye of the Oriental doll used in previous chapters shows the dot pattern created by a halftone screen.

coarse paper, 100 (40 cm) for smooth uncoated paper, 120 (48 cm) for coated paper, and 133 (54 cm) for quality coated paper. A line screen of 150 (60 cm) or more requires quality coated paper in the hands of a skilled offset printer. Figure 12.5 shows how choice of line screen affects reproduction of a halftone image on the paper used for this book.

Before specifying the line screen to be used on a particular project, know the press on which it will be run. A high-quality offset press reproduces an image well with a 150- or 166-line (60- or 66-cm) screen, while a coarser line screen is more satisfactory if a small, utilitarian press that tends to darken images is to be used. Choice of screen size also depends on the distance from which the printed piece will be viewed. A finer screen should be used for photographs in a small brochure, for example, than for photographs which are to appear on a large poster.

Halftone screens are available in crossline glass or contact form, the latter

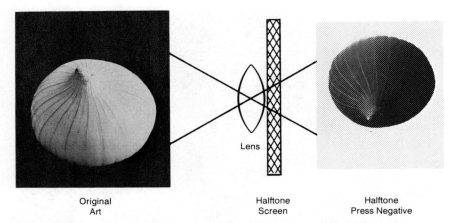

Original Art — Halftone Screen — Halftone Press Negative

FIGURE 12.3. The process of halftone photography used to create a halftone press negative.

Lines per Inch	Lines per Centimeter	Lines per Inch	Lines per Centimeter
50	20	120	48
60	24	133	54
65	26	150	60
70	28	166	66
75	30	175	70
80	32	200	80
85	34	250	100
100	40	300	120
110	44	400	160

FIGURE 12.4. Approximate equivalent values of halftone line screens in inches and centimeters.

being a flexible sheet of acetate containing vignetted dots with softened edges, as opposed to the sharply defined edges of a glass screen. Also used is film with a built-in 133-line (54-cm) screen ("Autoscreen") to bypass the need for a separate screen. As covered in Chapter 6, an array of special-effect screens may be substituted for a halftone screen for design purposes. Included in this category are shading screens to add shades of gray to line art.

In addition to size, halftone dot pattern influences reproduction of an image by shape of the dots. Common are the conventional square or an elliptical (or "chain dot") shape. Unless you specify to the contrary, your engraver will use a square pattern. The elliptical choice is better, however, when a more gradual change in midtones is desired (such as quality reproduction of a portrait).

A halftone screen is usually positioned so that the dot pattern is at a 45°

FIGURE 12.5. The effect of different line screens on halftone reproduction: 85 and 100 (top); 120 and 133 (bottom).

angle to the photograph edge rather than 90°. This procedure diminishes the visible dot pattern, softening the image we see reproduced on paper. When a second halftone negative is shot in preparation for printing a duotone (see Chapter 6), one is made at 45° and the other at 105° (or at least 30° difference) to avoid overlapping dots and creating a moiré pattern or muddy image. For this same reason, a prescreened halftone (one copied from an already printed piece) must be shot again at a different angle.

Ganging Halftones

Just as line art, halftone art can be grouped on the copyboard ("ganged") to save time and money. Ganged photographs or illustrations must be shot to the same size and require no individual attention. Once the halftone negative is developed, the separate pieces can be cut apart.

The drawback to this procedure is a noticeable loss of quality caused by the need to find a "happy medium" of exposure for dissimilar shots (such as those photographed with different lighting or film). Some images may look washed out while others appear muddy. Never gang halftones on a quality job and specifically prohibit others from doing so when you suspect your engraver might try to gang them to cut corners against a firm bid.

EVALUATING HALFTONE NEGATIVES

A well-prepared halftone negative should have a range of intensity from opaque to transparent that will produce a range of tones on the press from nearly white to nearly black. Pure transparent or pure opaque (except for offset) signals an unacceptable breakdown in the dot pattern which the press cannot accommodate. It should also produce a gradual range of midtones so that shades of gray in the original art are retained. Finally, halftone dots should be black not gray caused by faulty developing of the film.

To check the quality of a halftone negative made from a doubtful original, ask for an individual proof of it. Proofing it before negative assembly will prevent unnecessary delay and expense if you find that the quality is not acceptable. Ask your engraver to provide a "blackprint" (or "silverprint"), a proof made by contact printing the halftone negative onto photosensitive paper. Negatives can be ganged for quick, inexpensive proofing to produce a very close representation of what you'll see on the press.

COLOR NEGATIVES

Color in print production may be a flat color used as a spot accent or as a background tint. This type of color uses inks as individual colors—red and black, green and blue. Though the inks may touch, they do not create new combination colors. This section deals with the other kind of color in print production—process color. By using transparent inks side by side to create the illusion of several color variations, process color reproduces the full range of hues in original art (or nearly so).

Although the technique has limitations, it is the closest modern lithography has come to duplicating color photography or illustration on the press. Process color may also be used to reproduce line art appearing on the same page as full-color work by combining colors to approximate a flat color.

Process or full-color reproduction is employed in print projects for the same reasons that art of any kind is used. Color calls attention as an accent or helps establish a mood or image. Perhaps most important, however, color is effective when you need to explain or communicate color itself. For example, a color photograph of spring flowers communicates the range and intensity of color the viewer would see in real life, something a black-and-white image cannot do. Sales pieces rely heavily on color to show prospective customers the color of their products when it is a major selling point.

The communication, accent, and mood potential of color has significant cost implications. Color costs substantially more than black and white to prepare for the press and substantially more on the press because of the need for multiple inks and precise registration. A magazine using full color throughout may cost twice what it would if produced in black and white. This section will cover the ways negatives are made from full-color continuous-tone art, as contrasted with the procedures covered previously for making negatives from line or black-and-white continuous-tone art.

HOW COLORS ARE REPRODUCED

"Color separation" is the process of isolating each hue, value, and chroma in original art and making a series of negatives that will recreate those qualities on the press. "Hue" is a synonym for what we commonly perceive as color—the red of a cherry, the green of a tree. "Value" (or tone) is the lightness or darkness of a hue—the bright red of a cherry versus the pale pink of a cherry blossom. "Chroma" is the purity of a hue—the true red of a cherry versus the red-orange of a plum. These characteristics combine as the eye perceives color and are the foundation of color separation. The process divides the chroma of a color image into component primary hues, varying their values so that highlights, shadows, and midtones are retained.

Separation and reproduction make use of two different perceptions of color. The primary visible colors of red, green, and blue are additive colors. When added together in different combinations, they produce the secondary visible colors. Blue and red produce magenta (a purplish red called "process red"), red and green produce yellow, and green and blue produce cyan (a bluish green called "process blue"); all added together produce white. When art is separated to make color negatives, the primary colors are used as filters to achieve that separation.

Another name for secondary visible colors is subtractive. These are the principal colors used on the press to recreate a full-color image on paper because they are capable of reproducing the complete color spectrum while the three primary colors are not. When one primary visible color is subtracted from white, a combination of the other two remains. Magenta and yellow combine to produce red, yellow and cyan produce green, cyan and magenta produce blue, and all three combined produce black.

Three-color process printing, then, uses magenta, yellow, and cyan inks. Because their ability to produce a satisfactory intensity of black for detail and tone variations is weak, most process printing is four color: magenta, yellow, cyan, and black. Five-color process printing uses a fifth transparent ink to lend a different cast to the combinations, an opaque ink for a specific spot color, or varnish to give a high luster to photographs or other art. Even more colors may be used on a job requiring exact color matching of the original. At the other extreme, "two-color" (or "twin-color") process printing uses any two of the four basic colors in different screen values to create a range of colors.

HOW COLORS ARE SEPARATED

Though one of three methods may be used to separate colors for your print project, the basic process is the same. The primary visible colors of red, blue, and green are used in conjunction with light to filter out or separate the secondary visible colors. For example, a blue filter (composed of magenta and cyan) reflects those colors when subjected to light, creating opaque areas on the film in magenta and cyan portions of the image and nearly transparent areas in yellow portions. The positive press plate made from that film negative (the "yellow printer") will carry yellow ink. Similarly, a green filter (composed of cyan and yellow) leaves magenta, and a red filter (composed of yellow and magenta) leaves cyan. A special filter or a combination one absorbs all three colors to create a black printer. This process is illustrated in Figure 12.6. Pure black may be printed in shadow areas by etching out other colors from those areas on the negatives.

A halftone screen is employed in the process to convert the image into a dot pattern, just as for black-and-white continuous-tone negatives. A different screen angle is used for each color so that halftone dots lie side by side to create the visual effect of full color. For example, a yellow dot of ink next to a cyan dot gives the appearance of green. Typical screen angles are 45° for the black negative, 75° for magenta, 90° for yellow, and 105° for cyan. A 30° difference is needed between the strong colors (black, magenta, and cyan) to prevent a moiré pattern, but only a 15° difference between yellow and any other color. Figure 12.7 shows how these angles are achieved with a halftone screen. Yellow is also the exception in screen value; because it is a comparatively weak color, it may be reproduced using a line screen one value coarser than that used for the other colors.

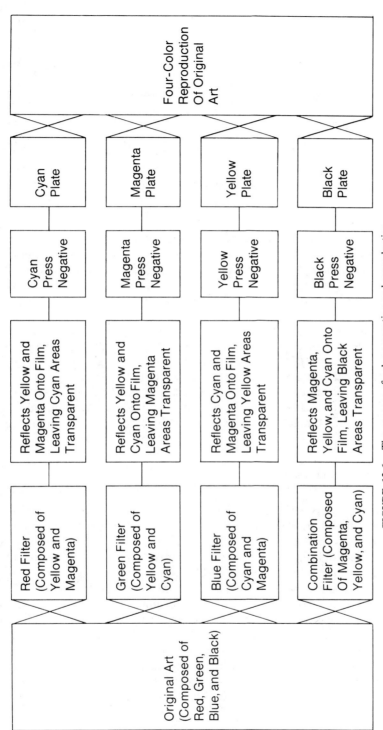

FIGURE 12.6. The process of color separation and reproduction.

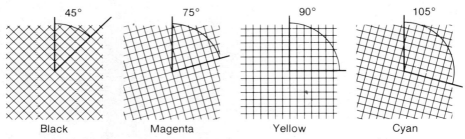

FIGURE 12.7. Halftone screen angles for reproduction of side-by-side colors.

Regardless of the separation method used, you can build in efficiencies so that the process goes smoothly and makes the best use of your budget. Try to schedule color separations before mechanical artwork is completed. Your engraver will require several days to do the separations (and possibly several more for adjustments). As soon as crop and size are final, this production step can be accomplished so the project isn't kept waiting.

When submitting art for separation, provide the finished size as well as the exact percent of enlargement or reduction. A mistake in sizing probably won't be noticed until the work is completed, and the engraver is entitled to charge full value for redoing a job because of customer error.

Plan ahead for multiple use of color separations. If you can use the same photograph in the same size in a brochure and again in an annual report, your production budget will be substantially extended, given the cost of separations.

Direct Color Separation

This method of separating colors exposes the image directly to the filter and halftone screen in one step. The illustration or photograph to be separated is placed on the copyboard of a process camera (or if a transparency, in a holder for backlighting) and the process camera set for the specified enlargement or reduction. Continuous-tone film is exposed through each filter and any color-correction mask in turn, but also through a halftone screen. Colors are separated in the resulting series of halftone negatives. Figure 12.8 diagrams this process which may vary somewhat by engraver. While fast and efficient, the whole process of direct color separation must be repeated for each size of separation you want from the original.

A form of direct color separation can be used to convert a photographic color negative into separate color printers. Intermediate black-and-white prints are made from the negative and then screened to produce the separate printers.

COLOR NEGATIVES

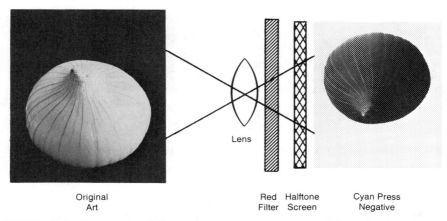

FIGURE 12.8. The process of direct color separation. It is repeated for each color negative.

Indirect Color Separation

As in direct color separation, art is placed on the copyboard or in the transparency holder of a process camera, and color filters are used in conjunction with light. The difference is that it creates intermediate separations that can be retouched for color correction before the halftone screen is added. Though more time consuming, this approach has the advantage of greater control over color quality on the final negatives. Also, different-sized separations may be made from the intermediate separations without having to repeat the first part of the process. Figure 12.9 diagrams this process.

Electronic Color Scanning

Technology has made possible the separation of colors by electronic scanning. This popular method employs a laser beam to "read" the color makeup of a print or transparency flexible enough to be wrapped around a revolving cylinder. (Some designs can take rigid originals.) This information is translated to photocells equipped with color filters and passed on to a common computer (or to separate computers for each color negative, depending on manufacturer) in the form of electrical impulses. Based on the paper, ink, and tonal qualities of the job, the computer controls how light beams expose scanner film according to the area scanned and add the halftone screen.

It can incorporate color correction or background tint as well as expand or condense an image to alter its proportions for design purposes. The resulting negatives can be closely controlled in quality and produced at a high rate of speed, making this separation method especially attractive. Figure 12.10

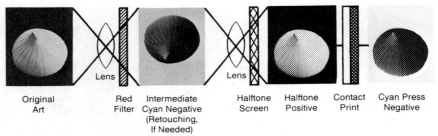

FIGURE 12.9. The process of indirect color separation. It is repeated for each color negative.

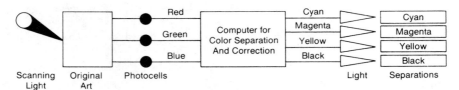

FIGURE 12.10. The process of color separation using an electronic scanner.

diagrams this process. Because scanner capability varies widely by manufacturer, ask your engraver for details if a scanner is to be used for your job.

Dupe and Assembly Separations

When highlight and shadow densities are comparable, several pieces of color art may be separated at the same time by a procedure called "dupe and assembly" ("pre-assembly" or "pre-positioning"). It differs from simple gang shooting (which may also be done, at a sacrifice in quality) in that shots are treated individually first and then ganged for actual separation.

Color shots are selected, cropped, scaled, and keyed to an exact layout. The engraver takes the art and specifications and makes a to-size duplicate of each shot, using filters to correct colors to your satisfaction in the process. The duplicates are then assembled in place according to the layout and gang separations made, resulting in a large negative for each color the full size of the spread. This procedure offers individual color correction and economical separation, and the savings in stripping time and cost is well worth the extra film required.

COLOR CORRECTION

To achieve the image you want on the press, the balance of colors may need to be altered so that flesh tones contain more red, for example, or so that grass is truly green. Such color correction is necessary when the wrong type of film or filter was used when an object was photographed, when processing was faulty, or when poor lighting conditions at the time a photograph was taken cast an unnatural color over the image. An unsatisfactory cast may also be created by the juxtaposition of colors; for example, a blue sofa next to a yellow chair may make the latter appear slightly green. Correction is necessary when the true color of an object is washed out and needs to be intensified on the press.

"Photo-masking" uses a color mask at the time of separation to correct for optical impurities in printing inks, improve tonal range from highlight to shadow, and sharpen detail. Weak positive or negative masks of original art are made from intermediate separation negatives using frosted acetate masking filters. These masks are then registered with the original image when final separations are made. An alternative is to use special masking film placed in contact with the original art during exposure. "Direct screening" uses the same special masking film, one per negative, in the back of the camera. Figure 12.11 diagrams how a correction mask is used during preparation of a cyan press negative by direct color separation.

Electronic scanning produces similar results using a computer to modify color balance. The computer can be set to incorporate color changes by varying exposure as negatives are made. This process does not allow correction of a particular area of the image or addition of localized detail.

Hand-retouching may correct color before or after negatives are made. A skilled photo laboratory or retouching specialist can retouch an enlarged duplicate of a transparency using dyes appropriate for the film, or your engraver might retouch the intermediate negative created during indirect color separation. A more predictable method is localized or overall dot etching in which halftone dots are chemically reduced. When done on the finished press negative, dot etching increases the amount of color that will print; when done on the film positive, it decreases the amount.

EVALUATING COLOR SEPARATIONS

Like other phases of print production, the process of color separation produces a proof for your approval before a job moves to the next phase. Colors are assembled for a proof in the order they will be laid down on the press.

Types of Proofs

The "overlay" (or "color transparent") system of pre-press proofing exposes each negative to a light-sensitive transparent sheet of acetate. The pretreated sheet is then developed to bring out the corresponding process color and the sheets assembled in register to create the composite image. Color-Key is a common brand name for this system. Because transparent sheets are used, separations may be evaluated over the paper on which a job will run, but they are not accurate enough to use for proofing color balance.

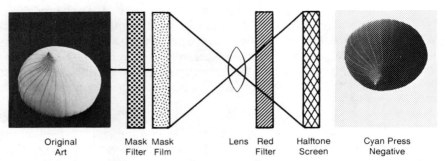

FIGURE 12.11. Use of a correction mask during direct color separation. The process is repeated for each color negative.

The "transfer" ("integral" or "color opaque") system of pre-press proofing exposes each negative to the same opaque sheet. Starting with a solid base sheet, a Mylar film is laminated to it and exposed to the first negative. The film is then removed and the base sheet dusted with a corresponding toner that is accepted by only the exposed areas. Excess toner is removed, the base sheet re-laminated, and the process repeated for the other negatives. Cromalin is a common brand name for this system which has the advantage of presenting the recreated image on a single sheet suitable for checking color balance.

Progressive proofs ("progressives" or "progs") are another proofing option, although they are much more expensive. Plates are made from color negatives and each application of ink checked in turn on a four-color proofing press or on the actual press to be used for the job. A color bar outside the image area provides an independent guide to the printer to check color strength and accuracy. In addition to tying up a press for proofing purposes, waiting to check separations until this phase of production can mean expensive replating if one or more negatives must be altered to meet your approval. The advantage is that the image can be proofed using the actual paper and ink specified for the project.

What To Check

A well-prepared color proof should be cropped and sized to your specifications so that the image will fit the space allotted for it on the layout. The proof should be clean, free of scratches or spots that might be transferred from negative to press plate to paper.

Colors should be as true to the original as can be achieved. When separating a photograph of an object that is portable, leave it with your engraver as a guide and refer to it as you evaluate a color proof (and later as you do a press check). Though you may not know how to correct a color problem, you should be able to describe the symptoms to your engraver or printer.

Too much yellow, for example, may be the fault of too little magenta, cyan, and/or black.

Okay the proof with your initials or initial the changes to be made, realizing that they may take several days to make if new separations will be required. Since the color proof will accompany your job to the press, ask to compare it (and the original art) with what you see during a press check (as explained in Chapter 14).

FACTORS INFLUENCING COLOR REPRODUCTION

Several variables affect the quality of a process-color project. Though some may be more important than others on a given job, they all have an influence. When even one has been compromised, the overall product is jeopardized.

Quality of Original Art

Art for color separation is classified as "reflective copy" (such as photographs or drawings) or "transmissive copy" (such as slides). Not every piece of reflective or transmissive copy will reproduce well on paper. As with black-and-white art, look for a contrast of colors with highlights, shadows, and a range of middle tones. A photograph with an overall blue or red cast lacks contrast and will not reproduce satisfactorily.

Art should be free of scratches, fingerprints, and other blemishes. It should be close to the size needed because it will lose quality if enlarged substantially. For this reason, a color print on textured paper should not be enlarged at all. A painting with areas of thick paint or other art heavily textured should first be photographed and the transparency or smooth-surfaced print separated, instead of trying to make separations directly from the original.

Color transparencies reproduce highlights better than color prints and may cost less to separate. Therefore, they are preferred for top-quality reproduction. When faced with a transparency that is lighter than you would like and one that is darker, choose the darker one to give your engraver the most image with which to work.

View transparencies in intense light (a light table or slide previewer), preferably surrounded by a frame of light. A shot viewed on a screen in total darkness will not be perceived under comparable lighting conditions to the printed page, nor will minute scratches be detected.

Check the effect of various color-correction options (to be discussed in detail later) on a transparency by viewing it through film patches commercially available for this purpose. By doing an advance check before a slide goes to an engraver for separation, you may achieve satisfactory separations on the first try.

Group art that will appear on the same spread to see the relationship of colors and the overall effect of their combination. The time to notice that a spread is too green is when selecting art, not when you see it on the press.

A painting in which different types or brands of paint have been used or one with certain color combinations may not separate satisfactorily. Review with your engraver the color art that you have or plan to secure for a job. Sound advice early in a project can avoid wasting time and money and contribute immeasurably to the quality of color reproduction.

Paper

Because the paper on which a job is run is the light source for the colors printed on it (thus the reason for using transparent ink), it has a significant influence on the quality of color reproduction. A rough-textured paper will scatter light, resulting in muted colors. A sheet that is especially absorbent will allow little light to reflect. The most intense, vivid colors are printed on smooth, white, bright, coated paper, all factors promoting light reflection.

Ink

Positioning two colors next to one another to create the illusion of a new third color works only if the inks used are transparent. It must be formulated to permit light to reflect off both areas of color so that the eye perceives the combination.

Opaque ink does not reflect and is used for flat-color reproduction. It is a constant color not strongly influenced by color of paper, an adjacent color, or a color that might have been printed under it. The color you see is the color you will get when printed on white paper.

Work out with your printer the specific inks to be used for reproducing your job on the press and advise your engraver. Process colors may vary by brand, and separations should be prepared accordingly for the best results.

Ink density is a related factor. When coverage is light or irregular as a job runs on the press, the image will be weak or spotty. If you have any doubt about your printer's attention to ink density, include a requirement for ample and consistent coverage in your specifications. Especially on a long press run, density may vary, and the printer may need to charge extra for close monitoring to keep it consistent.

Satisfactory color reproduction must take into account the sequence in which inks are laid down on the press. The order depends on brand of ink, paper, press, and the operator's experience. A common sequence is yellow, magenta, cyan, and black. When text or other line work is to be printed in black, that color may be first.

Ink laid down while the previous color is still wet will look different from wet-on-dry ink. Consult with your printer to understand fully what you can expect. Wet-on-wet inks are used sparingly to prevent blending together as they dry. To compensate, more black is used. Wet-on-dry ink is more intense, with only an accent of black.

Engraver Skill

Though you will see a proof of color separations, a skilled engraver can save you the time and frustration of trying to salvage poorly done work at that point in production. Interview an engraver as you would another supplier. Visit the shop, ask to see work in progress, and borrow samples of finished products. Get names of references and check them for reliability of work, estimating, and billing. Discuss your specific needs and timing, possibly using a past project as an example. Get acquainted on a small project if you can and communicate full specifications and schedule requirements in writing for each job.

Type of Press

Quality color reproduction can be done by a one-color press or the most sophisticated five-color. Any press designed for clean printing can produce process color in the hands of a skilled operator. Ask to see comparable color work done on the press your printer plans to use for your job and make sure your quality requirements are reflected in those samples. Though the price may look attractive, a large press designed for volume catalog work may not have the coverage and registration control capability of a smaller press that costs more to operate. Start with quality in mind and if budget is tight, see if you can save elsewhere before taking a chance on the press.

Color Perception

One person may perceive color quite differently from another. What looks satisfactory to you may appear off-color to someone else. While you can't account for this factor in your audience, you can build in some control during production. Look at original art, proofs, and press sheets in the same light. Fluorescent light and natural sunlight have distinctly different effects on color perception. Check your perception in comparable light to what you can anticipate viewers of the printed image will use. Also, one person should evaluate color during production to provide a consistent perception and direction to suppliers.

STRIPPING AND PROOFING

The process of "stripping" is an intermediate step to assemble line, halftone, and color negatives for platemaking. Also called "stripping-in," "film assembly," "flat assembly," "image assembly," or "negative assembly," it is part of engraving services and can be accomplished by an independent engraver or by your printer's engraving department. This section covers techniques used to strip negatives and points out production details that can be effectively done "on the negative" at the time of assembly. It also includes information on different assembly requirements when working with hot-metal composition and film positives for gravure.

STRIPPING TECHNIQUES

The technician doing negative assembly is called a "stripper," whether or not actual stripping of one type of negative into another is required on a given job. A stripper works with negatives and full knowledge of the kind and size of press on which a job will run, paper size, trim size, and how the image is to be positioned on the page. Working on a light or stripping table, the stripper prepares a negative for the press using cutting tools, a masking sheet on which the negative is to be mounted, opaquing liquid and tape, cleaning fluid, and any special tools necessary to complete a particular job.

Stripping of line negatives alone or of line negatives to which halftone or color negatives will be added is either single flat or multiple flat. Based on the specifications of your job, the stripper will determine which procedure to follow. Either type of assembly can be used for an individual project or for a group of small pieces to run on the same sheet. Ganging them onto a large flat saves stripper handling time and allows you to take advantage of any economies of using a large press for one run over a small press for multiple runs. To understand either approach, however, first requires an understanding of the basic assembly process, using as an example a simple line negative.

A stripper can start negative assembly with a full-sized negative made from mechanical artwork or can take negatives of individual elements (such as blocks of type, headlines, and art) and assemble them. Because the latter process is more complex and expensive, the following information will assume that line elements are already in place in the form of camera-ready artwork (as covered in the previous chapter) to make the most efficient use of stripper time.

Basic Assembly

To position the image accurately for the press and to mask out non-image areas, a stripper uses a commercial "masking sheet" ("goldenrod," "stripping base," or "stripping vinyl"). It is an orange or yellow opaque sheet of paper or plastic that blocks the passage of light. It is ruled with an overall grid and guidelines for positioning the image in relation to press requirements. Figure 12.12 shows a masking sheet with its heading and footing detail. Notice that the masking sheet has a top-of-paper line, providing an

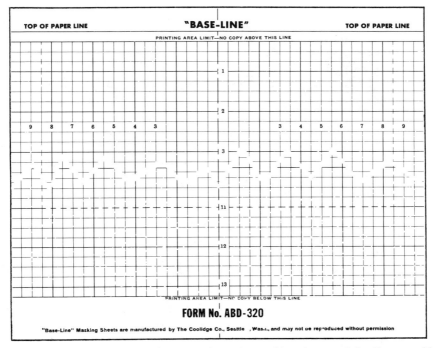

FIGURE 12.12. The upper and lower portions of a masking sheet used for film assembly. (Courtesy of Base-Line, Inc., Kent, WA.)

allowance for the press gripper (which grips a sheet to feed it through the press) and also for the plate clamp (which holds the plate on the press).

The stripper tapes the masking sheet upside down to a light table and tapes the negative emulsion-side up (wrong-reading) over it, lining up center lines and leaving specified margins. The negative and masking-sheet combination (a "flat") is then turned over so that the sheet is on top and the negative underneath is right-reading. Image areas are revealed by carefully cutting away the masking sheet.

After these holes are cut, the stripper closely examines the revealed negative for scratches, dust, or pinholes of light, spotting defects in opaque areas with liquid or tape and using a cleaning fluid on transparent areas. Finally, the stripper checks the flat to confirm that the image is properly positioned, fully revealed, and clean; that any cut edges of film are beveled to prevent shadows during plate exposure; and that the negative and masking sheet are securely taped together. Figure 12.13 illustrates this basic assembly process.

To add a screen to an image, it is converted into a film negative or positive, whichever form will make the image area transparent. A contact screen is then stripped in behind the transparent image and the press plate exposed.

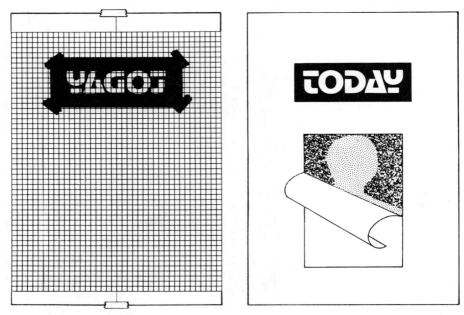

FIGURE 12.13. The basic stripping process, showing line work taped to a masking sheet and two segments of the sheet cut away to reveal type and a halftone.

Single-Flat Assembly

This method requires stripping of one flat only to produce one or more printing plates. Types of single-flat assembly are single plate, step-and-repeat, and keying.

Single-Plate Stripping. As illustrated in Figure 12.14, single-plate stripping results in a single negative intended to produce a single press plate for a one-color job. This method of assembly can be used for line work only (as shown) or for a project also involving halftone and/or color negatives stripped into the line flat.

Step-and-Repeat Stripping. The procedure for single-flat assembly is also followed for a negative intended to produce a single press plate with multiple, duplicate images. The "step-and-repeat" process shown in Figure 12.15 saves the time and expense of preparing multiple flats of the same line or a combination of line and halftone image. Instead, a press plate is exposed multiple times to the same image. Step-and-repeat stripping can be accomplished manually by using stripping materials especially designed for this purpose or mechanically by using a step-and-repeat machine. If your print project requires several duplicates of the same image (such as 20 identical mailing labels per press plate), discuss step-and-repeat stripping with your

STRIPPING AND PROOFING

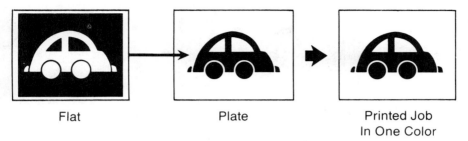

FIGURE 12.14. Single-plate stripping for an offset press.

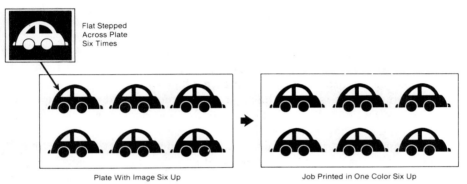

FIGURE 12.15. Step-and-repeat stripping for an offset press.

engraver; it may save substantial time and money, particularly if it can be done by machine.

Keying Stripping. One flat may also be prepared when the need is for multiple press plates, each to print only a portion of the total image. The stripper leaves one side of the masking sheet attached over each hole and keys these flaps to indicate to the platemaker which flap is to be lifted to expose which plate. An alternative is to attach a separate hinged flap over each hole. As shown in Figure 12.16, this procedure can be used to save stripping time and money if elements for a multicolor job are loosely registered. It is not effective if lap or hairline register must be used (see Chapter 11 for full information on types of register).

Multiple-Flat Assembly

This method of assembly requires stripping of more than one flat to produce one or more press plates. It is used when a single-color job has screens or other effects requiring individual treatment or when two or more colors must be printed in lap or hairline register. Types of multiple-flat assembly are complementary and multiple.

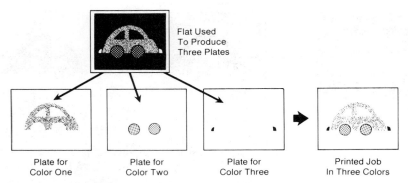

FIGURE 12.16. Keying stripping for an offset press.

Complementary Stripping. Two or more complementary flats are used to expose one press plate when a job is to run in one color but elements on the page must be exposed separately. Figure 12.17 shows how this stripping method would apply to a page having a gray screen around unscreened art. One flat is prepared for the screen and another for the art; then the two are used to expose a single press plate.

Complementary stripping is also used when one element requires a longer exposure time than others, when different screen values of the same color are specified, when type is to surprint over art, or when illustrations or photographs are too close to one another or to their cutlines to make taping them in place possible (see later information on stripping in halftones). An additional use is for reversing type out of art, in which case a film positive of the type (instead of a negative) is stripped into a flat.

Multiple Stripping. This assembly procedure is used when separate plates must be prepared to print separate colors. They may be flat colors in hairline or lap register (as illustrated in Figure 12.18) or full-color process colors in hairline register.

Halftone Assembly

Because line and continuous-tone elements are converted into negatives in two separate processes, the stripper must combine them so that one press plate can be made. The halftone negative is stripped into the line negative so that only one piece of film goes to the platemaker.

Once a line negative is fully assembled and checked, the stripper selects all the halftone negatives for that particular flat and trims them to within ¼ in. (0.63 cm) of the crop marks. With the flat upside down on the light table, the stripper tapes each halftone over its corresponding clear window on the line negative, securing it on at least two opposite sides outside the image area so the tape doesn't show on either the halftone or on adjacent type.

STRIPPING AND PROOFING

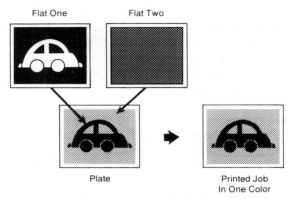

FIGURE 12.17. Complementary stripping for an offset press.

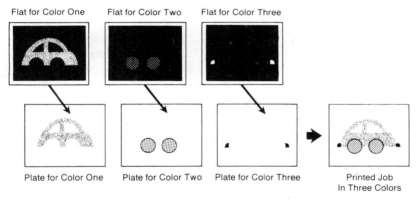

FIGURE 12.18. Multiple stripping for an offset press.

A more complicated procedure may be followed if masking film was not used on the mechanical to create a clear window on the negative. The mechanical artist can draw a precise box or designate on the tissue-paper overlay where a halftone is to go when art is an irregular shape or when placement in relation to a screen must be exact. If no window was drawn, the stripper scribes one into the line negative and makes a positive to create a black guideline on white. The window is then painted in fully and converted back into a negative, resulting in a transparent window for the halftone. A simpler alternative is for the stripper to mask a sheet of clear acetate to the appropriate window size.

Complementary stripping is used to avoid trimming the halftone negative. When preserved in its entirety, the full image is available, should you want to change its position in the window or use it on a subsequent project cropped differently. A sheet of clear acetate is used as the base for the halftone flat, and the stripper carefully outlines each window on the masking sheet with opaque tape so that edges and corners will be crisp.

Another instance in which complementary stripping offers greater design possibilities is when you want to place halftones closer together or closer to adjacent type than taping will allow. They can be assembled on a separate flat (or flats) and, if precisely registered to the line negative, will be properly positioned on the press plate.

Two halftone negatives may be butted if screen angles are matched to prevent a dark line or a space between them. After aligning screen angles and crop marks along their common edge, the stripper tapes them together, trims them simultaneously, and tapes them into the window on the flat.

Pin Registration

The pin-registration system can be used to assure accurate registration from paste-up through platemaking. Though standard registration marks are satisfactory for a small job, pin registration is effective when doing a full-color piece in hairline register or when doing a piece that requires accurate registration across a wide sheet of paper or film (such as a booklet that is to print eight pages up on the press).

During stripping, pin registration allows the stripper to align a series of punched holes on one negative with the same series on all others for that job using separate metal pins or a metal pin bar. When securely affixed to a light table, these pins precisely hold each negative in place as the stripper works. They serve the same purpose during exposure of press plates.

Imposition

Individual negatives must be assembled in the proper sequence and position for printing and binding. To avoid any confusion during stripping of a multipage or multipanel piece, send a project in for engraving accompanied by an imposition dummy provided by your printer. As shown in Figure 12.19, pages are clearly numbered on an imposition dummy, and it is notched to indicate the top of each page when the dummy is unfolded. Your stripper will refer to this dummy during assembly and when checking the job.

Possible impositions are as follows. Before deciding on use of a second color or giving instructions to a stripper, ask your printer which is preferred for the press to be used and length of run. Also confirm if two-sided sheets are to be printed "head-to-head" (tops of images correspond on front and back sides) or "head-to-foot" (the top of the image on the front corresponds with the bottom of the image on the back so that the sheet is flipped or rolled in use).

"Sheetwise" ("work-and-back," "print-and-back," or "front-and-back") imposition prints only the front image on one side of a sheet. It is

STRIPPING AND PROOFING

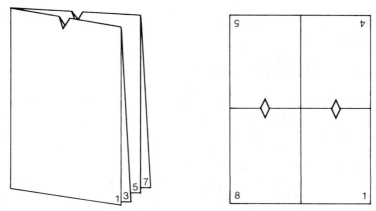

FIGURE 12.19. An imposition dummy notched to indicate the top edge of pages.

then dried and sent through the press again (using the same gripper edge) to print the back image. Two different printing plates are used.

"Work-and-turn" ("print-and-turn") imposition groups all pages of a signature on one side of the sheet for printing and is appropriate for a sheet to be cut apart crosswise. The sheet is then dried and turned over side to side, and the same press plate and gripper edge are used to back up the pages.

"Work-and-tumble" ("print-and-tumble," "work-and-flop," or "work-and-roll") imposition also groups all pages of a signature on one side of the sheet for printing but is appropriate for a sheet to be cut apart lengthwise. The sheet is printed, dried, and turned end to end, and then printed on the back using the same press plate but a different gripper edge.

"Work-and-twirl" ("work-and-twist") imposition prints two identical images on the front of a sheet, and the back is left blank. One image is printed, and then the sheet is twirled to print the second image or color. This approach may be preferred on vertically and horizontally ruled forms to be printed by letterpress. Half the plate may be vertical rules and the other half horizontal. After one impression, the sheet is twirled to complete each form. The same press plate is used, but no right-angle butting of metal rules needs to be done to make it, thus saving composition time.

On a project requiring more than one pass through the press for multiple colors, the stripper may need to compensate for paper expansion. A sheet may absorb moisture the first time through an offset press causing some expansion (more so across than with the grain). While offset paper or a skilled press operator can minimize such expansion, the stripper may also need to make slight alterations in placement of elements on the flat for subsequent impositions. Discuss this possibility with your printer so that it can be accounted for in the engraving estimate and schedule.

Paper may also expand because of too much tension on a web press. Though not enough to break the paper, excessive tension can cause it to stretch and distort the image when tension is released ("web gain"). A skilled operator working with quality paper can control tension to prevent such expansion.

CHANGES ON THE NEGATIVE

By the time a project is being stripped, it cannot undergo major changes without a serious impact on your budget and schedule. However, several minor changes can be made with little effort and/or expense. Also, certain production details that might have been done during paste-up can be executed during stripping instead.

If elements were not assembled onto a mechanical or if assembly was incomplete, that work can be done by the stripper. You can buy additional stripping time to position and align type and art based on a mechanical dummy or detailed sketch. In the face of a close deadline, you can also buy extra stripping time so that work can commence while you await a piece of missing type or art to be stripped in at the last minute.

A skilled stripper may scribe rules on the negative with a scribing tool that cuts through the film emulsion; ends are squared off with opaque tape. This procedure creates a crisper line than a ruling pen or transfer rules on the mechanical but can be used only on film with a soft emulsion, a type not popular today for general use because of its soft nature.

Unwanted elements can be opaqued on the negative so that they won't appear on the finished piece. A segment of a drawing, a paragraph of type, or a rule can easily be opaqued out with liquid or tape.

A halftone window can be reduced in one or both dimensions on a single flat with opaquing tape. When preparing complementary flats, the stripper can simply allow for a larger or smaller halftone as the image area is defined with opaque tape.

The overall image can be reduced or enlarged at the negative stage of production by making a duplicate negative to the new size before the flat is assembled. If you change your mind about size after the flat is assembled, the stripper must remask the new negative and, if halftones are involved, resize them accordingly and strip them in again.

If you want to reposition the overall image on the page, the stripper can do so as the flat is assembled. After the masking sheet is cut, however, repositioning requires cutting a new sheet.

A headline, block of type, or art can be straightened or moved slightly on the negative. The area is cut out, repositioned, and taped into place. The masking sheet isn't affected if such a change is minor.

Type corrections can be stripped in on the negative by resetting several lines of text that include the correction or by resetting an entire headline or cutline. A line negative is made of the new type and positioned over the old type. The stripper cuts through both layers of film, removes the old one, and

STRIPPING AND PROOFING

secures the correction in place with tape using opaquing liquid to seal any light cracks.

Art can be substituted on the negative with little trouble if the new art occupies the same space as what it replaces. When other elements must be repositioned to accommodate a larger substitute, considerable stripping work will have to be redone.

NEGATIVE PROOFING

The work of the stripper culminates in a proof of the completed flats before offset press plates are made. The proof that you see at this stage of production is your last look at a project before it is on the press. Thus your educated review of it is critical to the timely completion of the job within the quality standards and budget set.

Because making a proof of flats is a cost item, you will save some expense by not requesting a proof on a simple, routine job involving line negatives only. Such a job can be proofed when in artwork form. However, a proof of negatives is essential and the cost justified on any project that you consider complex or more than routine. Multiple panels or pages, multiple colors, a combination of line and halftone work, and a two-sided job to be precisely backed up are complexities that, individually or collectively, signal the need for a proof. A project that you classify as expensive also requires a proof, even if it is simply laid out and executed, to help assure that you will be getting what you paid for off the press. Finally, a proof is wise when you are working with a new engraver.

By approving a proof, you accept responsibility for the work that you okay (including any errors you overlook). Therefore, your engraver may insist on a proof to avoid disagreements later, even if you don't think one is necessary. When volunteered by your engraver, a proof should be furnished at no cost to you.

Proofing Techniques

A proof is made after the stripper has checked flats and found that they conform with specifications and are ready for the platemaker. Several options are available to make a photographic proof of negatives, depending on engraver equipment. All those mentioned are wet processes, similar to the printing of a photograph in that photosensitive paper is exposed to a negative and light, and then developed in liquid chemicals and allowed to dry. Therefore, these processes are not dimensionally stable; the size of the image area may be slightly altered from the original during wetting and drying. Though names may be used interchangeably, all photographic proofing processes are not identical.

A "blueline" is produced using blueprint or Dylux paper. The image appears as blue on off-white. While a blueline is a very common, inexpensive proof, it remains sensitive to light and should be kept in an envelope or folder until you return it to your engraver.

A "brownline" is produced using brownprint paper. Also called a "Vandyke" (or "Van Dyke"), it creates a brown-on-white image. It is not light sensitive after developing.

The proof can be made with the same vacuum frame as is used to expose press plates. The paper may be sensitized on one side or both. With the former, sheets must be glued or taped together for you to see how back-to-back panels or pages would be printed. With the latter, both sides are exposed so that the proof closely resembles the printed piece. A proof may be delivered to you trimmed or folded as the finished product will appear or it can be left as flat sheets that you will have to trim and fold according to trim and fold lines outside the image area.

For reliable proofs of halftones, request blackprints (as covered earlier). They more precisely present halftone images than either a blueline or brownline will.

Checking a Proof

Though a proof can't show you every detail of a job as it will look when printed, it is quite informative. When checking a proof, look for the following qualities. Figure 12.20 summarizes them in a checklist that could be completed for each job to help you proof thoroughly.

1. *Alignment.* All elements should be straight with the page and with one another (unless the design calls for diagonal or mixed alignment). Without the distraction of trimmed edges on a mechanical or the masking sheet on a flat, a proof accurately presents how well elements are aligned.

2. *Completeness.* All elements called for in the layout should be seen on the proof. If something is missing there, it will also be missing when the job is printed.

3. *Margins.* They should be as specified and consistent throughout the proof. When you hold a two-sided proof to the light, back-to-back margins should be close (if not identical). If they are noticeably off, point out the discrepancy; though attention to exact registration may have slipped when the proof was made, you need assurance that images are properly positioned on the flat so that they will line up on the press.

4. *Panel or Page Sequence.* The order you see on the proof should match the layout. If the piece folds across the top, panels or pages on the back as well as the front should be right-side up.

5. *Folds.* Vertical and horizontal folds should be as specified in both location and type of fold. If a proof done by your printer's engraving department comes to you folded differently, check that the printer has accurate

STRIPPING AND PROOFING

```
                          Proof Checklist

    For _____

    _____ Alignment              _____ Color Breaks
    _____ Completeness           _____ Screens
    _____ Margins                _____ Reverse/Surprint/Mortise
    _____ Panel or Page Sequence _____ Condition of Type
    _____ Folds                  _____ Scribed Rules
    _____ Art                    _____ Cleanliness
    _____ Windows                _____ Union Bug

    Artwork to be returned upon completion to _____

    Date _____  By _____
```

FIGURE 12.20. An example of a checklist for proofing a press negative.

specifications on the job ticket so that the printed piece will be folded later as you intended.

6. *Art.* Illustrations and photographs should appear on the proof in the position and size requested and be cropped properly. Each should also be right-side forward. A stripper not paying close attention can inadvertently flop a halftone so that it will print backwards (or you can have it flopped on purpose for design reasons).

7. *Windows.* Edges should be straight and free of irregularities, and corners should be crisp. Check that edges are also the size and shape specified, remembering that size may have been slightly altered as the proof was made.

8. *Color Breaks.* To see process color, you have to ask for a proof of color separations, as covered earlier in this chapter. A photographic proof can't show you the actual flat color(s) to be used or ink intensity, but it should show color breaks. To do so, the proof is fully exposed in the main color and underexposed in the second color so that it appears weaker. Your engraver may also mark those elements that will print in the second color. Check that these elements are as specified.

9. *Screens.* The proof should show screen placement and value. Because a proof isn't an exact prediction of what you will see on the press, you may want to confirm screen value with your engraver when you return the proof.

10. *Reverse, Surprint, and Mortise.* These special effects should appear on the proof and be sufficiently close to the finished product for you to judge their effectiveness. If you have difficulty reading reversed or sur-

printed type across a photograph on the proof, reconsider using it. It will not reproduce much better on the press.

11. *Condition of Type.* Text and headline type should be undamaged by the stripping process. Check for scratches or areas where it might have been opaqued out by mistake and circle any damage.

12. *Scribed Rules.* Any rules added by the engraver should be straight, crisp from end to end, and placed where specified. Corners should be clean and square on a scribed box.

13. *Cleanliness.* The proof should be free of spots, either light or dark. Circle any spots you see in both type and art areas so the engraver will check for dust or pinholes of light on the negative.

14. *Union Bug.* If printing is to be done in a union shop and you want the union symbol (or "bug") to appear on the piece, the printer's engraving department will add it to the artwork, and you will see it on the proof. Check that it's where you want it and, if it's been arbitrarily added, you can ask that it be deleted.

Return a proof to your engraver with any corrections or changes clearly marked. Initial it as "Okay to print as is" or "Okay to print as revised." Ask for a corrected proof if revisions are numerous or if you have any reason to doubt that those revisions will all be made before the job is on the press. Always return a proof so that it can be referred to by the press operator during printing and be used later as a binding dummy.

HOT-METAL AND GRAVURE ASSEMBLY

The assembly process is different from that just covered if you are using hot-metal composition for line and halftone elements or if the job is to be printed on a gravure press. Refer to later information on letterpress and gravure plates and to the following chapter on presses for a full understanding of how these types of assembly relate to other steps in production using these presses.

Hot-Metal Composition

The fact that type is set using hot metal doesn't necessarily mean that your engraver will be dealing with hot metal by the time images are assembled for press plates. The type could be set, repro proofs made, mechanical artwork prepared, and line and halftone negatives made as just explained for offset. When halftones are also to be made using hot metal, however, image assembly is different.

Based on a mechanical dummy or detailed sketch, hot metal is arranged in a form ("chase"). Instead of pasting down paper galleys and headlines, actual type is arranged to your specifications and a space left where continuous-tone art is to appear. To reproduce that art in metal, the engraver uses a halftone screen to make a negative with a dot pattern—the same process as was presented earlier in this chapter. The pattern is coarse and generally no finer than a 100-line (40-cm) screen. The negative must then be converted into a small metal plate that will fit into the chase in the designated space.

The negative is exposed to a photosensitive metal plate which is then etched so that halftone dots (image areas) are higher than surrounding non-image areas. (This process is covered in detail later.) For effective reproduction of highlights and shadows, the plate will be worked over by hand. To lighten an area, the engraver uses a chemical to reduce the size of dots by re-etching. To darken an area, the engraver uses a polishing tool to enlarge the size of dots by burnishing. The finished halftone plate is mounted ("blocked") to be as high as the type and locked into the chase. A full press plate will then be made or the actual hot metal used on a platen or flatbed cylinder letterpress.

Gravure Film Positives

Plates for a gravure press must be made from film positives instead of negatives. For conventional gravure, a page is first assembled in negative form, including stripping in of continuous-tone negatives of art. Halftone negatives are not used because the overall image will be given a halftone screen later. Negatives are retouched as needed and converted to continuous-tone film positives. For gravure by the Dultgen or direct-transfer process, the overall halftone screen is added at the time negatives are converted to produce a halftone film positive (see later information on gravure platemaking).

PLATEMAKING

The final step before your job is on the press is for plates to be made. The type of press plate depends on the printing process and press to be used, length of run, the quality you require in the finished product, and the paper to be used. Though this section will not present the platemaking process in full technical detail, it will describe the most common processes, the plates that result, and how they satisfy requirements for durability, quality, and responsiveness to paper. As the buyer of this service, you should understand the basic process; for additional information, refer to more technical handbooks on platemaking or manufacturers' manuals.

Because most commercial work involves platemaking for offset, that procedure will be emphasized here and will be followed by information related to other types of presses. The relative simplicity of offset platemaking means it is most often done by a printer, although an engraving specialist may make offset as well as more complicated letterpress and gravure plates.

OFFSET PLATEMAKING

An offset press plate is "planographic." The image to be printed and the surrounding area are on one plane (or nearly so), unlike the raised image of letterpress or the recessed image of gravure. A planographic plate works on the principle that oil (ink) and water don't mix. It makes the image area receptive to ink only and the surrounding area receptive to water only.

The Basic Process

The most common way of making an offset plate starts with the negative flat prepared by a stripper. It is placed over a thin, photosensitive metal plate (or one made of specially processed paper or plastic). The glass cover of the platemaking unit ("vacuum frame") is closed over the negative and plate. Air is extracted, drawing down the negative in full contact with the plate. When the exposure light is turned on, light passes through the transparent portion of the negative, exposing only image areas to the plate. The exposed areas harden and become waterproof; the emulsion remains water soluble on non-image areas and is washed away during developing. The plate is chemically developed by hand or machine and given a gum coating to minimize the amount of water needed during printing, before being ready for the press. Figure 12.21 diagrams this process.

Your printer may be able to remove an unwanted image from a plate using a special eraser or deletion fluid or add an image ("set-in") using a chemical treatment and re-exposing the affected area. This capability can be time and cost saving if, for example, you need 1000 copies of a brochure printed with a return address and 1000 without a return address, or if you want to add a telephone number on a subsequent printing using the original plate.

Proper preparation of an offset plate will prevent production delay and poor quality later on the press. A plate that wears out before the end of the run was poorly exposed or of a sort that was not durable. A worn plate causes blotches on the page as the distinction between image and non-image areas deteriorates. Blotches may also be caused by improper exposure or processing of the plate.

Trade customs assign ownership of press plates to your printer. They can be stored for you if you expect to run the job again. Your printer may prefer to save only the negatives, however. Customer changes are easier to make on negatives before rerunning jobs; plates have a limited shelf life, and storing both plates and negatives is usually not an efficient use of space.

Offset Plates

The printer selects a type of offset plate designed for the press on which it will run and for the length of the run. Though all have commercial applica-

FIGURE 12.21. The process of offset platemaking.

tion, some are particularly suited to the convenience and utilitarian quality standards of small offsets. Because of the ease and economy of making offset plates, only originals are used. Should a second plate be needed for a long press run, another original is made from the artwork.

Conventional Plates. A form of planographic plate called a "surface plate" is the most frequently selected variety for projects ranging from fewer than 1000 upwards to 300,000 copies. It may be sensitized on one or both sides by the manufacturer or by the printer.

A "deep-etch" plate made from a film positive of the negative flat is exposed to light and chemically etched in image areas so that it can carry more ink. It is especially effective for quality color work and high-speed press runs of 500,000 or more.

A "bimetal" plate is a deep-etch plate composed of more than one metal and used for high-speed runs of a million or more. The raised image area is made of an ink-receptive metal, and the recessed non-image area is made of a water-receptive metal.

"Photopolymer" is another form of deep-etch plate made of plastic. Though designed for letterpress (as covered later), it can also be used for dry offset printing, eliminating the need for water that can slow ink drying and expand the paper. This process is called "letterset," "dry offset," "indirect letterpress," "indirect relief," or "offset letterpress."

Direct-Image Plates. A non-photographic direct-image plate or master is made of paper, a plastic/paper or metal/paper combination, or acetate. It is used on a small office or commercial offset press when the image is typed or hand-lettered. A direct-image master is not photosensitive but is chemically treated to make what is written or drawn on it with a special ribbon or pen slightly greasy. It is ruled with nonrepro guidelines for proper placement of text and display type. Thus the image is ink receptive and the surrounding area water receptive. Type set by a hot-metal method may be reproduced on this form of offset plate by proofing the type directly onto the plate using a proofing press.

The image can be transferred photographically to a version of direct-image plate having a photosensitized emulsion. The photographic direct-image plate ("photo-direct" or "projection speed") is the basis for "quick printing" by offset. Typed, drawn, or typeset camera-ready, positive copy is placed on the copyboard of a camera unit designed for this form of plate and the lens set for same size, reduction, or enlargement. Light is reflected from the image area to expose the plate (stored in the unit as precut sheets or in a roll). The plate then passes through a developer and fixer and is dried, all within seconds of exposure. The low cost of this process makes it ideal for short runs of utilitarian quality that are entirely line copy.

Diffusion-Transfer Plates. The same process used to convert halftones into line copy (as described in Chapter 11) can be used to make an offset plate for a small press. Camera-ready, positive copy is exposed in a vacuum frame to special negative paper which, in turn, is fed into a processing unit in contact with the chemically treated diffusion-transfer paper plate. The plate is ready when developed, fixed, and dried.

Electrostatic Plates. This type of offset plate is produced using "xerography"—the process that is behind Xerox and other photocopiers. An electrical charge is given to the multilayer plate. Stored in the processor as single sheets or in roll form, the plate is exposed to camera-ready, positive copy. The charge is neutralized in non-image areas so that when toner is applied and heated, it will adhere only to charged image areas. The image may be the same size, reduced, or enlarged, and the plate made of paper or metal.

Facsimile Plates. The advent of lasers and communication satellites has inspired development of an offset platemaking system that is fast and can be simultaneous in several locations. Camera-ready, positive copy is scanned by a reading laser and translated by an exposing laser onto a presensitized surface plate, line by line. When processed, the plate is ready for the press.

PLATEMAKING

While reading and exposing lasers may be in the same room, they may also be linked by satellite around the world.

LETTERPRESS PLATEMAKING

Letterpress is relief printing, relying on a raised, inked surface for transfer of an image onto paper. Therefore the process of making plates for a letterpress differs substantially from that used for offset.

While actual hot-metal type can be locked onto a letterpress and inked, it is not the more common method used today for printing in any quantity. An original plate (or "cut") is photoengraved, starting with a repro proof of hot type made into pages or cold type pasted up into camera-ready artwork. Another form of letterpress plate is a duplicate of hot type or of a metal engraved plate.

Original Plates

The first part of the photoengraving process for an original letterpress plate resembles that used for conventional offset plates. The flat is assembled and the negative exposed in a vacuum frame to a photosensitized metal plate, hardening image areas. It is developed and then chemically etched and routed to make surrounding metal lower than image areas. When cleaned and blocked on a piece of wood, metal, or plastic, the resulting plate is "type-high"—the 0.918 in. (2.33 cm) from feet to face required by a letterpress. Figure 12.22 diagrams this platemaking process.

"Photopolymer" plates are made from photosensitive plastic bonded to a film or metal base instead of all metal. It is hardened after exposure, chemically etched, cleaned, and blocked. This form of plate can be made quickly and inexpensively from cold type. Though capable of a range of halftone-screen reproduction, it is used primarily for newspapers and other work of utilitarian quality.

Duplicate Plates

If the plate is to be mounted on a rotary letterpress, it must be curved rather than used flat. To do so, a duplicate must be made directly from hot-metal forms or from the metal engraved plate. Duplicate plates are also made for simultaneous printing of a job or quality printing throughout a lengthy press run using a series of duplicates.

An "electrotype" is a mold (usually plastic) of a metal plate made under pressure. It is sprayed with a layer of silver that makes it receptive to thin coatings of pure metals or alloys when subjected to an electrical current (electrolysis). The shell is then removed from the mold, backed with metal for support, blocked for a platen or flatbed cylinder letterpress, or curved for

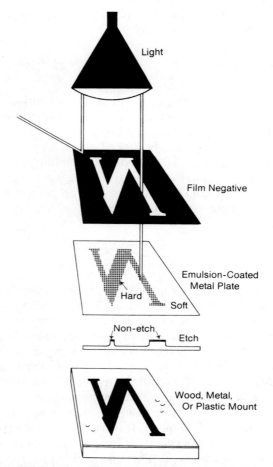

FIGURE 12.22. The process of letterpress platemaking.

a rotary type. This form of duplicate is effective with halftone screens as fine as 150-line (60-cm) and is capable of high-quality reproduction.

A duplicate lightweight plastic plate may be quickly made using a similar process. A plastic mold must first be made of the metal plate using pressure and heat. This mold is then attached to a platemaker press that automatically extrudes a layer of plastic that is squeezed up against the mold, making a duplicate plate. It can be strengthened by electrolysis to increase its life on the press.

A "stereotype" is commonly used by newspapers when the halftone screen is 85-line (34-cm) or less. Using high pressure, a paper or plastic mold ("matrix" or "mat") is made of the metal image area. This mat is then dried, curved, and a metal plate cast from it, corresponding with press requirements. This same duplicating process is used by advertisers to supply ads in appropriate form to publications printed by letterpress. An ad is converted to a mat that can be mailed inexpensively to publications for easy conversion back into metal.

Flexible duplicate plates can be prepared for high-volume printing on plastic or other flexible materials ("flexography"). Pressure and heat are used to mold a sheet of plastic matrix material to the original metal plate. When cool and hard, the mold is repositioned with a layer of rubber or plastic and more pressure and heat applied. When the assembly is cooled, the flexible duplicate is peeled away and mounted on a supportive backing material.

GRAVURE PLATEMAKING

The gravure process of printing is based on the opposite principle from letterpress. The image area is recessed or cut into the plate and the surrounding area wiped clean. When inked and subjected to pressure, highly absorbent paper withdraws the image from etched portions of the plate. The result is a soft appearance with subtle tones and a subdued or invisible dot pattern, at a cost reflecting the use of relatively economical paper.

To print commercially by gravure, the entire image—including type and other line copy—must be screened so that dots, not solid lines or areas, are etched into the plate. Because gravure presses are either sheet-fed rotary or web-fed rotary (rotogravure), any type of gravure plate is curved. The plate *is* the printing cylinder and must hold up for a lengthy press run at high speed.

A conventional gravure plate is screened so that all dot cells are the same size but are etched to varying depths. The darker the tone to be printed, the deeper the etch so more ink will be present for the paper to absorb. A variable process produces dot cells of varying size as well as depth. Direct transfer results in dot cells of varying size but the same depth. Gravure plates of any type are expensive and the process best reserved for a project with a very long press run and a need for exceptional halftone quality.

Film prepared for gravure platemaking is positive not negative. The first step in the conventional process is to expose a halftone screen (commonly 150-line or 60-cm) in a vacuum frame to a "resist"—a photosensitive gelatin ("carbon tissue") or silver ("Rotofilm") stripping base. It is so named because it becomes acid resistant when exposed to light. Strong light makes the resist hard throughout; less intense light hardens it to lesser depths.

The screened resist is then exposed to the film positive, hardening the resist to varying depths. It is wrapped around a copper cylinder face down and developed in hot water as the cylinder rotates, removing the outside layer of the stripping base. Soluble portions of the resist are washed away. Chemical etching is applied next to eat cells of the same size into the metal of the cylinder, the depth varying according to how much resist remains to restrict it. Figure 12.23 diagrams this process.

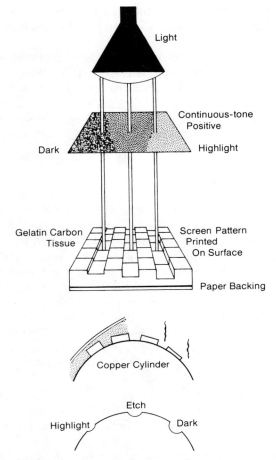

FIGURE 12.23. The process of gravure platemaking. When carbon tissue and soft gelatin are washed away, hardened gelatin is left to resist acid etching. The thicker the gelatin, the less etching occurs.

A halftone quality between conventional gravure and letterpress or offset is achieved by starting with a continuous-tone positive superimposed over a halftone positive instead of a line/continuous-tone combination. Known as the Dultgen process, it results in ink cells that vary in both size and depth and in a visible dot pattern on the finished product.

The direct-transfer method exposes a halftone film positive to a photosensitive copper plate. Ink cells vary in size only, not in depth. The result is less subtle reproduction than either the conventional or Dultgen method, but it is suitable for the printing of packaging materials.

Gravure doesn't provide a means for making duplicate plates. The process can be shortened somewhat by using duplicate continuous-tone positives or duplicate resists.

SCREEN STENCILS

Screen printing uses a stencil through which ink is forced to create an image on paper, metal, fabric, or other material. The stencil can be prepared by hand or by a photographic process. Both produce original stencils, starting with artwork done in type, ink, pencil, or paint. It must be camera-ready for a photographic stencil. If to be hand-cut, solid images need only be outlined because the technician cuts away the centers.

Hand-Cut Stencils

A two-layer film is used in this conventional method of making a stencil (also called "knife-cut"). The top layer is a colored transparent film and the bottom layer a clear plastic backing. The film is placed over the artwork and a stencil knife used to outline image areas, cutting through the top layer only. Cutout film is removed, leaving right-reading holes in the top layer. The stencil is placed on the under side of the printing screen so that it is right-reading on the top and adhered by softening the film with thinner. When the backing is peeled off, the stencil is ready for inking and printing.

Photographic Stencils

A presensitized film on a temporary backing is used in this method. A film positive is made of camera-ready artwork and contact printed onto the film. When developed and washed, right-reading image areas are etched out. The film is dried on the screen to adhere it and the backing peeled off. Other forms of photographic stencils may be used, depending on supplier preference.

CHAPTER THIRTEEN

Printing Methods and Printers

OFFSET LITHOGRAPHY   385
   The Basic Process, 385
      Feeding System, 386
      Cylinder System, 386
      Dampening and Inking Systems, 387
      Delivery System, 387
   Advantages and Disadvantages, 388
      Advantages, 388
      Disadvantages, 389
   Special-Purpose Offset, 389
      Offset Duplication, 389
      Quick Printing, 389
      Waterless Offset, 390

LETTERPRESS   390
   Types of Letterpresses, 390
      Platen, 391
      Flatbed Cylinder, 391
      Rotary, 392
   Advantages and Disadvantages, 393
      Advantages, 393

      Disadvantages, 394
  Special Relief Printing, 394
      Flexography, 394
      Letterset, 394

GRAVURE 395
  The Gravure Process, 395
  Gravure Presses and Quality, 395

OTHER PRINTING METHODS 396
  Screen Printing, 396
      The Screening Process, 397
      Screening Presses, 397
  Line Engraving and Thermography, 398
  Collotype, 398
  Xerography, Electronic, and Ink-Jet Printing, 398
      Xerography, 399
      Electronic Printing, 399
      Ink-Jet Printing, 399
  Mimeography and Ditto, 399

SELECTING AND WORKING WITH PRINTERS 400
  Types of Printers, 400
      Generalists, 400
      Specialists, 401
  Print-Shop Organization, 402
      Sales, 403
      Production, 403
      Accounting, 403
  Evaluating a Printer, 404
      Competence, 404
      Reliability, 405
      Cost, 405
      Convenience, 406
  Working With a Printer, 406
      Trade Customs, 406
      What a Printer Needs, 407
      The Press Check, 408
      Unfair Practices, 410

OFFSET LITHOGRAPHY   385

Though the time and complexity involved in planning and preparing a project for the press far outweigh that of the actual printing, this production step is the focus of all preceding effort. The press proves the quality of preparatory work, putting elements together in final form on the printed page—or provides indelible evidence of their shortcomings. Printing may also be the single most expensive phase of production, given the combination of press charges and the cost of paper if supplied by the printer.

This chapter presents each of the major printing processes and information on others for which you might have occasional use, followed by guidelines on selecting and working with printers. If you need more technical information on the various types of presses, consult specialized publications or observe presses in operation when the press operator or your print salesperson can conveniently answer questions.

Because of its dominance in the commercial-printing industry, offset lithography is emphasized in this chapter. In the United States, some 28,000 commercial shops were doing business in 1983. Of the revenue they generated, more than 60% was from offset printing. Letterpress accounted for 25%, gravure and screen for 10%, and other forms for the remaining 5%.

OFFSET LITHOGRAPHY

Printing by offset is a relative newcomer to the industry, dating from the turn of the nineteenth century when it was literally writing ("graphy") on a stone ("litho"). The printer wrote backwards on a stone with a grease pencil, wet the stone with water to repel the ink in non-image areas, inked it, and pressed paper over the stone to transfer the writing. The stone has given way to metal and even paper plates; but the principle of lithography remains the same. It relies on oil-based ink in image areas and a water solution in non-image areas.

Whether done by a quality offset press or an offset duplicator, offset lithography is "planographic printing" (or "chemical printing"). Image and non-image areas are on the same (or nearly the same) plane. The term "photo offset" underscores the photographic process that is used to make offset press plates. "Offset" explains how the image is transferred from plate to paper. It is offset from the plate onto a rubber cylinder or blanket that, in turn, prints it onto the paper. The flexibility afforded by this intermediate step is at the heart of offset lithography and its appeal today for a wide variety of projects ranging from simple brochures to magazines and continuous forms.

THE BASIC PROCESS

All offset printing is done on a rotary press made up of five components, as diagramed in Figure 13.1. Though the specific configuration varies by press design and manufacturer, each component is integral to the offset process.

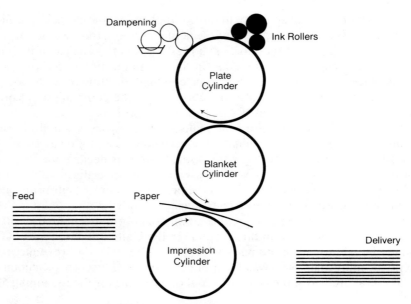

FIGURE 13.1. How an offset press operates.

Feeding System

Paper gets to the plate and ink positioned for accurate registration via a feeding system. A sheet-fed offset uses individual sheets of paper cut to the appropriate size for the job and stacked at the feed end of the press. A mechanism sends them one at a time into the press and an elevator automatically raises the stack to assure a constant flow of paper.

A web-fed offset press uses a roll of paper the appropriate width. It can be paired with trimming and folding equipment for a continuous flow through to the finished piece (referred to as "in-line" capability). The extra work needed to make it ready for printing is justified on a long press run by its increased speed (two to four times faster than a sheet-fed offset).

An attachment modifies the standard feeding system to accommodate envelopes or already-printed pieces to be "imprinted." For example, sales brochures printed at a company's headquarters can be imprinted with the local address and phone number. Ideally, such pieces are imprinted flat and then folded. Pre-folded brochures can be imprinted, although the process is necessarily slow. A short run is efficiently imprinted on a letterpress using a rubber stamp-like plate. Offset, however, is a more practical choice for a medium to long press run.

Cylinder System

A three-cylinder (or roller) system is the impression component of an offset press. The plate is clamped to the plate cylinder, dampened, and inked. The image is transferred to the offset cylinder and, as the paper passes between, the offset and impression cylinders impress the image. Packing material may be used on one of these cylinders to establish proper impression contact. The offset cylinder maintains image quality on the plate throughout the run. Because it is made of rubber, it also provides a flexible contact point with the paper, allowing successful printing of textured and uncoated stock. An offset press may accept an attachment for imprinting or special tape on the blanket for scoring or perforating, although an offset press won't score or perforate as well as a letterpress or special finishing equipment.

Dampening and Inking Systems

A series of small rollers wets non-image areas of the plate. A fountain holds the dampening solution—a mixture of water with chemicals added to make it especially wet and keep moisture on the paper to a minimum. Because true offset is a wet process, offset papers have been developed that will remain dimensionally stable, despite the presence of moisture. Their thin coating of sizing isn't noticeable to the eye or hand, but it prevents the dampening solution from penetrating the paper and causing it to change size.

Another series of rollers distributes ink from the ink fountain to the dampening plate. The operator adjusts ink flow in relation to the paper being used. An offset press may be designed to ink one side of the paper at a time or both sides at once (a "perfecting press" or "perfector"), using two blanket cylinders pressing against one another with the paper between them as substitutes for an impression cylinder. A web press may have a turning bar to turn the paper over during the run for printing on the other side. A non-perfecting sheet-fed offset requires turning the stack of sheets over for another pass through the press using the same edge ("work-and-turn") or the opposite edge ("work-and-tumble") for the press gripper.

It may print one to four or more colors at a time using a common impression cylinder for two or more blanket cylinders. By dividing the ink fountain with a dam, a one-color press can print two or more colors if they are widely spaced on the sheet. This "split-fountain printing" requires a skilled press operator and is best limited to a short press run to avoid mixing of colors. When a fusing of colors is desired, dams are removed for "rainbow-fountain printing."

Multiple rollers assure uniform distribution of dampening solution and ink, the number depending on manufacturer and press quality. These rollers are thoroughly cleaned between press runs or between printing of different colors on the same piece. A print bill includes the cost of this "washup."

Delivery System

An offset press has some means for collecting the printed paper. It may be a movable tray similar to the feeding system on a sheet-fed offset or a conveyor belt leading from a web offset's in-line folding unit. A mechanism may spray a fine drying powder onto sheets as they come off the press to prevent "set-off" ("setting-off," "off-set," or "off-setting")—spotting of adjacent sheets because of wet ink. A web press may have a heated drying and remoisturizing chamber through which the paper passes before it is folded.

ADVANTAGES AND DISADVANTAGES

The popularity of offset is evidence of the long list of advantages it has over its closest competitors: letterpress and gravure. It also has disadvantages, however, and is not necessarily the best choice for a given print project. A skilled operator using a quality press and well-prepared plates can capitalize on the former and minimize the latter.

Advantages

A sheet- or web-fed offset press can be operated at a speed faster than a flatbed cylinder letterpress. Makeready requires less time with an offset than with any type of letterpress. The platemaking process is also faster for offset than letterpress and gravure. A web offset that folds the finished piece saves time over any type of sheet-fed press when delivery to the bindery and preparation time at the bindery are taken into account.

In addition to the cost implications of time, offset saves in several areas. Cold-type composition is a natural for a job to be printed by offset because it can so easily be converted into press negatives and plates. The material used for offset plates is comparatively inexpensive, especially if a form of paper plate is appropriate. When printing multiple flat colors or process color, plate savings add up. Offset also saves during engraving if numerous continuous-tone illustrations or photographs are to be given a halftone screen.

Art and type reproduce particularly well on an offset press. It retains delicate lines better than the overall screened image of gravure and reproduces a pleasing tonal range in halftones. In skilled hands, process-color work can be better on an offset than on a letterpress.

The rubber cylinder integral to offset allows quality printing on a variety of papers or other materials (such as plastic, cloth, or metal). It can be highly textured, coated or uncoated, lightweight or very heavy and bulky. The flexible rubber compensates for these variables to impress the entire image onto paper. A range of sheet or roll sizes is also possible with the adjustment capability of offset feeding and delivery systems.

Offset has few restrictions for the designer. It can accommodate virtually any layout—and offset's use of a positive-reading plate makes that layout easy to see on the press to spot minor corrections.

Offset plates and negatives are convenient to store if you expect to need them again. Plates are not heavy and bulky as are those for letterpress, and metal plates can easily be preserved for months in anticipation of their subsequent reuse.

Disadvantages

Paper waste can be high during the process of adjusting an offset press before a run. Unless a sized offset paper is used, it can expand and curl when dampened, thus distorting the image. Inking is not as intense as with letterpress or gravure, and more operator attention is needed to assure consistent quality throughout an offset run.

SPECIAL-PURPOSE OFFSET

The offset process just described applies to a press designed and operated to produce a quality printed piece as well as to offset presses intended for more utilitarian work. The most common uses of the latter are for in-house offset duplication and quick or instant printing. In addition to letterset covered later in this chapter, other waterless methods are variations on offset.

Offset Duplication

Offset presses designed and priced for doing simple jobs in an office reprographics department or small print shop are broadly classified as "duplicators." They are usually restricted to paper no larger than 11" × 17" (279 × 432 mm) and to one color. Though web-fed models are available, sheet-fed duplicators are more popular. Common brands are Multilith, A.B. Dick, and Chief. Paper or metal plates can be used, and makeready is minimal.

A duplicator is inexpensive for press runs under 10,000, easy to operate, and relatively fast for simple newsletters, office forms, postcards, and envelopes. However, it lacks the flexibility of a more substantial offset press. It is not capable of close registration or (because it has fewer ink rollers) of consistent inking throughout a run without constant operator attention. Also, halftone reproduction is utilitarian. A duplicator can accommodate light to heavy paper and may accept attachments for such functions as imprinting.

Automatic duplicators have been developed to use electrostatic or diffusion-transfer plates and require almost no operator attention. With some, all the operator must do is insert original copy. The platemaking/duplicating unit automatically makes a plate, attaches it to a cylinder, prints the required number of copies, and stacks them. It can repeat this process for multiple pages, all automatically.

Quick Printing

This adaptation of the offset press has established quick printing in the industry between office duplication and quality commercial work. The "quick" comes about primarily because inexpensive paper plates are made within seconds from camera-ready artwork using a photo-direct unit (as explained in Chapter 12). Paired with a small offset press adjusted for standard sizes and papers, the cost savings can be substantial for the customer wanting utilitarian printing done "while you wait" without the overhead of an in-house reprographics department.

A quick printer uses single-color duplicators and, depending on demand, may also have an offset press capable of doing more quality reproduction, although still in a small, one-color format (preferably black) with a run under 10,000. A quick-print shop that is an extension of a larger plant may accept multicolor, long-run work; but it will be done in the plant and priced accordingly, not within the equipment and price-list confines of the quick-print shop.

Waterless Offset

"Direct lithography" or ("dilitho") uses inexpensive metal planographic plates and letterpress inks. Unlike letterset, it has no blanket but rather prints directly onto paper from a modified letterpress. "Driography" eliminates the need for water by using a silicone-treated metal planographic plate. (See later information on a third waterless process, letterset, presented in the section on letterpress because of its use of a relief instead of planographic plate.)

LETTERPRESS

Letterpress, the oldest form of image transfer, is "relief printing." Image areas are raised above non-image areas and inked. A rubber stamp, woodcut, or the humble potato block uses this same principle. The image to be printed is a reverse or mirror image of what you see on the resultant printed page.

Letterpress quality ranges from utilitarian to excellent. At its best, letterpress still sets the standard for all other automated print processes. Depending on the type of letterpress used, it is suitable for very small jobs and projects with press runs in the hundreds of thousands. Letterpress printing can be recognized by the sharp, crisp image it reproduces, by a noticeable spread of ink when the printed page is viewed with a magnifying glass, and possibly by a slight embossing of the paper.

TYPES OF LETTERPRESSES

Relief printing by letterpress is done on three types of presses: platen, flatbed cylinder, and rotary. They will be explained in historical order; however, their popularity today is in the reverse order. For that reason, more information will be presented on the rotary letterpress. Each type uses a multiple-roller system to distribute ink evenly across the printing plate or type.

Platen

A platen letterpress contacts plate (or actual metal type) and paper on flat surfaces. A flatbed platen lowers the paper directly down onto the inked plate, while a more modern clamshell platen operates as its name implies. These two types are diagramed in Figure 13.2. Whether hand-fed or automatic, a platen press uses sheets of paper and prints one color at a time. The paper can range from lightweight bond to heavy bristol.

A platen press is restricted today to simple jobs with very short press runs ("job printing") because it is slow. It is also used when a small job requires stamping, die-cutting, embossing, or scoring not more efficiently done (or possible) on special bindery equipment.

Flatbed Cylinder

This development in letterpress technology speeded reproduction and provided a more even impression on the paper. As diagramed in Figure 13.3, the plate (or actual metal type) remains on a flat surface as with a platen, but the paper is caught by an impression cylinder and rolled over it by the back-and-forth motion of the bed. The plate may be mounted horizontally or vertically, depending on manufacturer. An example of this variety is a press used to proof galleys of hot-metal type. It is sheet-fed and can accommodate light to heavy paper. A flatbed cylinder may be designed with two pairs of plates and cylinders to print both sides of the paper at once to make it a perfecting press or allow it to print two colors.

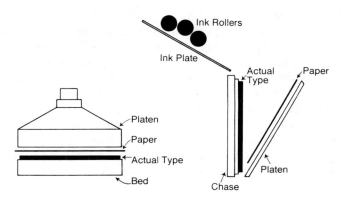

FIGURE 13.2. Flatbed (left) and clamshell (right) platen letterpresses. They use actual type instead of a plate.

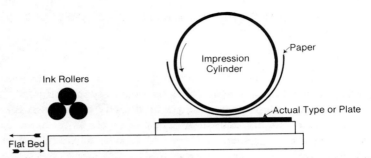

FIGURE 13.3. How a flatbed-cylinder letterpress operates. It can use actual type or a plate.

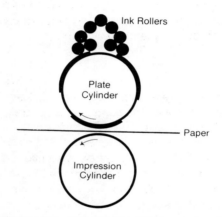

FIGURE 13.4. How a rotary letterpress operates.

Rotary

Though manufactured in sheet-fed form, a rotary letterpress is more commonly web-fed, accounting for its continued favored position in the industry for long-run projects requiring high-speed production. The plate is curved around a cylinder that rotates in the opposite direction from an impression cylinder, printing the paper as it passes between them (as diagramed in Figure 13.4). On a sheet-fed model, the paper is carried by the impression cylinder, as with the flatbed-cylinder letterpress.

It can print one to six colors at a time and may be designed as a perfector. The press may have multiple pairs of cylinders or one impression cylinder serving multiple plate cylinders. Because the plates are curved, they must be duplicates; actual metal type cannot be used directly on this form of letterpress.

For effective operation of a rotary letterpress (or either of the other types), the image area must be type-high: 0.918 in. or 2.33 cm from feet to face. The impression cylinder must be packed to assure that the proper pressure is applied when the paper contacts the plate. Too much pressure results in a "sock impression" with a noticeable embossing of the paper; too little pressure results in a weak image. The ideal is a "kiss impression" so that the full image is transferred without embossing the paper. During makeready, the operator adds packing material where the image is to be intense (such as type and halftone-shadow areas). This process is one reason that letterpress makeready is time consuming in relation to other printing methods.

Though a range of papers can be used on letterpress, a coated stock is the most effective (thus the preponderance of enamel paper when letterpress dominated the industry). Its hard, non-absorbent surface assures a kiss impression when the press is properly adjusted. Textured paper will not print in recessed areas on a letterpress, resulting in a spotty image. Newsprint can be used by covering the impression cylinder to make it somewhat resilient, although don't expect quality reproduction.

A web-fed rotary letterpress is often the choice of newspapers and magazines needing fast printing in high volume or customers requiring printed continuous forms. It also has the advantage of being able to incorporate folding on the delivery end so that the newspaper is completely ready for the stands in one quick operation. Book sections printed on a web can be folded in-line into "signatures" (sections) for easy binding. Web paper passes through a heated drying or curing unit during this process to prevent set-off, followed by treatment to replace moisture in the paper to prevent cracking.

ADVANTAGES AND DISADVANTAGES

Letterpress still claims a significant portion of the print market because of certain advantages over other processes. The list of disadvantages to be presented here, however, explains why that portion has become smaller with improvements in offset.

Advantages

When coated paper is used, ink is intense on a letterpress job and halftone reproduction excellent for color as well as for black and white. Type, rules, and lines in illustrations are well defined. Letterpress can produce consistent quality throughout a lengthy press run, assuming plates are not used beyond their lifetime. Because it's the only printing process that can print directly from hot-metal type, it may preclude platemaking per se and allow you to make corrections easily as well as leave the type standing for subsequent reuse.

Rotary letterpress is fast, especially web-fed presses. It can't match the speed of rotogravure but can be a fast alternative to offset on a job not requiring the halftone-reproduction quality of gravure. Small jobs can be done quickly on a platen or flatbed-cylinder letterpress directly from hot-metal forms.

Finally, letterpress can readily accommodate numbering, scoring, and other attachments and a variety of paper sizes and weights. Because it involves no water solution as with offset, letterpress doesn't distort paper size during impression.

Disadvantages

Because of the extensive makeready work required by letterpress, it can be a slower process than offset before the press is actually running. Engraving letterpress plates is also time consuming and costly, especially if numerous halftones are involved. When printing directly from hot-metal forms, positioning of elements on the page is limited, unlike on a mechanical for photoengraving in which type and art can be placed anywhere on the page.

A letterpress cannot print pure white or pure black in halftones because of its reliance on halftone dots at intervals to carry ink on the press. Depending on press quality, reproduction may be restricted to four colors only to prevent overlapping of inks in a moiré pattern.

SPECIAL RELIEF PRINTING

Variations on traditional letterpress have been developed for special production needs. Each is a form of relief printing.

Flexography

This variation uses a flexible rubber or soft plastic plate and very fluid, fast-drying ink to print images on plastic, cellophane, foil, or non-absorbent paper. The plate is made from a mold of hot metal or a letterpress plate and mounted on a rotary letterpress, usually a web.

Flexography (or "flexo") cannot produce sharp images so is generally limited to bold line work only and type larger than 6 points. Flat-color reproduction is excellent, however, making this process attractive for wrappers, gift wrap, and many other packaging needs. Also, toilet tissue and paper towels are printed by flexography because the lightweight paper used can be fed through a web press. Since their designs don't require crisp reproduction, this process is acceptable even though the paper used is highly absorbent.

Letterset

This process is a combination of letterpress and offset. Also called "indirect relief," "indirect letterpress," "offset letterpress," "dry offset," or "waterless lithography," it involves making a press plate in low relief and running it on a type of letterpress or on an offset press without dampening solution. The offset plate cylinder must be changed to allow the press to accept such a plate.

It lays down a heavy deposit of ink characteristic of letterpress while retaining the intermediate blanket of offset. Plates are right-reading made from negatives or right-reading hot metal made from Linotype "patrices" instead of matrices. By eliminating the need for dampening solution, this form of relief printing increases the range of papers that can be used. Plate cost is between traditional letterpress and offset, though possibly expanded plate life may compensate for the price difference. The lack of moisture and the ability to lay down a thick layer of ink make letterset a choice for printing on metal or plastic packaging materials and on paper especially sensitive to dampening solution.

GRAVURE

You may see a product of gravure printing at least once a week in your Sunday newspaper. The pictorial or magazine supplement ("roto section") is probably produced by this method because it offers high-speed, quality reproduction of art on low-grade paper. Gravure is an intaglio process— meaning that the image to be printed is below the surface of the plate instead of even with it as in offset or above it as in letterpress. The original process was called "photogravure" and accomplished by hand. "Rotogravure" refers to printing on a web-fed rotary gravure press.

THE GRAVURE PROCESS

A gravure plate is prepared by one of the three methods presented in the previous chapter. It is wrong-reading and assembled directly on the printing or plate cylinder of the press. As diagramed in Figure 13.5, the plate rotates in a tray of ink or against an inking roller; another design sprays on the ink. The plate is scraped by a squeegee ("doctor blade") to remove ink from surface, non-image areas. Pressure is exerted by an impression cylinder so that paper passing between the two cylinders is printed under high pressure. Because the transparent ink used for gravure dries instantly, no powder or heat is needed, and full-color work is wet-on-dry ink. Folding equipment can be paired with a rotogravure to produce a fully finished product in-line.

GRAVURE PRESSES AND QUALITY

Whether sheet-fed or web-fed, a gravure press is rotary. Most are web-fed with application for projects requiring economical high-speed reproduction in high volume (in excess of 100,000 copies). Though plate preparation is

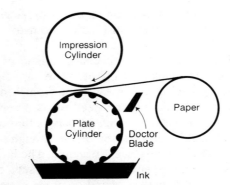

FIGURE 13.5. How a gravure press operates.

more expensive than for offset or letterpress, the gravure plate has a much longer life.

Gravure can accommodate inexpensive wide paper as well as quality stock and plastics, making it even more financially attractive. Its ability to accomplish close registration at the same time, the range of ink intensity and tones, and its multiple-color capability encourage gravure's use on process-color reproduction of illustrations and photographs. Sheet-fed gravure is especially suited to fine-art reproduction.

Because of the screen added over the entire image area, a job destined for gravure should have sans serif or very simple serif type larger than 8 points and be of medium to bold weight so that the detail lost by screening won't be noticeable. Reverses should be avoided except with simple display type.

Work done by gravure can be recognized under a magnifying glass by faintly ragged edges on type and other line work caused by the screen. In addition to newspaper supplements, the process is used commercially for catalogs, packaging, and posters.

OTHER PRINTING METHODS

Several other printing processes are available for your use on appropriate projects. Though not nearly as often used for volume work as those already covered, they do have their applications, either commercially or in in-house reprographic departments.

SCREEN PRINTING

Also called "silkscreening" or "silkscreen printing," this method of transferring an image has commercial as well as artistic application ("serigraphy"). It is a simple process used for posters, display materials, packaging, and even street signs. Though not suited to work involving much body type or to reproduction of full-color art, it is ideal when you require heavy, intense inking to produce a brilliant, opaque image, especially if the ink is lighter than the paper. When viewed with a magnifying glass, a screened image will have crisp edges and corners (if skillfully done) and the hatched lines from the screen may be faintly visible.

Screen printing can be on paper (ranging from light to heavy and smooth to rough), glass, metal, plastic, wood, leather, or fabric. The thick ink used for commercial printing can be opaque, fluorescent, or metallic, glossy or matte. The image area can even be flocked or tinseled while wet as seen on wallpaper, T-shirts, or greeting cards.

When done by hand, screen printing is ideal for very short runs (such as 10 posters or multiple panels for a billboard). Automated screening allows the use of this method for several hundred copies, although the price may not be competitive with offset or other processes.

The Screening Process

The design to be printed is reproduced on a stencil by hand-cutting or photographically, as covered in the previous chapter. This stencil is open in image areas and blocked in non-image areas. It is adhered to a porous screen stretched taut over a frame. This screen may be silk, polyester, organdy, nylon, or wire mesh. The non-image area around the stencil is masked with paper or liquid so that only live matter will print.

The screen is lowered onto the paper and inked using a rubber squeegee to distribute the ink evenly and force it through the screen onto the paper in areas that are open on the stencil. The resulting print is then dried. The process is repeated if a second color is involved using another stencil for those image areas.

Screen printing is ideal for bold line work. It may also be used for halftones, although not as satisfactorily. Start with a coarsely screened halftone (such as 65-line or 26-cm). Specify that the technician use an especially closely woven mesh for fine and uniform ink coverage. A close mesh is also preferred when screening detailed line work.

Screening Presses

A few posters or lobby cards will be screened by hand and air-dried; but a volume project in which you want the look of screening may be done on a semi- or fully automatic press, paired with an automatic dryer. The semi-automatic version requires hand-feeding of paper. An automatic web screen-printing press prints rolls of paper or fabric by passing it between a screen cylinder (with ink and a squeegee inside) and an impression cylinder to apply pressure. A cylindrical press is designed for screen-printing cans and other round objects by moving the screen. Other automated presses have been developed to speed production of specialty items (such as T-shirts).

LINE ENGRAVING AND THERMOGRAPHY

Line (or metal) engraving, an intaglio process similar to gravure, involves cutting a wrong-reading image into a polished copper ("copperplate") or steel ("steel die") plate. Engraving may be by hand or machine, working freehand or from a design photographically transferred to the plate. A matching right-reading die is made so that, when the plate is inked and the surface wiped clean, paper can be positioned between the plate and the die, firmly printing the image.

Recognized by a slightly raised image on the front and indentation on the back caused by this pressure, true steel engraving is considered the highest-quality printing because of its ability to reproduce crisp lines and fine detail. Though an expensive choice today, it is still used for fine stationery and announcements.

"Thermography" ("fake engraving" or "imagic printing") attempts to duplicate the raised lettering of real engraving by dusting wet ink printed by one of the standard methods with a resin and applying heat. The result is a shiny raised image that, unlike real engraving, can be readily scraped away.

COLLOTYPE

Also known as "screenless offset," "photogelatin," or "blackbox" printing, collotype was developed in the mid-1800s as a screenless process. A photosensitive gelatin on a metal base is exposed to a film negative, tanning or hardening the gelatin. When washed, the resulting plate accepts more ink in darker, harder areas than in lighter, softer areas. Thus it can transfer a range of tones without requiring the dot pattern of a halftone screen.

A conventional collotype plate must run on a direct-rotary collotype press using a very smooth, uncoated paper. It costs considerably less than a letterpress or gravure plate to engrave. A collotype plate prepared by the Optak or Aquatone process uses a very fine halftone screen and the same principle. It can be run on an offset press because the dot pattern established by the screen holds up better than a conventional collotype plate under the speed of an offset.

Art reproduces especially well with this process, although it cannot be proofed before it's on the press. New plates must be made instead of duplicates for a press run in excess of some 5000 impressions—the life of one collotype plate. This process is occasionally used for posters, packaging, and images on plastic to be backlighted.

XEROGRAPHY, ELECTRONIC, AND INK-JET PRINTING

These processes are pressureless and plateless methods of reproducing an image. Xerography is the most common, yet the others have a place in an increasingly computerized industry.

Xerography

This popular reproduction process uses intense light to expose typed, typeset, or drawn camera-ready artwork to electrostatically charged paper. The paper retains its charge in image areas, losing it in all others. A powder or liquid toner then develops those image areas; when heat is applied, the image is set. This process is what happens when you make a photocopy on a Xerox or other office copier. The name literally means "dry writing."

These units have been improved in recent years to be an economical alternative to office offset duplicators, quick printing, or other methods that are more primitive. While a basic copier may be the most cost-efficient choice for a few copies, larger, more sophisticated, faster units are efficient for a hundred or more copies, without the time-consuming task of platemaking. Copiers have been designed to reduce or enlarge images or even to reproduce in color and continuous tone on one or two sides of xerographic or bond paper. They can count, collate, staple, and make address labels on special paper. They can be operated by anyone in an office and are frequently seen in small print shops for economical use on short-run, simple jobs.

Electronic Printing

Similar to xerography, this process is photographic. The image, however, is converted (or created) by computer into a pattern projected by laser beam onto a photoconductor drum. It is reproduced with toner onto plain paper. Electronic printing allows all-electronic production of forms, bills, and other pieces generated by computer. Each copy, however, must be constructed individually; so some slight variation in image can occur.

Ink-Jet Printing

Also individually reproduced, ink-jet copies are created using very fine jets of ink. A computer controls the flow of electrically charged droplets of ink. Hundreds of thousands of characters can be reproduced per second from computer-generated information, making ink-jet printing ("non-impact printing" or "digitalized-press printing") a choice for personalized computer letters and related projects.

MIMEOGRAPHY AND DITTO

Although replaced to a great extent by xerography and offset duplicators, stencil and spirit duplicating remain important processes in schools, churches, and small offices without sufficient volume to make other methods financially feasible. Both are planographic processes and best limited to fewer than 100 copies.

Mimeography ("mimeo" or "stencil duplicating") uses a stencil made from a thin plastic sheet with a waxy coating over a backing sheet. The image is typed onto the plastic without using a ribbon or drawn using a stylus to cut through the coating. When mounted on a mimeograph machine and the backing peeled away, the stencil allows passage of ink through to the paper in cut areas only, similar to screen printing. Mimeo or other uncoated, utilitarian, medium-weight paper can be used in standard sizes or heavier paper for cards.

Another form of mimeograph uses an electronic scanner to "read" a typed or drawn original, transferring the image onto a stencil as small holes in the coating. Either approach is limited to line work only.

Ditto ("spirit duplicating" or "hectography") is characterized by its purple color (usually) and noticeable smell from the chemical used. A line-only image is typed or drawn on special paper, backed by a sheet of colored carbon paper. A wrong-reading image is created on the back of the paper. When mounted on a ditto machine and the backing carbon peeled away, the master is brought into contact with special printing paper. Wet with a solvent, the paper causes some of the image to transfer from the master. Copies can be made until all the ink from the master is used up. This same principle is applied when children learn the rudiments of printing using a gelatin slab. A right-reading master is turned over and pressed down onto the gelatin. Paper is then pressed onto the slab until the ink is used up.

SELECTING AND WORKING WITH PRINTERS

The printing methods just presented are used by several types of commercial printers. Knowing the processes is one thing; but selecting and working with printers is quite another. Like other suppliers, printers have a range of equipment and skill capabilities, and expectations of you as a customer. This section provides information on the types of commercial printers available to you and extensive guidelines for evaluating and getting along with those you select for your projects.

TYPES OF PRINTERS

Though specific terminology may differ geographically, commercial printers can be categorized into generalists and specialists. They work for the public, while in-house printers serve the needs of specific companies, associations, and publishers.

Generalists

A general shop (or general job printer) accepts a wide variety of work, limited only by equipment, skill, and staffing level. It will have an assort-

SELECTING AND WORKING WITH PRINTERS

ment of presses and may include typesetting, paste-up, and binding among its services. This type of printer may also offer comprehensive image assembly (especially if presses are offset) or subcontract that aspect of production to an engraver. Single-color or multicolor, short or long run, small or large format, excellent or utilitarian quality are all within the scope of a general shop.

Lettershop. This type of generalist provides similar services but on a more limited scale. The neighborhood job printer who does utilitarian to medium work on business stationery, small brochures, and simple newsletters is in this subcategory. Services are most likely limited to printing, photostats, and basic binding but possibly include some typesetting, paste-up, and engraving. The onus is on the customer of a lettershop to do much of the preliminary work before bringing in the job.

A lettershop takes walk-in and scheduled business but may also do work "for the trade." By doing subcontract work for larger printers, this kind of shop expands their capabilities, volume potential, and/or price appeal on a package of projects.

Quick Printers. This type of shop does a range of work, all of which has in common utilitarian quality of simple design produced quickly to standard specifications. A limited choice of standard papers is offered with the best pricing on jobs printed in one color (black). You must supply camera-ready artwork from which a paper offset plate is made on the premises. A quick printer might accept a job with halftones, multiple colors, folding, or binding requirements to be printed on your choice of paper; however, this type of shop isn't geared to such complexities, as may be reflected in resultant lower quality, slower turnaround, and higher pricing. Quick printing has carved out a niche in the market for standard, no-frills production at a low cost "while you wait" and performs predictably (and satisfactorily) when customers recognize its limitations.

Specialists

This broad category encompasses printers who concentrate on meeting the needs of particular types of customers. They may offer a full range of services from typesetting through binding, but to a specific market.

Advertising. These printers focus on direct-mail brochures, newspaper inserts, tabloids, and other sales pieces. They print in high volume and are equipped to meet demanding color-reproduction requirements of advertisers and advertising agencies. Expect an advertising printer to have at least one web press, probably a rotogravure if high-volume quality color work is required.

Publishing. Other specialists concentrate on printing magazines, newspapers, and books. They may be full service and further specialize in quality magazines or utilitarian newspapers. In these shops are found large multicolor, high-speed web presses. They can offer attractive prices on short-run publications as well if you agree to use the same paper and ink as the job ahead of you on a web press.

Catalogs and directories are also printed by publishing specialists, although they serve a different type of publishing market. The emphasis is on high volume at high speed, at a quality level to match customer demand.

Packaging. These printers take on high-volume production of simple one-color boxes, multicolor cans, and other needs of the packaging industry. They may do die-cutting, folding, and gluing in addition to printing—whatever corresponds with end uses. Printing may be on paper, metal, or plastic for all manner of packaging and labeling needs. Their equipment may include automated screen printers and flexographic letterpresses.

Business. Specifically serving business and financial customers are printers who specialize in forms, financial booklets, insurance policies, contracts, and other business-related projects. They may offer specialty typesetting, proofreading, format assistance, and binding. Those producing continuous forms for computers have web presses dedicated to this purpose and equipped with in-line perforating and/or folding units to produce a finished product off the press.

Novelty. This broad specialty includes printers who produce stickers, campaign buttons, T-shirts, pencils, balloons, billboards, tickets, and other novelty items imprinted with company logos, sales slogans, and so on. They use screen printing, offset, letterpress—whatever process will achieve a particular result. Because this type of printer is so specialized, you may need to go out of town to find someone to do a particular novelty job. Look for full-service capability to ensure that artwork meets all specifications for printing. Make certain your requirements are fully communicated because you probably won't know exactly what your job will look like until it's off the press.

PRINT-SHOP ORGANIZATION

Although very large printing plants are scattered throughout North America, they are not the norm. Most employ fewer than a dozen people, including the owner or general manager. While print-shop organization may vary by size, location, and type of printing, it does incorporate certain standard functions. These functions may be performed by just a few people or formally divided into departments, each with several employees. As the buyer of printing services, you should become familiar with the structure of the

SELECTING AND WORKING WITH PRINTERS

shops you use so that production and the customer/supplier relationship run smoothly.

Sales

Under the direction of a sales manager, prospective and past customers are contacted in search of new business. The sales manager may be an administrator with little customer contact or a salesperson, estimator, and administrator all in one.

The salesperson is your most frequent contact with a print shop, giving you an estimate or bid, picking up your job, bringing proofs, and, in general, seeing that the project goes smoothly by staying in close contact with you and the shop's production manager.

A salesperson may call on you on a regular basis or, if you are a frequent customer, see you only when you call for the next project. Print salespeople may feel entitled to show up unannounced to see production managers for advertising agencies and large companies. However, if you're charged with much more than production management, don't hesitate to ask for at least a few hours' notice. If you're busy, put the salesperson off until a more convenient time. If you never expect to have any work for that shop, ask to be dropped from the prospect list.

Production

Responsibilities of the production manager encompass purchasing materials and supervising all production services offered by the shop. These may include typesetting and proofing, designing, engraving, printing, finishing, binding, and shipping—summarized as prep (preparatory), press, and bindery. The production manager schedules a job from the time it comes in until it is delivered (or picked up) and sees that specifications and quality are to your satisfaction.

On a major project or one with production complexities, you may work directly with the production manager who, in turn, may ask you to confer with the stripper, press operator, or other specialist working on your job. However, the production manager must be kept fully informed of any changes or problems that might affect the overall schedule and/or specifications.

Accounting

Invoices and bills are generated by the accounting or office side of a print shop. The manager makes certain jobs are billed soon after completion, that all charges are included, that the cost is as estimated or bid, and that payments are duly recorded. Again, your salesperson is your liaison unless complications arise. Communicate any special billing requirements (such as needing extra copies of invoices or wanting overtime charges listed).

When a job costs more than estimated or bid, sales, production, and accounting managers talk it over to come up with an appropriate surcharge, depending on responsibility for the additional cost as substantiated by production records. Your salesperson should discuss the situation with you in advance of billing. You may concur or choose to argue the point.

EVALUATING A PRINTER

The process for selecting a printer is much like that of other suppliers discussed elsewhere in this book. Look at competence, reliability, cost, and convenience through a combination of questioning, visiting the shop, reviewing samples, and checking the shop's references.

Unless you are buying print services in a small town with limited or no options or need to print only one type of project, you should evaluate and work with more than one print shop, either in town or out of town. Having a backup is always wise in case your regular shop can't accommodate your schedule. Also, a variety of projects requires a variety of capabilities, possibly even a range of quality. Printers respect loyalty, and being able to call regularly on familiar printers makes your job easier. However, they also recognize that two projects are seldom alike and that not everyone fits their capabilities and/or interests.

The emphasis here is on evaluating printing. See earlier chapters and Chapter 15 for guidelines on selecting related services such as typesetting, finishing, and binding, though they may all be provided under the same roof.

Competence

A qualified printer should have the equipment and skill to print your work in the volume you require, by your deadline. Visit the shop to see the types and sizes of presses available and ask for this information in writing for your files. Observe press operators and other employees in action to see if they appear to know what they're doing and are attentive. Talk over a forthcoming job with the salesperson and production manager. Listen for a knowledge of other stages of production and a "work-with" attitude, a philosophy of doing business with—not just for—customers to help assure smooth production and satisfactory results.

Ask to have samples for review of work similar to what you expect to produce. If you can borrow or keep them, check them thoroughly for quality of press work (as covered later regarding a press check). Also ask for names of current customers to check as references.

Printer competence should extend to timely and realistic estimating of jobs. It may be done by hand by the sales manager in a small shop or by computer by an estimator in a large operation. Discuss the procedure used

SELECTING AND WORKING WITH PRINTERS

by this printer and ask the shop's references about their experiences with the validity and timeliness of estimates. In pricing and all other areas, a competent printer will place a high priority on satisfactory customer relations.

Competence should also extend to subcontractors the printer may use. Discuss this practice thoroughly so that you understand under what circumstances subcontracting is done. Make certain that the printer will stand behind what is delivered, whether from that shop or from a subcontractor.

Reliability

A reliable printer meets commitments, both in preliminary work and in delivery of work to the bindery or of the finished job. You should have assurances that you will get what you specify—that is, the printer will not substitute different paper from that specified, use off-color ink, or ignore corrections you had asked to be made on the proof or during a press check. The printer's quality standards should meet or exceed your own.

Though you should discuss these reliability points with the printer, you will glean more objective information from the printer's references. Another source is a related supplier (such as a trusted typographer) who has experience working indirectly with that shop. Listening to the word "on the street" can reassure you about the prospects for a fruitful association with a printer or steer you away from a potentially hazardous association.

Cost

A printer should price services competitively and be willing to compete with other comparable shops for your work. If you are in the habit of asking for estimates from dissimilar shops whose pricing can't possibly be competitive because of different standards, a quality printer may rightly balk at participating in the competition. Union and non-union shops may not charge at similar rates. If for internal reasons you need to use union labor, make certain that all those competing for a job meet this criterion.

A job that is produced per the specifications under which it was priced should be billed accordingly. Check with the printer's references to see if the printer does so and if extra charges are discussed in advance. Also ask if invoices and/or statements are sent in a timely manner. A printer who delays billing you can make budgeting for subsequent jobs difficult.

Determine what the printer charges for overtime and rush work and how they are defined. A shop with two or three shifts shouldn't need to impose an overtime charge but may bill services at increased rates to compensate for expanded hours. "Rush" should be defined by activity (such as faster than two working days for negatives or faster than five working days off the press).

Printers calculate charges based on hourly rates for each of the services provided that include a percentage for profit. When they supply paper, they customarily add a percentage as a handling fee. If you expect to need to incur overtime or rush charges to meet your schedule, ask about individual rates so that you can designate the work to be done under surcharge for the least impact on your budget.

Convenience

Unless you plan to do your own pick-up and delivery, ask what service the printer supplies, either as part of the total price of a job or for a per-trip fee. You might pick up and return the proof yourself, for example, but not want to be responsible for getting the printed piece to a bindery or to your office. When researching an out-of-town shop, determine how long-distance phone calls would be billed and if a surcharge would be added for long-distance pick-up and delivery.

Even if you are after press work only on a forthcoming project, ascertain what other services the printer offers. They might include typesetting, design, paste-up, engraving, finishing, and/or binding. A shop offering skilled execution of these related steps may be able to price them lower than independent suppliers because of reduced overhead.

A printer who can store quantities of paper for you offers both convenience and the cost advantage to you of allowing bulk purchases. Humidity-controlled storage may be on premises for immediate needs and in an off-site warehouse for more long-range use. If you expect to print masters for imprinting later, make sure the printer can store the quantity you will need to make using masters an efficient option.

WORKING WITH A PRINTER

If you and your printer both look upon print production as a cooperative venture, the process will be smoother and more rewarding for all concerned than approaching it in any other frame of mind. Attitude is not enough, however. As the buyer, you need to know what a printer expects of you, practices to avoid as the customer, and how to recognize when you are being duped.

Trade Customs

Accepted practices of the industry are presented in Appendix A and may be listed on the back of your printer's quotation sheet. They are used as guidelines, and actual practices in a given shop may vary somewhat. The following points bear special mention.

Proofs. The onus is on you as the customer, not on the printer, to correct errors. Any mistake that you overlook on the proof is your problem. Unless the printer happens to catch it, the job will be printed, delivered, and billed as satisfactory. The cost of a rerun is your responsibility. This same practice is followed if you don't want to see a proof or if you fail to return it before giving the go-ahead to print.

If you keep the press waiting for a press check, standby time will be charged as an extra cost beyond the quoted price. The printer is responsible for notifying you of the approximate time your job will be ready for a press check. You are responsible for being there or for paying for your tardiness.

Quantity. Delivery of plus or minus 10% is recognized in the industry as responsive on any job of 10,000 copies or less. You may negotiate a different tolerance in larger quantities or specify a minimum amount acceptable, in which case the overrun tolerance may be higher. You will be billed for what is delivered. For instance, if you ask for 1000 copies and receive 900 or 1100, you will be billed accordingly. If you require *not less than* 1000 copies, you may receive and be billed for even more than 1100. Always save the delivery receipt until a job is billed to confirm the number received.

Delivery. A printer will have a defined geographic area within which jobs will be delivered at no extra cost or at a standard delivery charge. If you need delivery (or pick-up) outside that territory, the printer is entitled to charge extra or to ask you to make the arrangements. The price is also based on complete, one-time delivery unless at the time of quotation you stipulated a split delivery.

Quick-Print Customs. The nature of a quick-print operation has produced some differences in the practices of this type of shop versus a conventional printer. Pricing is based on receiving camera-ready artwork and on your claiming the finished job at the counter. A quick printer will certainly charge extra to do any typesetting and/or paste-up but is also entitled to add a surcharge for cleaning up your artwork.

A "reasonable" variation in color is considered acceptable because of the minimal control inherent in this type of operation. A 5% tolerance is allowed for delivery if the job requires any kind of binding work. Should you provide your own paper, it must be in sufficient quantity to account for waste and be in acceptable condition.

What a Printer Needs

To produce a satisfactory project, your printer needs elements of the job, full information, and your participation. The elements depend on what you are asking the printer to do, whether just the press work or an entire project from typesetting through binding. For just printing, you would provide the press negatives or other appropriate copy from which plates are to be made (as covered in the previous chapter). If the printer is to do the engraving, you need to provide camera-ready artwork and an imposition dummy to define the sequence of pages or panels. This dummy would be used by the printer as a guide to binding or sent on with the job to an outside bindery.

Figure 13.6 could be adapted to any project requiring one or more of the services listed of your printer or, with the addition of a deadline for each, be used for separate suppliers of these services. Should you also ask your printer to do the typesetting, use the form shown in Figure 10.13. When such information is clearly presented in writing at the onset of work, it makes the printer's job (and yours) much easier. Notice that, in the case of a rerun, the original job number is called for so that the printer can quickly locate the old negative or plate. Confirm press run and delivery requirements now and reconfirm them when you do a press check to prevent oversights.

The Press Check

On a first project with a new printer, a piece with full-color reproduction, or any other job about which you have concerns, do a press check. The printer will appreciate your participation (and may insist on it) to help ensure that the work is satisfactory before the run is completed. You should be notified when the press is to start in time for you to get to the shop to see early pages. Plan to check both sides of a job unless the reverse side is uncomplicated and you're willing to trust it solely to the printer. Figure 13.7 summarizes common things to look for when you check a job on the press. Recognize that few adjustments can be made at this stage of production without severe cost and schedule implications. Now is not the time to catch a typographical error and expect your job to be out on schedule as estimated.

Color. The color should be what you want, matched to the original art as you perceive it. Color matching is not a job for a committee; the same person should approve color reproduction on the press as approved the color separations. Colors should be where you want them; a rule you specified as red should be red, not black. Ink intensity on a full-color job should be balanced so that colors are also balanced on the printed sheet. You may have had colors corrected on separations; but that image can be distorted by too intense magenta ink or another color, thus requiring the press operator to adjust for color balance.

Inking. Ink should cover the image area, leaving no part of type or art unprinted. Coverage should be of consistent intensity from edge to edge and top to bottom to avoid a mottled look. Solid areas should be solid, with no "hickies" (spots) caused by dirt in the ink or paper fibers sticking to the offset blanket. Ink should be confined to the image area with no "fugitive color" showing up as spots or streaks in margins or other white space. Ink may also stray in the form of "ghosting" (or "fade-out") in which color left over on the blanket from a heavily inked area creates a shadow effect in a large area of adjacent white space. Little can be done about ghosting at this

Printing Specifications Checklist

For _Conference Flyer_

Engraving Number of Photographs _one, black and white_
Special Treatment Required _—_
Line Screen _133_
Gray/Tint Screen % _10% black_
Tint Screen Color _—_
Proof/No. of Copies _one_
Return Artwork _with negative_
Other _—_

Printing Finished Size Flat/Folded _8½ x 11 flat_
Color(s) - Standard/Mixed _black_
Press Run _300_
Paper Name/Color/Weight _Hammermill Offset Opaque, Cream White, 70#_
Rerun of Previous Job Number _—_
Press Check Wanted _no_
Other _—_

Binding Type of Fold _—_
Type of Binding _—_
Packaging _wrapped or boxed_
Delivery Location _Pacific Bldg, 8th floor, reception desk_
Completed by _July 12_
Other _—_

FIGURE 13.6. An example of a form for organizing printing specifications.

Press-check Checklist

For _____

_____ Color Accuracy   _____ Registration
_____ Color Break   _____ Alignment of Pages/Panels
_____ Ink Coverage   _____ Picking/Embossing of Paper
_____ Ink Consistency   _____ Press Run Confirmed
_____ Spots/Streaks/Ghosting   _____ Delivery Confirmed

Date _____ By _____

FIGURE 13.7. An example of a checklist to be used during a press check.

stage of production. The possibility needs to be realized when a piece is designed to avoid the juxtaposition of intense and uninked areas.

Registration. While elements that make up one press plate can be checked for registration on a proof, any job that involves more than one plate must be checked for registration on the press. Elements in different colors should align exactly as specified. Also, pages or panels printed on the front should register with those printed on the back, regardless of how many plates are used.

A related concern is the registration of halftone screens so that no moiré pattern is formed. The image should be sharp, ranging from almost black to almost white. A moiré pattern indicates that the screens used to make halftone negatives overlap because of improper angling.

Paper Surface. Ink that is too tacky or paper that is weak may cause picking away of fibers from the paper, creating a fuzzy appearance. Short of selecting a stronger paper, the ink can be modified or the press slowed. If picking has occurred on an offset press, the blanket must be cleaned to avoid hickies in image areas caused by fibers stuck to it.

A press set to make a heavier impression than necessary may emboss the paper. The problem can be remedied by adjusting the pressure to lighten the impression. Check the result, however, to see that no image is lost by adjusting it too far.

Unfair Practices

A successful customer/printer relationship is based on mutual respect, cooperation, and fair dealings. However, certain unfair practices on the part of customers and printers can frustrate the relationship or result in the customer's being duped.

Printer Beware. If you habitually engage in any or all of the following unfair practices, you may find yourself without a printer. Since word travels fast in the industry, your reputation may discourage printers all over town from doing business with you.

Avoid price squeezing—trying to beat down a printer's price to cost or below cost. To stay in operation, a shop must make a profit. You might shave back that margin on a big or initial job; but the printer will not seek your business if you try it time after time.

Avoid rush work as standard procedure. To accommodate a rush project, a printer must juggle the schedule and put jobs for other customers in jeopardy. More than an occasional rush request betrays your lack of organization and appreciation for the printer's situation. If you habitually need faster turnaround, go to a printer equipped and staffed to provide it and expect to pay more.

Avoid piecemeal delivery of material when at all possible. Production can be more predictably scheduled and accomplished when all the artwork comes in at once for engraving or when all the plates can be made at the same time. Similarly, a printer can work more efficiently if all copies of a piece can be printed in a continuous run rather than some one day and the rest the next. Each time a job must be put aside and picked up again, efficiency decreases and cost increases.

Avoid continual changes, either in specifications or content. Work can proceed smoothly only if changes are kept to a minimum—and if you make them as soon as possible instead of waiting until the last minute. A change in schedule caused by an earlier change in specifications or content or by your change of mind disrupts the flow of work for you and other customers.

Avoid continual requests for donated printing. Only occasionally and only for a worthy cause should a printer be asked to contribute labor and/or materials. If your church needs a brochure or you need personal business cards, present the need. If your printer volunteers to contribute fully or partially on the strength of your past and future business, accept the offer. If not, be prepared to pay.

Customer Beware. Not every printer is reputable. To avoid becoming the victim of a printing scam, be aware of the following unfair practices.

Extra work done at the printer's option should not be reflected in your bill. If your printer chooses to rerun a job without consulting you to achieve better results or to make a proof when none was requested, you should not pay a surcharge. A printer who agrees to meet a certain quality standard should price services accordingly in the first place and then perform, instead of trying to pass on inefficiency or incompetence to you.

A job should be printed, priced, and delivered to your specifications and be no less in quality or turnaround time than you requested. You don't have to accept (and pay for) a lower grade of paper, noticeably off-color ink, or weak excuses for delays. Know what you specified and examine what you get and when you get it. Accept reduced quality and performance only if the bill is reduced accordingly. Beware of any print estimate or bid that is substantially lower than that of other comparable shops. An intentionally low initial price could be followed by a much higher price on subsequent jobs once the printer has your business. You will be better off working each time with a reputable shop and paying a fair price than trying to budget in a situation in which prices are either too low or too high and not reflective of the true costs of particular projects.

CHAPTER FOURTEEN

Paper and Ink

SELECTING AND USING PAPER   414

  Paper Manufacture, 414
    Paper Formation, 414
    Paper Textures and Finishes, 415
  Types of Printing Paper, 416
    Bond, 416
    Book or Text, 417
    Cover, 418
    Bristol or Board, 418
    Specialty, 418
  Buying Paper, 419
    Paper Characteristics, 419
    Saving on Paper, 424
    Envelopes, 427

SELECTING AND USING INK   431

  Ink Properties, 432
    Composition, 432
    Drying Methods, 433
    Matching Ink to Press, 434
  Color Selection, 435

The physical ingredients of a printed piece are the paper on which it is printed and the ink that is used. Although printing can be done on other materials, paper continues to be the most common surface, and its selection and use are a major consideration in print production. Ink, too, is available in different types, each variety meeting a particular need in the industry.

This chapter presents information on selecting and using paper and ink. While these decisions might be left to your graphic designer or printer, an educated approval is your responsibility, especially since paper alone may represent up to half the cost of producing a printed piece.

SELECTING AND USING PAPER

The discovery by early Egyptians that a drawing surface could be manufactured from the papyrus plant and the development in China of true paper initiated today's plethora of papers. They are available in many weights, finishes, and colors, changing with consumer requirements and tastes.

The papermaking process determines what is available to the print buyer and will be summarized in this section. Types of paper, their qualities, and their uses will then be presented, followed by specific information on sizes and weights to assist in their wise purchase. The section closes with common styles and sizes of envelopes. Because of the complexity of the subject, information here is necessarily limited to what the print buyer should know to complete most projects. Expect your paper merchant or printer to provide more specialized information—and recognize that its interpretation may differ among sources in the industry.

PAPER MANUFACTURE

In addition to making paper from papyrus (from which "paper" is derived), early civilizations used fine calfskin ("vellum"), sheepskin ("parchment"), or paper made from cloth or rags. Paper came into prominence in Europe about the time Columbus sailed to North America coincident with the invention of movable type, making the printed word available to the masses. Paper manufacture was mechanized in the early 1800s, relegating papermaking by hand to an art form and its commercial manufacture to a major component of the forest-products industry.

Paper Formation

All paper today is made from vegetable fibers (cellulose). The composition may be all wood, all cotton (in the form of new, undyed or bleached rags or fine, raw cotton fibers), a combination of the two, or recycled paper from which the ink has been removed. Other sources of cellulose are sometimes tapped, but these three are the most common. Fine writing papers are high in rag content to make them strong and long lasting, although comparatively expensive. Wood pulp is used throughout the range of papers available, including a strong competitor in quality to all-rag bond. Although made into a more limited range of papers, recycled paper is also available in many varieties.

The first step in papermaking is converting of wood into chips and rags into small pieces. A coarse paper (such as newsprint) results when whole logs are debarked and chipped, while finer grades have resins and other impurities removed. Leaf-bearing hardwoods produce short fibers, resulting in smooth, pliable, bulky papers. Needle-bearing softwoods produce long fibers, resulting in coarse, strong, compact papers. Rags produce even longer fibers free of impurities, resulting in strong, bright papers.

The chips are further refined mechanically or chemically into pulp which then goes through a process of washing and screening to extract resin and other impurities, leaving pure cellulose fibers. The more cleaning, the higher the quality of the resulting paper. It is also bleached to achieve maximum natural brightness. The pulp is then mixed and beaten with a large quantity of water and dyes (for color), sizing (for water and ink resistance), fillers (for opacity and smoothness), and/or a fixative (for stability).

The milklike solution (or "stock") is released at high speed to flow over a fine, long, wire screen, the weight of paper determined by the rate of release and the speed of the forward motion of the continuous wire. The screen's side-to-side vibrating action interweaves and mats the pulp fibers, aligning them in a predominately lengthwise direction to create a grain. At the same time, sufficient water is extracted to form a sheet of paper. This wide sheet is dried further through a series of heated rollers and finished to the desired texture before being cut into sheets or slit into rolls.

The "wire" side of paper (possibly evident by close examination) was in contact with the wire screen, and the "felt" side was on top. Paper coated on only one side is coated on the felt side, and any watermark is right-reading on the felt side, considered to be the more printable.

Paper Textures and Finishes

While the paper is still wet, it can be given one of many textures. A "dandy roller" pushes fibers into an overall texture or a genuine watermark, a subtle image visible when a sheet is held to the light. A true laid finish with texture in parallel columns is made using a special wire screen. A comparable texture is more commonly achieved today using a dandy roller. When the same wire mesh is used on the dandy roller as on the screen, the paper is referred to as "wove"; both its top and bottom appearance are determined by the wire mesh. Paper may also be embossed to create a pronounced texture. A "deckle" edge is made by feathering wet edges with a stream of water. (When such paper is cut, the deckle edge will always be with the grain.)

Once it has been through the dandy roller, paper passes from the screen to wool felt that carries it between pressing rollers to squeeze out water and then between drying rollers to extract most of the moisture. Some moisture is retained because completely dry paper would be too brittle for printing. (For this reason, paper that is heated during printing to dry the ink is also remoisturized.)

To give it the desired final finish, paper passes between highly polished calendering rollers. The smoother the finish and the higher the gloss desired, the more calendering is done. "Supercalendering" uses pressure, steam, and friction for extra gloss and smoothness. An artificial watermark may be added at this stage of production, as may surface sizing for smoothness.

The paper may be given a layer of coating during calendering to produce an enamel finish. A mirrorlike coating ("cast coating") is achieved by pressing the wet coated paper against a highly polished drying roller to achieve the most glazed enamel finish possible. In increasing order of smoothness, uncoated finishes are antique, eggshell, vellum, machine finish (MF), English finish (EF), and supercalendered (or "super"). In increasing order of gloss, coated finishes are matte, dull coated, gloss coated (textured coated or embossed), and cast coated.

The final production step is cutting the paper to market size. It may be slit and sold on a roll for web-fed presses or cut into sheets for sheet-fed presses.

TYPES OF PRINTING PAPER

As is evident from the summary of paper manufacture, several types (or "grades") of paper can be produced by varying content and/or processing. Although the number and makeup of classifications may also vary by your source of reference, five are used in this handbook: bond, book or text, cover, bristol, and specialty. Each will be presented in turn, with particular emphasis on those papers most commonly specified for print projects.

Bond

These high-quality printing papers are used for correspondence and record-keeping and are sometimes referred to as business or writing papers. They are lighter in weight than most book papers but often manufactured with identical finishes so that pieces printed for different purposes will all match. In addition to the popular laid texture, other common textures for papers broadly defined as bond are "cockle" (wrinkled), "pebble" (stippled), "ripple" (wavy), and "linen" (woven).

Originally used for printing of stocks and bonds, bond paper itself is now a staple in offices for stationery and related uses. It has a smooth, hard, durable finish and may be made entirely of wood fiber ("flat" or "sulfite" bond) or contain some rag fibers. The highest-grade bond is all rag. (Rag bonds are almost always watermarked with their content.) The sizing added to make bond paper especially suited to writing and typing also gives it a finish acceptable for offset printing.

Erasable bond has been specially coated to make erasing easy, although its susceptibility to smudging renders it inappropriate for printing. Ledger (or record) paper is also easily erasable. It is strong and nonglaring, making it especially suited to financial forms. Safety (or cheque) paper is treated to achieve an opposite objective; erasures on checks or forms printed on it are difficult to make and easy to detect.

The lightweights in this category are onionskin, manifold, and bible papers. They have hard surfaces and are much thinner than other bond papers. In addition to being used to make carbon copies of correspondence, unerasable or uncoated onionskin is also printed as letterhead to save on international postage. Manifold is an all-wood variation on onionskin often printed and sandwiched with carbon paper to create snap-out, multiple-copy forms. Bible (or India) paper can be printed on a web press to produce a relatively thin, durable, lightweight book of many pages (such as a directory).

Parchment is related to these types of writing papers in that it is finished to a hard, almost translucent surface. Its added weight, however, and mottled appearance make it resemble real parchment made from sheepskin, and it is a frequent choice for certificates.

Mimeo paper is classified as a bond paper, although it is not as finely processed as other varieties in this category. It is manufactured to be more porous than offset paper to correspond with the less sophisticated inking and drying system of the mimeograph.

Book or Text

This general-purpose paper is used extensively today for brochures, booklets, newsletters, and many other corporate and association printed pieces. Papers with a smooth surface are called "vellum" or simply "book," while those with a textured surface are referred to as "text." Wove or laid are antique textured finishes; wove has a subtle, clothlike texture and laid a distinct texture in horizontal and vertical rows resembling a venetian blind. Supercalendered book paper has undergone extra calendering to give it a very smooth, glossy surface with or without the addition of a coating.

Coated book papers are produced primarily for quality reproduction of halftones. They may have a dull (matte) finish or have different degrees of gloss (enamel). Paper coated on one side (C1S) is used for posters, book covers, and labels; paper coated on two sides (C2S) is used for magazines, catalogs, annual reports, and other high-image pieces requiring quality halftone reproduction on both sides. An English ("plater" or "plate") finish using a clay additive creates a suedelike surface noticeably soft to the touch. A mirrorlike gloss and the smoothest surface are achieved by cast coating.

The natural opacity of book papers may be augmented with fillers to produce very opaque sheets that will prevent showthrough from front to back, even with heavy inking. Prolonged cleaning and bleaching increase brightness of both white and colored sheets, as do fluorescent dyes.

Offset paper is specially treated to make it stable during printing. A sizing is applied to each side to prevent the water solution used on an offset press from dampening the fibers and causing the paper to expand and to prevent fibers from sticking to the offset blanket. It is manufactured in coated and uncoated versions in a variety of textures.

Cover

Heavier than bond or book papers, cover stock is used extensively for brochures, magazine and report covers, presentation folders, announcements, business cards, and other projects requiring sturdy paper. It is produced with many finishes and in an array of colors to match bond and book papers and may be coated or uncoated. Cover papers must be scored (preferably with the grain) to prevent breaking of fibers when folded. They are heavy enough to hold up well under embossing, debossing, or die-cutting.

"Duplex" cover paper is two thin sheets (or "plies") laminated together to make a cover-weight sheet. One version is color-on-white, while color-on-color provides color on both plies. The darker color can be used for the outside cover of a booklet, for example, and inside pages matched to the lighter color of the inside cover.

Bristol or Board

This category contains the heaviest papers of all. They may be minimally calendered to leave them bulky, manufactured thick and extensively calendered to make them heavy, or two or more plies of lighter-weight paper laminated together for strength. These papers are used for file folders, notebook dividers, postcards, posters, packaging, and other projects requiring substantial body. They must be scored to fold without breaking. They can be die-cut and, in the lighter weights, embossed or debossed. This category includes tag, pressboard, cardboard, chipboard, index bristol, and printing bristol. Because the look, feel, and use of these papers are so closely related, study samples from different suppliers to determine exactly which one will be suitable for a given project.

Specialty

This catch-all classification contains papers not appropriate to group in other categories. Roughly processed papers with a high pulp content include newsprint and a more calendered variation called rotopaper for use on a rotogravure press.

Carbonless paper (NCR for "no carbon required") is coated so that a typed or written image will transfer from one sheet to the next. The first

SELECTING AND USING PAPER

sheet of three, for example, is coated on the back only (CB), the second on the front and back (CFB), and the third on the front only (CF).

Kraft paper has the longest paper fibers of any type and, because of the resulting strength, is used for grocery bags and wrapping paper. It is also available for commercial printing in a range of colors for brochures, magazine inserts, and other projects for which its smooth surface and pulpy appearance are suited.

Another specialty sheet is label or gummed paper. It is coated with a moisture-sensitive or pressure-sensitive adhesive on one side and used for mailing and other labels. The pressure-sensitive version has a peel-off backing and is classified as either removable or permanent. The former is used for name tags (though it can damage leather, sequins, and silk), bumper stickers, and other temporary signage. The latter is used for permanent stickers and signage.

Metallic papers in cover weights are finished to have a silver or gold luster on one side. Other specialty papers come and go on the market; for the most current selection, consult paper-company sample books.

BUYING PAPER

The easiest way to buy paper for a print job is to leave the selection to your graphic designer and/or printer. However, a general understanding of the qualities, sizes, and weights to be considered will make you an educated buyer much more likely to be satisfied with your purchase—with the paper itself and with the bill. This section builds on the previous explanation of paper types by covering general characteristics to look for, paper weights, common sizes, bulk, and ways to save money on your purchase.

Consumer demand has prompted the proliferation of an array of papers that undergoes continual change. Unless you buy sufficient printing to justify a close, detailed study, find trustworthy paper merchants and printers who can keep you up to date on what is available and how sheets can be used effectively. Paper merchants stock the products of several manufacturers and can supply unprinted or printed samples as well as a bulking dummy to show you how a multipage piece you are planning will "bulk out" (how thick it will be) using a given paper.

Paper Characteristics

The characteristics readily apparent when evaluating papers are color, texture, and finish. They are made in a rainbow of colors and in a wide assortment of textures and finishes, each with a role to play in commercial printing. Beyond these obvious characteristics, however, are factors that influence how well a paper will perform for your intended use. When you can't find the ideal sheet, look for one exhibiting the characteristics most essential for your job.

Grain. In the production process, paper fibers are aligned to give the finished sheet a grain. "Long grain" refers to grain running with the long dimension of a sheet. "Short grain" refers to grain running with the short dimension of a sheet. The fibers that establish grain may also be long or short. Paper made from short fibers is pliable, while paper made from long fibers is strong.

In addition to these characteristics created by grain, grain also determines how a piece that is to be folded should be cut out of a sheet. As explained in detail in Chapter 6, a piece to be folded is usually printed so that the fold is with the grain. Such a fold is crisper than one made across the grain, and the printed page will lay flatter. A cross-grain fold may break the fibers and crack (unless the fibers are very short), producing a ragged edge and causing the page to curl. An exception is a table tent (or tent card); the fold is made across the grain so that grain strength helps the card stand. For an illustration of ways to identify paper grain, see Figure 6.2.

Grain is noted in paper-sample books and on paper labels by underlining the dimension that is with the grain. For example, "17 × 22" means that the grain runs in the 22-in. direction. Grain direction may also be indicated by printing that dimension in boldface type or by simply listing it last.

Paper Weight. The weight designation of paper is used along with paper name and color when you or your designer specifies the sheet on which a job is to be printed. Weight designations have evolved with the industry, leading to confusing terminology and subtle differences that can easily confuse both novice and experienced buyers. This section will present basic guidelines. For more detailed information, consult your printer, paper merchant, or technical publications.

"Basis weight" (or "basic weight") is the weight of a single sheet of a particular paper based in the U.S. system on the weight of one "ream" (500 sheets) of its "basis size"—the most common size in which it is produced. "Substance" is another term for basis weight. In the metric system, it is calculated using the weight of one square meter of paper expressed as grams per square meter (g/m^2). Figure 14.1 lists the metric equivalents of basis weights for book and cover sheets. To find g/m^2 of weights not given, multiply the basis weight by 1406.5 and then divide that figure by the number of square inches in the basis size.

For example, the basis size of book paper is 25" × 38" (635 × 965 mm). One ream of a particular book paper in that size might weigh 70 lb. It is said, then, to be a "70-lb book," "basis 70 book," or "substance 70 book." In metric grammage, one square meter of that paper would weight 104 g and be referred to as a "104-g book."

"M weight" is the weight of 1000 sheets of any given size. This approach to calculating weight is simpler than basis weight because it doesn't require knowing the basis size. It can be used for any size of sheet and is what your printer uses to order from a paper house. Figure 14.2 is a label from a ream

SELECTING AND USING PAPER

Paper	Basis Size	Basis Weight	Grams per Square Meter (g/m^2)
Book	25" × 38"	30	44
	or	45	67
	635 × 965 mm	50	74
		60	89
		70	104
		80	118
		90	133
		100	148
		120	178
Cover	20" × 26"	55	148
	or	60	162
	508 × 660 mm	63	170
		65	176
		68	184
		80	216
		90	143
		110	298
		120	325

FIGURE 14.1. Metric equivalents of basis weights of book and cover papers.

8½×11–10M SUB. 20	**MIMEOGRAPH**	CANARY 500 SHEETS GRAIN LONG

FIGURE 14.2. Information on the label of a ream of precut paper.

of precut paper. Type of paper is given in the center. On the right are listed color, number of sheets, and grain direction. On the left are sheet size and the M weight, followed by the substance or basis weight.

A sheet of book paper cut to the same size as a sheet of cover paper will not have the same M weight. A thousand sheets of the heavier cover paper will have a higher M weight. They could, however, have similar basis weights, given different standard sheet sizes. This similarity is a prime contributor to the confusion of paper selection.

Figure 14.3 presents equivalent basis weights of five types of printing papers. For example, a 55-lb cover sheet is equivalent in weight to a 100-lb book or a 40-lb bond paper. If you can't find an appropriate sheet in one

Bond 17" × 22"	Book 25" × 38"	Cover 20" × 26"	Printing Bristol 22½" × 28½"	Index Bristol 25½" × 30½"
—	30	—	—	—
13	—	—	—	—
16	40	—	—	—
—	45	—	—	—
20	50	—	—	—
24	60	—	—	—
28	70	—	—	—
32	80	—	—	—
36	90	—	—	—
40	100	55	67	—
—	—	60	—	90
—	—	63	—	—
—	120	65	80	—
—	—	68	—	—
—	—	—	—	110
—	—	80	—	—
—	—	—	100	—
—	—	90	—	—

FIGURE 14.3. Approximate equivalent weights of common printing papers in their basis sizes.

category, check to see if you can substitute. Other characteristics (such as bulk), however, will not be comparable.

Paper Size. Different types of papers are manufactured in different basis sizes. This practice stems from how the various papers are most often used and on long-established custom. Following are sheet basis sizes for frequently used printing papers (metric paper sizes are always given in millimeters):

	Inches	Millimeters
Bond	17 × 22	432 × 559
Book or Text	25 × 38	635 × 965
	17½ × 22½	445 × 572
Cover	20 × 26	508 × 660
Printing Bristol	22½ × 28½	572 × 724
	22½ × 35	572 × 889
Index Bristol	25½ × 30½	648 × 775
Newsprint	24 × 36	610 × 914

SELECTING AND USING PAPER

Basic sizes of general printing papers in the metrically based International Standards Organization (ISO) system are grouped in an A series; a B series of sizes between A sizes identifies sheets used for such specialized projects as maps and charts. Note that each number in the A series is the previous sheet cut in two, starting with a sheet equal in area to 1 m². A4 in the ISO system corresponds most closely with the U.S. standard $8\frac{1}{2}'' \times 11''$.

	Millimeters	Inches
A0	841 × 1189	$33\frac{1}{4} \times 46\frac{3}{4}$
A1	594 × 841	$23\frac{3}{8} \times 33\frac{1}{8}$
A2	420 × 594	$16\frac{1}{2} \times 23\frac{3}{8}$
A3	297 × 420	$11\frac{3}{4} \times 16\frac{1}{2}$
A4	210 × 297	$8\frac{1}{4} \times 11\frac{3}{4}$
A5	148 × 210	$5\frac{7}{8} \times 8\frac{1}{4}$
A6	105 × 148	$4\frac{1}{8} \times 5\frac{7}{8}$

Many more sizes are available, although they may involve a special order at a commensurately higher cost. Precut sizes are also offered, especially in bond papers and book stock heavily used in standard smaller sizes. Roll-paper sizes correspond in width with cross-grain standard sizes.

Paper Bulk. A sheet not as finely finished as another is said to have more bulk. Although the two may weigh the same, a 16-page magazine printed on the bulkier sheet will be thicker and probably more opaque. Choose a bulky paper (or one manufactured to be opaque) when you want to increase opacity without adding weight or choose a more calendered paper when you want to keep down the bulk of a multipage document.

Bulk is determined by caliper measure, calculated in thousands of an inch or in micrometers using a caliper or micrometer instrument. Caliper may also be expressed in points; for example, a 20-lb bond paper measuring 0.003 in. (76 μm) is said to be a "3-point book."

Runnability. Paper must be able to pass through the press without causing expensive delays. Runnable paper is flat, clean, trimmed accurately, and balanced in moisture content with that of the pressroom. Roll paper is evenly and tightly wound with no or few splices and of sufficient strength not to break during operation of the web press on which it will be used. Ink affinity corresponds with the ink and type of press to be used. A quality offset press will accommodate offset or bond papers having virtually any texture or finish. The best result off a quality letterpress occurs when a smooth, coated paper is used. Gravure favors a smooth, absorbent paper, preferably a high-grade newsprint manufactured especially for gravure ("rotopaper").

Opacity. A paper that prevents ink on the front from showing through on the back is said to be opaque. The degree of printed opacity depends on how it is manufactured. A sheet can be opaque because of its color, weight, and/or bulk. Fillers also contribute; they can make a difference in opacity when all other characteristics are comparable.

To do a rough check for opacity, hold a sheet to a printed page with inking similar to the piece you are preparing. The more opaque it is, the less you will be able to see through to the sheet beneath it. If your project is to be heavily inked on one or both sides, the sheet must be more opaque than one for a job that is to be lightly inked.

Brightness. A related quality is the brightness or brilliance of a sheet. If you look at several types of white paper, you will notice that some have a gray, blue, or yellow cast that creates a dull white, while others are a snappy white. The difference is achieved by the degree of cleaning and bleaching given the pulp as it is prepared for papermaking, by fluorescent dyes, and by other additives. A bright white or color will reflect more light than a dull one and create a greater contrast between the inked image and its background.

Other Factors. On a given project, one or more other factors may assume more importance than those just mentioned. They include strength, durability, permanence, and feel. The paper selected should also correspond with the halftone line screen used when printing by letterpress. The finer the line screen, the finer the paper should be because fine halftone detail will be lost if the paper is coarse. For example, newsprint is appropriate for a 65-line screen (26 cm), while a 133-line screen (54 cm) will reproduce best on highly calendered or coated stock.

Saving on Paper

Even if you buy paper only a few times a year and in small quantities, you can make each of those purchases to your advantage by applying the following money-saving tips. The more paper you need for a job, the more your budget will benefit.

Standard Sizes. The most economical size of paper is that produced as a standard. Paper is cut to standard sizes in response to market demand; the greater the demand for a given size, the more is produced and the lower its unit cost. A special size is a special order destined to cost more unless your order is large enough to offset the extra overhead incurred at the mill.

Basis sizes are always standard, although a manufacturer may also offer others as standard. Any size listed in a sample book is considered standard; however, ask if some may be more economically priced than others because of their greater demand.

SELECTING AND USING PAPER

Economical Use. The way a printed piece is to be cut out of a sheet has a lot to do with the total cost of paper. The less that is wasted, the less you will have to buy. Standard sizes for letterhead and business cards in North America are as follows:

	Inches	Millimeters
Business card, flat	$3\frac{1}{2} \times 2$	89×50
Business card, folded	$3\frac{1}{2} \times 4$	89×100
Letterhead, standard	$8\frac{1}{2} \times 11$	216×279
Letterhead, executive	$7\frac{1}{2} \times 10\frac{1}{2}$	191×267

Figure 14.4 illustrates economical ways to cut larger printed pieces from standard-sized paper. Those that fold are cut so that the first and primary fold is with the grain. Also note that more than one piece is cut from the cover paper shown in the illustration. When you can print a small job at the same time as a large one, you will save on both paper and printing. An alternative is to retain the runnable leftover portion for a subsequent job, thereby at least saving on paper.

Bulk Purchasing. Manufacturers offer price breaks at pre-set amounts (usually by the carton) so that the more paper you order at a time, the less it will cost per unit. You must be able to use it, however, before it can be damaged by changes in humidity or by improper handling that renders it unsuitable for the press.

You can take advantage of bulk rates without actually buying all the order yourself by using "skid paper." Printers will often keep on hand paper that is so commonly called for that they can buy it by the skid and be assured that it will be used up in a short amount of time. White book paper suitable for booklet pages is an example. Buying a share of skid paper will cost less than independently specifying another sheet.

A different way to take advantage of a bulk order is to piggyback your job onto one running ahead of it on the press. This practice is common for work done on a web press. If you can print your newspaper off what remains of the roll that is already on the press (using the same ink), you will save in both paper cost and makeready charges.

A variation on this plan is to coordinate your paper purchase with that of another buyer using the same printer. Both of you will save on monthly magazines, for instance, that are on the press close enough together to use up the paper within a few days, if not in immediate succession.

Buying Direct. A printer who buys paper for you will add a markup for the service. This extra cost is usually offset by the convenience of not having to be responsible for handling the paper yourself and by the assurance of having runnable paper on hand when it's needed. On a large project, how-

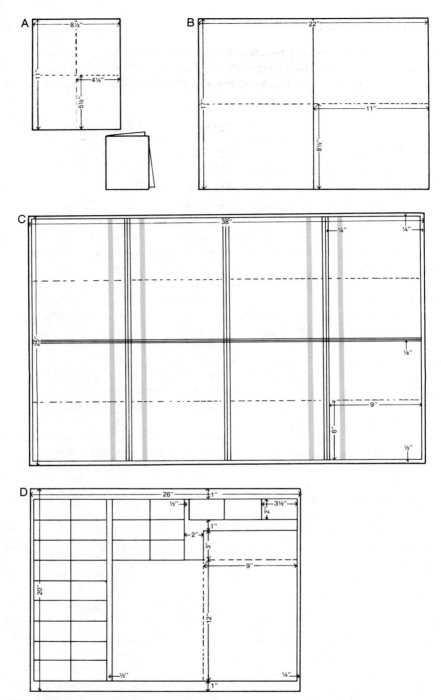

FIGURE 14.4. Economical cuts out of standard-sized paper for an announcement (A), a newsletter (B), a brochure (C), and a presentation folder (D). Note on D that otherwise wasted portions of the sheet are used to print sets of business cards.

SELECTING AND USING PAPER

ever, you may want to consider saving the markup expense by buying direct. Check first with your printer; previous experience may have led to a shop policy against this practice.

When you buy direct, you are responsible for making certain the paper is the specified weight and size and that it can be run on the press to be used. Never buy direct without first consulting your printer to confirm specifications. You must also guarantee that it is delivered in runnable condition in time to meet your press date and that any excess is removed after the work is completed. You may buy direct from a major paper house, from a large stationer's, or from a paper house specializing in close-out or job-lot papers. When shopping at the latter, be especially alert to runnability of the paper.

Envelopes

Several styles and sizes of envelopes are manufactured to match or complement papers, as shown in Figure 14.5. A paper-sample book indicates sheets with matching envelopes and the sizes and quantities available. Your printer may also keep on hand a supply of basic white envelopes in common sizes for utilitarian work. Special sizes and styles may be ordered, although the quantity needs to be well into the thousands to justify the extra cost.

When ordering a standard, determine the size you need. The paper house or envelope company will tell you which standard size most closely corresponds. You may also order by number, although the numbering system presented in the illustration has fallen out of use except in commercial sizes, and expansion-envelope sizes have no corresponding number.

Any envelope may be printed on the outside. Some styles are printed after assembly, while others must be printed before a sheet is die-cut, folded, and glued to make an envelope. Any envelope with a logo or other overall design on the inside to prevent the contents from being seen must be printed before assembly.

Closed-sided envelopes are delivered fully assembled to your printer for imprinting on the front using a press equipped with feed and delivery systems of the appropriate size. Open-sided envelopes are printed first on a sheet and then die-cut, folded, and glued. An envelope with printing on the flap may be delivered unfolded or printed before assembly. A design calling for halftones, very close registration, or a large solid or screened area may result in a surcharge.

Sizes given in Figure 14.5 are in inches. In the metrically based ISO system, a C series is used to identify sizes. Following are those dimensions; each is designed to hold the corresponding A-series sheet, as presented earlier in this chapter. The DL size is the standard commercial envelope, taking an A4 sheet folded into thirds.

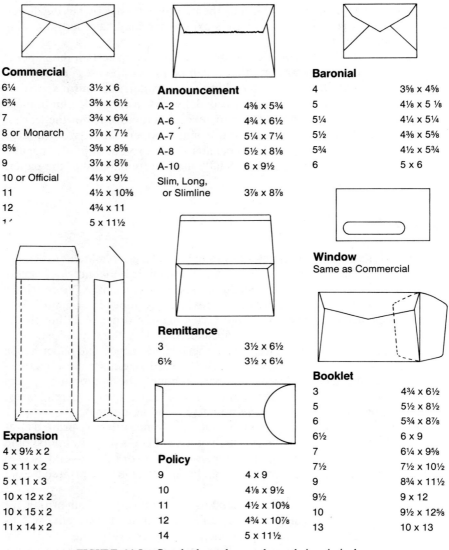

FIGURE 14.5. Standard envelope styles and sizes in inches.

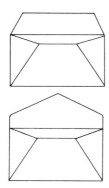

**Wallet Flap
or Bankers Flap**
Same as Commercial

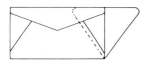

Postal - Saver
Same as Commercial

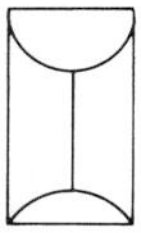

Coin

1	2¼ x 3½
3	2½ x 4¼
4	3 x 4½
4½	3 x 4⅞
5	2⅞ x 5¼
5½	3⅛ x 5½
6	3⅛ x 6
7	3½ x 6½

**Clasp or
String and Button**

10	3⅜ x 6
11	4½ x 10⅜
14	5 x 11½
15	4 x 6⅜
25	4⅝ x 6¾
35	5 x 7½
50	5½ x 8¼
55	6 x 9
63	6½ x 9½
68	7 x 10
75	7½ x 10
80	8 x 11
83	8½ x 11½
87	8¾ x 11¼
90	9 x 12
93	9½ x 12½
94	9¼ x 14½
95	10 x 12
97	10 x 13
98	10 x 15
105	11½ x 14½
110	12 x 15½

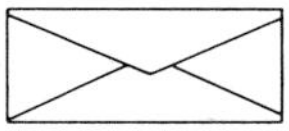

Ticket

3	1¹⁵⁄₁₆ x 4⁷⁄₁₆

Self-Sealing

6¾	3⅝ x 6½
7¾	3⅞ x 7½
10	4⅛ x 9½

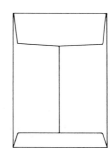

Catalog

1	6 x 9
1¾	6½ x 9½
2	6½ x 10
3	7 x 10
6	7½ x 10½
7	8 x 11
8	8¼ x 11½
9½	8½ x 10½
9¾	8¾ x 11¼
10½	9 x 12
12½	9½ x 12½
13½	10 x 13
14¼	11¼ x 14¼
14½	11½ x 14½

FIGURE 14.5. (Continued)

	Millimeters	Inches
C0	917 × 1297	$36\frac{1}{8} \times 51$
C1	648 × 917	$25\frac{1}{2} \times 36\frac{1}{8}$
C2	458 × 648	$18 \times 25\frac{1}{2}$
C3	324 × 458	$12\frac{3}{4} \times 18$
C4	229 × 324	$9 \times 12\frac{3}{4}$
C5	162 × 229	$6\frac{3}{8} \times 9$
C6	114 × 162	$4\frac{1}{2} \times 6\frac{3}{8}$
DL	110 × 220	$4\frac{5}{16} \times 8\frac{5}{8}$

Consult paper-sample books for bond and book papers that can be converted into matching envelopes. For larger sizes, open-sided styles, and those with protective padding, consult envelope companies specializing in their manufacture for the widest selection and best pricing.

Plan a print job to allow at least $\frac{1}{4}$ in. (0.64 cm) clearance in both directions for easy insertion into an envelope and more for a bulky mailing. The minimum envelope size for mailing in the United States and in Canada is $3\frac{1}{2}'' \times 5''$ (90 × 140 mm). In both countries, a single postcard or the combined front and back of an envelope must bulk to at least 7 points (178 μm).

Common types of envelopes are as follows. Each is illustrated in Figure 14.5.

1. Commercial (or regular) envelopes are used for correspondence and are made with a pointed flap from bond and book papers as well as kraft. The most common for business stationery is the No. 10 or official size and the No. 8 monarch or executive size. The No. 10 and the $6\frac{3}{4}$ are available in airmail form.

2. Announcement envelopes have a square flap (possibly with a deckle edge). They are used for formal announcements and invitations.

3. Baronial envelopes are similar in use to the announcement style. The flap is deep and pointed.

4. Window envelopes are styled the same as commercial but have a clear cellophane (glassine) or open (see through) window on the front for the address to show through. They are used for invoices, statements, and direct-mail solicitations.

5. Bankers flap (pointed) and wallet (or square) flap are heavy-duty envelopes designed for bulky correspondence. Although they look like the standard correspondence style, they are much stronger, making them a frequent choice for reusable interoffice envelopes.

6. Self-sealing envelopes have a pair of square flaps; the lower one folds up to adhere to the upper one without moisture. They are used for business papers when ease of handlng is required and may be made with a window.

7. Remittance envelopes are similar to self-sealing envelopes, although they have a deep, square flap that is not self-sealing. It prevents show-through of the contents and provides a large surface on which information may be printed.

8. Booklet envelopes are designed for mailing booklets and direct-mail pieces. They are an open-sided style, printed flat overall and then glued on the sides to present an unseamed back for advertising purposes. They may be made with a window on one or both sides.

9. Postal-saver envelopes are truly open-sided. They look like a commercial envelope but have one side that is only spot-glued so that they can be opened and closed for postal inspection.

10. Catalog (or open-ended) envelopes have wide seams and exposed, square top and bottom flaps. They are usually made from manila-colored kraft paper for strength and used to mail reports, magazines, catalogs, and other oversized documents. This same style made of Tyvek (a tough, water-resistant blend of synthetic fibers) is used for shipping purposes.

11. Clasp and string-and-button envelopes have a similar design with the addition of a metal clasp or a pair of paper or metal buttons with a string attached. The clasp version (sometimes called Columbian clasp) may have a gummed flap as well. These envelopes are strong and used for mailing large documents. The reusable nature of the string design makes it popular for interoffice mailing; holes can be added for this use to detect at a glance if all papers have been removed.

12. Expansion envelopes are three dimensional. They are constructed and scored to accommodate thick directories and other bulky pieces and have a gummed square flap or a "peel-and-seal" pressure-sensitive flap.

13. Policy envelopes are narrow with a rounded gummed flap on one end and an exposed, square bottom flap. They are used primarily for insurance policies and other legal papers.

14. Coin envelopes are small and used for paper currency and coins. They have an exposed, rounded bottom flap and a gummed, rounded top flap.

15. Ticket envelopes are the smallest style. They are made with the same open-sided design as booklet envelopes so they can be printed overall and are used for theater and other tickets.

SELECTING AND USING INK

The ink used in print production is much more than a means of creating an image of a particular color. The scores of colors available can obscure the importance of other ink properties that contribute to successful completion of a print project, both in its appearance and in its performance. The axiom, "Beauty is more than skin deep," can be applied to printing inks.

Expect to rely heavily on your printer to choose the most appropriate ink for a given job once the paper is selected. You will need to choose the color and any special qualities (such as fluorescence or high gloss). This section presents an overview of ink properties and color selection so you can make informed decisions. For more details on ink chemistry, consult your printer, technical publications, or manufacturers' literature.

INK PROPERTIES

Printing ink has a different composition from that used for writing ink, watercolors, or other liquids employed to create images. It is formulated to have properties particularly suited to the print medium and to the various printing methods used. Inks that are the workhorses of the industry share a similar "recipe" with specialty inks, although specific compositions differ.

Composition

Ink is made up of two primary components: pigment and a vehicle. Other ingredients are added to give it specific properties for a particular use.

Pigment. This component gives ink color—the black, green, or blue that we see on the printed page. Pigments are derived from organic and inorganic sources. In addition to contributing color, pigments determine if an ink is opaque (for flat-color work), transparent (for process-color work), or semi-transparent (for specialty work requiring that the ink beneath it be altered in color but not hidden). Other characteristics determined by pigment are permanence, the size of color particles (e.g., fine to extend plate life), and chemical resistance (such as to an overcoating of lacquer or to a drying process).

Fluorescent is a specialty ink created by using pigments that impart bright, intense color. Although two passes through the press are usually required to achieve the full effect, fluorescent inks are popular for posters, bumper stickers, and other display materials.

Magnetic ink allows the image to be "read" by an electronic scanner. It is printed and then magnetized for use on bank checks and packaging.

Vehicle. Usually a blend of synthetic and petroleum resins with a solvent, the vehicle gives ink its mobility, gloss, and abrasion resistance. It may be stiff and thick with high resistance to flow (high viscosity) or fluid and thin with low viscosity.

Depending on the vehicle, an ink may be termed long or short. A long ink flows freely in a long filament, while a short ink is thick and slow flowing, forming a short filament.

"Tack" is also determined by the vehicle. It is the stickiness of an ink that affects its direct transfer onto paper or onto an offset blanket. An ink must

SELECTING AND USING INK 433

be tacky enough to achieve the image transfer without picking fibers off the paper surface and depositing them on the blanket.

Other Components. Any ink that relies on oxidation in the drying process requires a "drying agent." Also called a "drier," the agent helps prevent ink from smudging caused by slow or incomplete drying.

Other ingredients ("additives" or "modifiers") prevent set-off and scuffing, add body, help control tackiness, and aid penetration. They can make ink rainproof and/or pasteproof for outdoor advertising or yard signs and allow printing on special surfaces (such as metal or plastic).

High-gloss inks are formulated for minimal penetration so that most of the ink remains on the surface. They are especially effective with coated papers. Quick-setting inks also have a high gloss but are made to penetrate quickly for very fast setting of a job on a high-speed press or a job that must be printed in quick succession on both sides.

Metallic inks have fine particles of bronze, aluminum, or copper added and usually require two passes through the press to achieve their full potential. They are most effectively used in solid areas on coated paper.

Varnish and Lacquer. Although not an ink, varnish is treated as an ink and applied on the press. Lacquer is applied later in the bindery using a coating machine that applies lacquer (or "Mar" coating) over the entire surface much as paper is coated during manufacture. Both add gloss (lacquer more than varnish) and protection. Varnish may be added in spots, while lacquer must be overall. Lacquer must also be applied over a resistant ink so that the ink won't bleed through and contaminate the image.

Drying Methods

The first phase of the drying process is setting. The solvent wicks out of the vehicle into the paper to bond the ink to the sheet, leaving the pigment on the surface to create the image. The sheet can be handled carefully once the ink is set but should not undergo any finishing or be folded or bound until the pigment hardens, thoroughly drying the ink.

Drying is usually achieved by one or more of the following major processes. Ink is formulated so that the method of drying matches the press and paper to be used.

Oxidation. Also called "oxidative polymerization," this method is the one used predominately with offset and letterpress inks. The oil-based vehicle absorbs oxygen, causing the pigment to gel and harden. The process can be speeded by applying heat.

Evaporation. This method is used for gravure, flexographic, and screen-printing inks as well as for stamping. The solvent-based vehicle evaporates rapidly so that wet inks don't run together or smudge on metal or plastic surfaces. The process is also referred to as "volatilization" since inks that dry by evaporation are very volatile. A fast-drying, heat-set ink employs both evaporation and heat on a lengthy, high-speed, quality run.

Absorption. Also called "penetration," the vehicle penetrates into paper fibers but may never dry fully. The ink is very thin and suitable for high-speed printing, making it a popular choice for newspapers, even though the still-damp ink comes off when the newspaper is handled.

Precipitation. The printed piece is exposed to moisture, precipitating out the solvent and allowing penetration into the paper. This process is used with two types of ink favored for food and other packaging: moisture-set that is relatively odor free and wax-set that requires the printed sheet to be dipped in wax to create waxed paper.

Matching Ink to Press

Just as a particular press is suited to a particular project, inks are formulated to have properties especially suited to particular presses. Your printer is responsible for selecting the best ink to achieve optimum results on a given press. For your information, however, the following are general guidelines.

Offset. A pasty, viscous, heavy-bodied ink is required. Because a relatively thin layer is applied, the ink must have an intense color. The specific intensity depends on the number of colors to be printed during one pass and in what order they will be laid down. Web-offset ink has a lower viscosity and less tack than sheet-fed ink to correspond with the higher speed of a web press.

Letterpress. Ink should be moderately tacky and less viscous than offset inks to achieve the best results from the raised surface of a letterpress plate or actual type. Specific qualities depend on the design and speed of letterpress to be used.

Gravure. Ink for this press must be very fluid, fine, and quick drying. The speed of a rotogravure requires the ink to dry instantly.

Letterset. This combination of letterpress and offset-press qualities requires a combination of their ink qualities also. Its thickness is similar to that of offset ink and its intensity less than offset but more than letterpress.

Flexography. Ink should be especially bright, intense, and opaque to achieve the best results. When used for packaging materials, light and abrasion resistance are also important. It must be very fluid and fast drying.

SELECTING AND USING INK

Screen Printing. Ink for this process must be thick. The type of project establishes other qualities; paint may even be used instead of ink.

COLOR SELECTION

You must specify what color of ink is to be used on any work you submit to a printer. Black is the easiest to specify, although even that color may vary by manufacturer. When a job requires heavy black inking, consult your printer to make sure the ink used is true black without a blue or purple cast.

One way to indicate ink color is to attach a sample and ask your printer to match it as closely as possible. The sample can be a tearsheet from a previous print job, for example, a piece of colored paper, or an actual object.

Another way that is more exact and requires less production time is to use the PANTONE®* MATCHING SYSTEM. Although other systems have been used in the industry, this one is now the most universally recognized. It is a formula system used as a guide to duplication by your printer of the color you specify, regardless of the particular type of ink to be used.

PANTONE Color samples are available in two forms of reference manual. A narrow book (the *PANTONE Color Formula Guide*) is used by printers and designers for quick reference. A larger, designer format (the *PANTONE Color Specifier*) has perforated squares of each color that can be torn out and attached to artwork. Some 500 different PANTONE colors are shown, each on white coated and uncoated papers so you can see how color intensity varies with paper finish. Both PANTONE color reference manuals may be purchased from an art-supply dealer or you can refer to your printer's copy if you don't need one frequently enough to justify the cost.

Using the formula corresponding with the ink number you specify, your printer will mix ink to match. Certain colors are basic and used to create the many possibilities offered. In addition to black and transparent white, they are PANTONE Yellow, PANTONE Warm Red, PANTONE Rubine Red, PANTONE Rhodamine Red, PANTONE Process Blue, PANTONE Reflex Blue, PANTONE Purple, and PANTONE Green. Any of these basic colors can be used at less cost than one your printer has to mix or have the manufacturer mix if you need a large amount.

Also available from an art-supply dealer is the PANTONE Color Tint Selector showing screen values of colors and PANTONE Color Markers identified by number for use on designer comps. Offered commercially by Pantone and by major paper companies are guides to show how PANTONE Colors look on colored papers or sheets of PANTONE Color Overlay for the same purpose. Such a guide is invaluable if you must frequently use combinations of colors. An alternative is to create your own guide of the inks and papers you most often use by keeping on file samples with ink numbers, paper colors, and screen values noted.

*Pantone, Inc.'s check-standard trademark for color reproduction and color reproduction materials.

CHAPTER FIFTEEN

Finishing, Binding, and Documenting the Job

BINDERY OPERATIONS 438
    Types of Binderies, 439
    Planning for the Bindery, 440

FORMS OF FINISHING 440
    Collating and Folding, 441
    Cutting, Trimming, and Slitting, 443
    Other Finishing Techniques, 443
        Drilling, Punching, and Cornering, 443
        Debossing, Embossing, and Stamping, 443
        Die-Cutting and Perforating, 444
        Scoring, 444
        Numbering, 444
        Tipping-In and Inserting, 445
        Laminating, 445

BINDING METHODS 445
    Looseleaf Binding, 445
        Plastic Comb, 445

　　　　Spiral, 447
　　　　Double-Wire Band, 447
　　　　Rings, 447
　　　　Screw-and-Post, 447
　　　　Friction, 448
　　　　Velo-Bind, 448
　　Stitching, 448
　　Gluing, 450
　　Perfect Binding, 450
　　Case Binding, 451

PACKAGING AND DELIVERY 452

DOCUMENTING THE JOB 453

The final touches are applied to a printed piece in what is collectively referred to as a binding operation. Finishing and binding encompass any additional work beyond printing necessary to prepare a piece for the reader. Simple jobs such as letterhead or flyers may not require additional work, while reports, periodicals, brochures, and other more complex jobs invariably do. Finishing includes collating, folding, trimming, and several other techniques that may be necessary to complete a project or prepare it for binding—the assembling and securing together of pages in order.

This chapter presents information on finishing and binding services commonly performed by a commercial bindery, a printer's bindery department, or in-house office equipment. An overview of bindery operations sets the stage for the details that follow on the many finishing techniques available and types of binding. The chapter closes with guidelines on documenting the job, without which your work would not be responsibly completed.

BINDERY OPERATIONS

Binding and finishing may be accomplished by a commercial bindery devoted exclusively to this phase of production, by a printer's bindery department, or by in-house reprographic staff using office equipment. This section covers the various types of binderies available and how to select from among them as well as how to plan a project with finishing, binding, and trade customs in mind.

TYPES OF BINDERIES

Several types of binderies are available to the print buyer depending on complexity of the job, quantity to be processed, quality required, and production schedule. Expect to rely heavily on your printer's recommendation, based on familiarity with your project and with bindery equipment, pricing, performance, and workload. These factors fluctuate in the industry so that the bindery you use this month may not be the preferred choice six months from now when you rerun the same job or next week when you have a different kind of project to be processed.

Like other subcontracted services secured by your printer, binding is billed to the printer who, in turn, adds a surcharge as compensation for handling and assuming responsibility for the work. You may, however, avoid this extra charge by purchasing bindery services direct. You will then be responsible for getting the printed piece to and from the bindery and for approving the work. Another instance when you might buy direct is when finishing and/or binding is all you need, for example, if a job already photocopied in-house or quick-printed must be collated and bound. When selecting a bindery with which to deal direct, tap the advice of trusted printers or ask a bindery for references and check them. Describe your project fully and get estimates of both cost and timing.

A trade or job bindery does work "for the trade," finishing and binding jobs beyond the scope or schedule of a printer's bindery department. It has versatile equipment and performs some hand work. A trade bindery is capable of doing virtually any medium to large project of a general nature.

A book or edition bindery specializes in medium to large runs of books. It does both hard- and soft-cover binding and is fully automated. Consider this type of bindery for a directory, catalog, or other substantial document.

A library bindery specializes in handcrafted work involving binding, rebinding, and repairing primarily books and periodicals. Turn to this specialist to bind back issues of a magazine, for example, or to salvage historic company records.

A print-shop bindery department does machine and manual work within the capability of equipment and staff. Generally described as pamphlet work, projects range from folding a brochure to collating, stitching, and trimming an annual report. A printer equips and staffs a bindery department to the extent that is economically practical, given workload and the capability and competitive pricing of trade binderies in the area. Therefore, one printer may almost never send out work for finishing and binding, while another almost always does.

Comprehensive graphic-supply dealers may also offer limited binding services. A business that does extensive photocopying may sell some of the mechanical bindings covered later in this chapter (such as comb or Velo-Bind) or have photocopy equipment that copies, collates, and binds in one in-line operation. Check the Yellow Pages when you have a small job for which such binding is appropriate.

Your company's reprographic department may offer similar binding capability for a small job or a large one if you can afford to wait several days for completion. Equipment may include a photocopier with in-line binding and a tabletop folder for simple folding of letters or lightweight brochures.

PLANNING FOR THE BINDERY

The time to make certain that a job can be finished and/or bound without difficulty is when the project is initiated, not after it's on the press. As you plan how a piece will be laid out, consider bindery requirements. Design folds that can be made easily by machine (unless your budget can sustain manual folding). Determine imposition so that the order of bound pages will be accurate and in sequence with folding equipment. Also make sure adequate margins will be left for trimming and binding. A bindery will provide a binding or imposition dummy folded and assembled to the exact size and shape of the finished piece to help assure that a project is planned to correspond with bindery capability.

Become familiar with trade customs of the industry (Appendix A). Especially note requirements for delivery to the bindery of a project to be finished and/or bound. Any work that is damp or that has been otherwise mishandled may be rejected as impossible or too difficult to process. The gripper side of sheets must be marked, along with the side used during printing as a guide to exact image placement on the sheet. These two sides are used by the bindery to set up and maintain registration during assembly.

The bindery is responsible for performing services to specifications. A job not properly finished and/or bound must be reprinted and reprocessed with the cost borne by the bindery if the fault lies there, by the customer, or by both. Make certain the bindery has full instructions and a dummy showing exactly what you want (and that the bindery has agreed that it can be done).

When in doubt about bindery performance, ask to be present when processing starts to check trim, alignment, completeness, and all specifications of the job (as suggested in Figure 15.1). Quality binding work includes all pages and is even around the edges with no overhang or short pages. Each page is securely bound so that the document won't fall apart under normal use. Page margins are even (or close) across the top and along the side throughout.

FORMS OF FINISHING

Several services may be performed by a bindery to add finishing touches beyond the press or to prepare a piece for binding. Some may be done manually, while others are semi- or fully automatic.

FORMS OF FINISHING

```
┌─────────────────────────────────────────────────────────────────────┐
│                    Finishing and Binding Checklist                   │
│                                                                      │
│   For _____ │
│                                                                      │
│      _____ Finishing per Specifications    _____ Sufficient Quantity Processed │
│      _____ All Pages Included              _____ Packaging Method Confirmed    │
│      _____ Trim Accurate                   _____ Delivery Date and Place Confirmed │
│      _____ Pages Aligned Top and Side      _____ Overrun Instructions Given   │
│      _____ No Short Pages                  _____ Other _____ │
│      _____ Binding per Specifications      _____ Other _____ │
│      _____ Binding Secure                                                       │
│                                                                                  │
│   Date _____    By _____  │
└─────────────────────────────────────────────────────────────────────┘
```

FIGURE 15.1. An example of a checklist for evaluating finishing and binding.

COLLATING AND FOLDING

The most frequently used finishing services are collating and folding. Collating usually precedes a form of binding, while folding may be an end in itself.

When sheets or signatures are gathered into the proper order, the process is called "collating." (Technically, single sheets are collated, while signatures are gathered.) When done by hand, printed pages are laid out in sequence on a table or in bins. They are picked up one at a time until the document is fully assembled.

This process is duplicated mechanically at higher speed by collating equipment. It may collate flat sheets only or signatures. A "full-service" flat-sheet collator gathers the sheets and folds, saddle-stitches, and trims them into the finished piece, all in one operation. Collators are designed with up to eight or more "stations" (slots) to accommodate many pages or folded signatures in preparation for perfect or case binding.

All folds made manually or by machine in a printed sheet are either parallel or a combination of parallel and right angle. Parallel folds are what the name implies; all folds are parallel with one another and most probably parallel with the paper edge. (A novelty fold may run diagonally.) A right-angle fold is made perpendicular with a parallel fold. Examples of these two types of folds are shown in Figure 15.2. Because a fold with the grain is crisper than one made across the grain, parallel folds are made with the grain; then the piece may be given a right-angle, cross-grain fold.

Automated folders have either a knife or buckle mechanism. A knife folder uses a blade to crease the paper and force it between folding rollers. A buckle folder uses a stop to buckle the paper and send it between the rollers. The knife design is more accurate for right-angle folds, but the buckle version has greater versatility, making it the more commonly used. Office tabletop folders are the buckle variety. More sophisticated "jobber" folders may

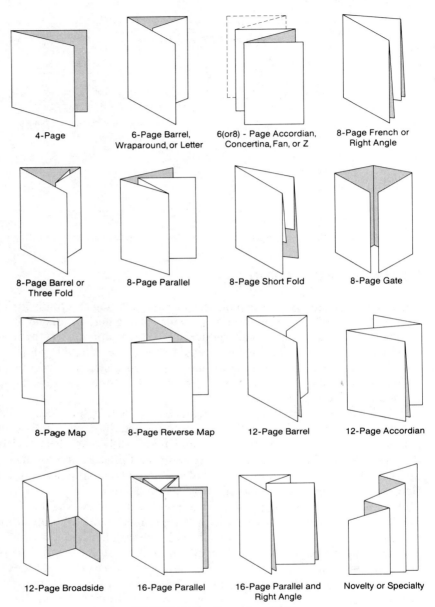

FIGURE 15.2. Examples of folds.

FORMS OF FINISHING 443

also score, slit, glue, and trim work in one pass or fold large sheets into signatures for perfect and case binding. The more folds a job requires, the higher its cost.

CUTTING, TRIMMING, AND SLITTING

Each of these procedures cuts paper, yet they are slightly different. "Cutting" in the bindery is the severing of paper to prepare it for some type of finishing and/or binding. Cutting may also occur before printing to size sheets for the press. Labels printed 12-up on a sheet illustrate a job for which cutting may be the only bindery activity. Cuts are usually parallel to the sheet edge, but angles can also be cut with special equipment.

"Trimming" is the severing of a strip of paper along the edge of a printed piece after other processing has occurred. For instance, a magazine is folded and bound and then the top, bottom, and outside edges are trimmed to achieve the desired finished size. Another example is a poster with art that is to bleed to the edge of the paper. That art is printed beyond the edge and the excess trimmed away to create the bleed effect.

"Slitting" is the severing of paper during the binding process itself. Binding equipment that takes a large sheet may fold, stitch, slit, and trim in one operation. Sheets printed 2-up with a newsletter, for example, may be folded and stitched down the center and then slit apart for individual trimming.

OTHER FINISHING TECHNIQUES

Several other finishing techniques may be considered as you plan a print project. Check with your printer or directly with a bindery to determine if equipment is available to do those techniques you want and at what cost.

Drilling, Punching, and Cornering

These procedures involve the cutting out of areas on a printed sheet. "Drilling" is the cutting out of round holes using a rotating drill, while "punching" (or "slotting") is the cutting out of rectangular holes (or any other shape) by a punch. "Cornering" is the punching away of square corners to make them rounded. When specifying drilling or punching, indicate the exact size, shape, and placement of holes to correspond with the method of binding. When specifying cornering, indicate the degree of rounding. If in doubt, include a past project as an illustration or the binder to be used with the job, or ask the bindery to provide examples from which to choose.

Debossing, Embossing, and Stamping

Although debossing and embossing may be done at the time a piece is printed by letterpress, they must be done later during finishing if another type of press was used. "Debossing" is the impressing of an image so that it is recessed on the front of the sheet, while "embossing" is the impressing of an image so that it is raised on the front. The area to be debossed or embossed may be inked or uninked ("blind debossing" or "blind embossing").

A metal die resembling a type character is cut into the shape to be debossed or embossed. It is mounted on an embossing press and registered exactly to the area to be impressed. A job to receive a deep or large impression is heated at the time of impression to soften the paper to prevent cracking.

"Stamping" ("hot-stamping," "die-stamping," or "blocking") uses a metal die and heat to create an image on book covers. A sheet of gold or silver leaf, imitation leaf, other metallic foil, or pigment is laid over the area to be stamped, and pressure and heat are applied. Metal type is used when letters, numbers, and/or words are to be stamped.

Die-Cutting and Perforating

Both of these finishing techniques cut the paper. They may be done at the time of printing by letterpress or offset (in the case of perforation) or in the bindery if a different printing method was used. A sharp metal die is made (or a standard die selected) in the appropriate shape of a die-cut design and mounted on a letterpress reserved for this purpose. The die for perforating may be designed to make a single row of slits ("slit perforation"), round holes ("pin perforation"), or a box of slits or holes. Pin perforation must be done on bindery equipment. Slit perforation may be done on a rotary perforating machine.

Scoring

Also called "creasing," this procedure makes an impression on heavy paper so that it will fold easily and evenly. It may be done by string scoring on a letterpress at the time of printing, by special scoring tape on an offset, or later in the bindery. Scoring may be a simple crease so that, for example, the cover of a perfect-bound directory will fold back easily. However, scoring of very heavy, long-grained paper requires a different approach to achieve a crisp fold. Parallel creases that almost touch are made on the front, creating two small bends in the paper in a W shape that share the stretch when the sides are folded back, rather than a single large bend.

Numbering

Tickets, receipts, and ballots are print projects commonly numbered in sequence. Numbers may be printed on a letterpress simultaneous with other printing or in the bindery. A numbering machine has five or more wheels, each with metal type characters from 0 to 9. The action of the press advances the numbers and prints one number per copy. The numbering machine may occasionally skip a number; when requesting this service, specify that skipped numbers are acceptable if a record of them is kept or that numbers must be in absolute sequence (expect to pay more for such a guarantee).

Tipping-In and Inserting

Tipping-in is the gluing of a single page into a document. Inserting adds loose advertising and reply cards to a magazine or other publication. Tipping-in occurs during assembly, while inserting occurs after binding. When done by machine, inserting uses a jet of air to insert the card between pages of a finished publication. The weight of pages and the insert's proximity to the binding edge hold it in place until the publication is thumbed by the reader.

Laminating

To protect covers of books and menus, maps, and other printed pieces that must sustain continual handling, a bindery may laminate them. A preheated film is applied to both sides of the sheet by a special machine. Heat and pressure from rollers bond the film to the paper. Any excess is then trimmed away.

BINDING METHODS

Binding is the securing of pages so they can be read in their proper order. Several methods are used today, each with characteristics making it appropriate for a particular type of project and/or budget.

LOOSELEAF BINDING

This method secures pages into some kind of mechanical binder. They are not secured to one another but rather rely on a binder to keep them together. Sheets can be assembled in any order desired, in any assortment of weights, colors, or sizes, and in virtually any even or odd number. They are then jogged by hand or machine to even the edges and bound. Pages may be removed at any time and, in some cases, added later, using the same binder. Each of the following binders is shown in Figure 15.3.

Plastic Comb

A popular choice of mechanical binder is a plastic comb—a series of broad, flat teeth on a solid backbone, curled inward more than full circle. Pages are assembled and, using a simple desktop binding machine, slots are punched

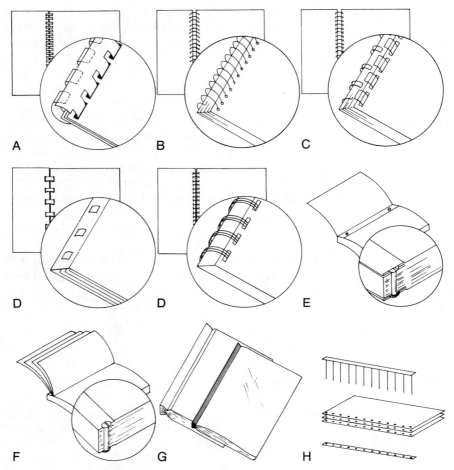

FIGURE 15.3. Types of mechanical binders: plastic comb (A), spiral (B), double-wire band (C), rings (D), screw-and-post with tongue cover (E), exposed screw-and-post (F), friction (G), and Velo-Bind (H).

along the binding edge. A comb is placed on the machine, opened, and the pages fitted over the teeth. When the comb is released, it recurls to secure the pages and the backbone becomes the document's spine.

This type of binding is very economical and easy to do in an office. It may also be done by a bindery using semi-automatic equipment. Combs are available in several colors, lengths, and widths and may be printed along the backbone with a title or company name. Also possible is "side comb binding" (or "hidden comb binding") in which a second or false cover wraps around the comb to hide it and present a solid spine much like a case-bound book.

BINDING METHODS

Spiral

Similar to the comb is the spiral (or "single-wire"), a continuous strand of wire or plastic. This form of looseleaf binding is done by a bindery using semi-automatic equipment. Pages are assembled and a row of small holes is punched along the edge to be bound. The spiral is twisted through them and the ends turned under to keep the spiral from unwinding.

Although not as convenient as the do-it-yourself comb, spiral binding does hold pages securely and allows a document to lay completely flat when opened up (although because of the spiral, facing pages are slightly out of line). Spirals are made in several lengths. They do not, however, offer a spine for printing.

Double-Wire Band

Much like a plastic comb but more secure, this mechanical binder is a row of double-wire teeth attached to a wire base. Pages are assembled and a row of slots punched along the edge to be bound. The teeth are inserted and curled to hold the pages in place. The work must be done by a bindery. Bands are fully exposed on the spine when separate front and back covers are used or partially exposed when a wraparound cover is used. Pages lay completely flat when the document is opened.

Rings

Ring mechanical binders may be loose or stationary. Loose plastic rings are tight coils threaded through slots on assembled pages. A few rings might be done by hand but any volume work is efficiently done by a bindery. Each ring is independent of the others, forming no solid spine. Pages lay flat when the document is opened.

The popular ring binder is related to the binders just mentioned. Styles range from the common three ring to those with several more rings in various colors, coverings, sizes, and widths of spine. Pages lay flat and may be assembled in any order, reassembled, removed, or added to. The cover and spine may be printed with a title, company name, or graphics.

Screw-and-Post

Done manually, a screw-and-post (or "binding post") binder holds pages together with a screw mechanism. Pages are assembled and drilled with as many holes as posts to be used, usually two or three for a standard-sized page. Front and back covers have corresponding holes through an inside flap or tongue to conceal the posts. Simple front and back covers may also be used. A metal post the appropriate height is inserted through all layers and a screw tightened on top to hold pages in place. Screw and post aren't visible from the outside. It is an economical binding solution; however, a wide margin must be left to accommodate the mechanism and account for the fact that pages won't lay flat when the document is opened or stay open unless pages are wide.

Friction

Two types of mechanical binders rely on friction to secure pages. Both are available at a stationer's and are for manual use only. A spring-back binder is made to be tightfitting near the spine so that, when pages are inserted, they are clamped securely in place.

A simpler manual approach is the plastic strip with similar built-in friction. Pages are assembled and the strip slipped over the edge to be bound. Strips are available in several colors and may be used in conjunction with a transparent vinyl cover or a printed one. With either friction style, extra margin must be left because a document so bound won't lay flat when opened or stay open unless pages are wide.

Velo-Bind

A patented process popular because of its economy and performance, Velo-Bind is identified by a thin strip of plastic along the spine edge of front and back covers. Pages are assembled and punched with a row of small holes along the binding edge. A pronged plastic strip resembling a rake is threaded through and sheared off flush with the back cover. A plain plastic strip is then riveted onto the back using heat. Pages are held securely and lay almost flat when the document is opened. The work must be done by a bindery or by office equipment purchased for this purpose.

STITCHING

This method of binding uses staples to secure pages. The process is called stitching because, when done by machine, the staples are actually segments of wire. The mechanical stitcher takes a strand of wire off a spool and in rapid motion pierces the pages, cuts the wire, and bends in the ends. The wire may be regular staple weight, heavier for thick publications, or coated for a piece to be exposed to moisture. Stitching may also be done manually using regular or heavy-duty staples. A piece bound by stitching may be either side-stitched or saddle-stitched, as illustrated in Figure 15.4. Pages are held together permanently and cannot easily be removed without weakening the binding.

"Side-stitching" (or "side-wiring") is the more simple method and is called "side-sewing" when thread is used instead of wire. Pages are assembled and two or more stitches made close to the edge to be bound (or only one stitch in an upper corner). An even or odd number of sheets may be

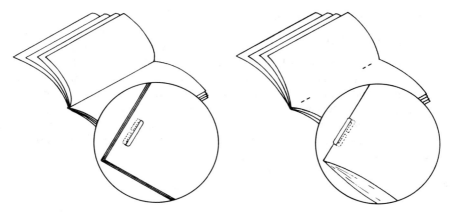

FIGURE 15.4. Side-stitching (left) and saddle-stitching (right).

bound in any assortment of weights, colors, or sizes. The process is easy to do by hand or, for a quantity, by a bindery on automated equipment. A wide margin must be left on each page to account for the stitching and the fact that pages won't lay flat when the document is opened or stay open unless pages are wide. Stitches should be tight so that wire ends on the back are firmly turned in. This process may be used with single pages or double pages folded vertically down the middle and with separate front and back covers or a wraparound cover.

"Saddle-stitching" (or "saddle-wiring") creates a more finished document and is popular for booklets, reports, magazines, and brochures. Sheets to be bound are double wide and, therefore, the process is for an even number of pages only. Included with regular pages may be inserts on different paper or of different sizes (such as advertising reply cards). Small inserts may be positioned to be held by a staple or to fall between stitches and be held by the fold only.

Assembled sheets are folded and opened across a saddle. Two or more stitches are made along the fold with wire ends on the inside. (Equipment may also stitch first and then fold, depending on design.) The piece is next trimmed on the top, bottom, and outside edges. The process may be done by hand but is more efficiently done by bindery equipment that folds, stitches, and trims in one rapid pass. Saddle-stitching allows a document to be opened completely flat and to stay open as it is read. The cover may be heavier than inside pages or the same paper ("self-cover").

As few as two sheets may be saddle-stitched; the process has an upper limit, however, depending on paper weight. A standard stitcher can secure a document up to $\frac{3}{8}$ in. (0.95 cm) thick, while heavy-duty equipment can do more pages. A document that is too thick will tax the stitching, starting with pulling away of the cover and center sheet. Also, pages of a thick document that is saddle-stitched fan out; to trim the outside edge evenly, the outside margin will have to vary from wide on outside pages to narrow at the center or the gutter will have to be shingled to accommodate. A bulking dummy supplied by your printer or paper house will show how thick the piece will be so the bindery can advise you about the appropriateness of saddle-stitching.

"Smyth sewing" is a related method used today primarily on signatures to be bound by the case method. It involves sewing through the gutter of double-wide pages with thread, similar to saddle-stitching. When all signatures are assembled in this way, case binding can proceed. Smyth sewing with thread may also be used on small documents in lieu of stitching with wire; the process may be manual or automatic.

GLUING

Two gluing procedures are used to bind pages. One is "pasted form," an alternative to saddle-stitching when lower cost is desired or when a document has an uneven number of pages. It is commonly used on advertising brochures and tabloids not expected to receive heavy wear. A ribbon of glue is laid down the middle of a full signature and then the sheet is folded over and trimmed. Pages lay almost flat with this method, although one side may lose more inside margin than the other because of unequal distribution of the glue. Center or outside pages can be removed, although the binding will be weakened.

"Padding" is a gluing procedure used primarily to make tablets of forms or scratch paper. Pages to be padded are assembled and compressed along the edge to be glued. Traditional padding is done with a water-soluble glue, while forms printed on pressure-sensitive paper are padded with a glue formulated for this purpose. Pages can be torn off easily and a cardboard back added for support. Padding may also be done in-line by a specially equipped photocopier, resulting in a smooth bead of glue along the spine.

PERFECT BINDING

A document that is too thick for saddle-stitching, that must sustain considerable wear, and that must present a quality appearance may be "perfect" (or "adhesive") bound. This permanent method (as shown in Figure 15.5) is commonly used on directories, catalogs, and large reports. Pages to be bound are assembled and the spine edge trimmed and roughed up. That edge is then glued, the cover wrapped around to secure pages and cover together, and outer edges trimmed.

Perfect binding is recognized by an evident spine which may be printed with the document title. A soft, cover-weight paper is used, as opposed to a hard cover (to be discussed next). Pages lay almost flat when the document is opened. Confirm the type of paper and quality of gluing, however, to make certain it will hold up under expected use.

A type of perfect binding may be done by office equipment or a bindery.

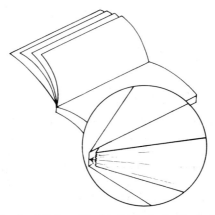

FIGURE 15.5. Perfect binding. The binding edge is glued directly into the spine.

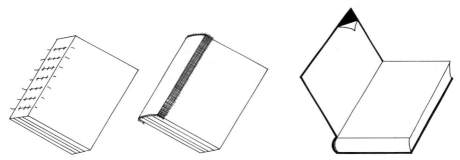

FIGURE 15.6. Case binding. The binding edge is glued to endpapers before the unit is glued into the cover.

Instead of a wraparound cover, single pages and separate front and back covers are glued along the edge. A narrow strip of paper is then wrapped around the spine and glued securely to the covers.

CASE BINDING

Books and other substantial documents that must withstand heavy wear on a permanent basis are "case" (or "edition") bound. As illustrated in Figure 15.6, this process is identified by a hard cover not glued directly to pages at the flat or rounded spine. Signatures are gathered and compressed along the edge to be bound. They are sewn together with heavy-duty thread and glued to a strip of cloth; then signatures are glued to the endpapers of a hard cover, usually a heavy board covered with cloth. The cover is creased as heat and pressure are applied to extract air. Pages lay somewhat flat when the book is opened, although the curve into the binding becomes more pronounced as thickness of the book increases.

PACKAGING AND DELIVERY

The final step in actual production is to package and deliver the finished piece. Several types of packaging may be done by your printer or bindery. You need to specify which one you want to correspond with how the piece will be distributed and/or stored. You also need to arrange for delivery. Both decisions should be specified at the time a job is priced.

Finished pieces may be bound with string or tape into a bundle, banded with a strip of paper into a bundle, or simply left on a skid. Each of these packaging methods leaves pieces exposed but should be adequate if they are to be distributed immediately by a mailing bureau or mailroom. Protection from air, dirt, and mishandling is afforded by wrapping pieces in kraft paper and taping the package securely, boxing them, or having them shrink-wrapped. Wrapping is less expensive than boxing; you pay for the boxes used to package your job, whether you or the bindery supplies them. Shrink-wrapping is a semi-automatic process using a special machine that envelopes the printed piece in a thin film of plastic and applies sufficient heat to seal the edges. This procedure provides the added protection of being moisture resistant.

With any method of packaging, a job can be slipsheeted to speed counting for distribution. A heavy or colored piece of paper may be inserted every 100 copies or whatever convenient number you designate. A project may also be packaged by count (such as 500 copies per box or bundle). Pieces to be shipped can be packaged to correspond with weight limits of the postal service or other carrier to be used.

Delivery may be to your office, to your company's loading dock, or to any other location within the service area of the print shop or bindery. If you need delivery to an outlying area or out of town, the extra charge will be added to your bill or you will have to make independent arrangements. You may also elect to pick up your job rather than pay a supplier's delivery charge either directly or indirectly as part of the overall cost. In such a case, supplier responsibility ends when you pick up the job. Whether a job is delivered or picked up, you should receive a packing slip or bill of lading indicating how many copies are in the order. Retain this slip to check the number against bindery charges.

According to trade customs, a 10% variance constitutes acceptable delivery. If you need no less than a certain number of finished copies, your printer will provide enough to the bindery to assure adequate delivery, taking into account higher waste in the bindery as finishing and binding requirements increase in number and complexity. You will be charged for whatever amount must be processed to assure delivery of your "not-less-than" figure. Should your printer provide the bindery with many more than you need, you may save money by asking that the excess not be processed. Instruct the bindery to dispose of the overrun or to deliver it to you.

DOCUMENTING THE JOB

Although delivery of a completed job may fulfill your responsibility for production of that particular project, it should not close your interest in it. The production process and costs should be documented for your reference as you plan specifications, budget, and schedule for future projects.

Information in production files may be of a short- or long-term nature. "Old copy" is the manuscript from which type was set as well as interview notes, other materials used to develop the copy, initialed approval forms, and production checklists suggested throughout this book. Old copy should be held for as long as questions might arise about your source of information or who approved it. On a regular publication, for example, you might retain old copy for the March issue until you do that issue again next year. Simply put it all into a marked envelope; if it hasn't proved of value during the intervening year, throw it away.

File illustrations and photographs separately from old copy since they require careful handling to avoid damage (covered in Chapter 7) and may be used again in another printed piece, display, or audiovisual presentation. As suggested in Figure 8.1, information can be included on past use to guide later decisions.

Mechanical artwork may be part of the production file if it is small or filed elsewhere if larger than your file drawer. It should be protected by a tissue-paper overlay or placed in an envelope. Whether stored in a file or laid flat in a drawer, it should be kept away from light and extremes in temperature and humidity if you expect to use it again in whole or in part.

A summary of production specifications is a valuable reference when doing the same project again or one with some characteristics in common. If you have used the specification forms suggested in this book, simply file them with at least five copies of the finished piece. (You may want to keep more copies in a separate file that can't be touched except for historical purposes.)

You might, instead, develop a summary form just for documentation purposes. It should include all type, print, finishing, and binding specifications and details on paper and ink. Also included should be the names and phone numbers of all suppliers and their job numbers for the project in case you re-order. Costs should be documented by supplier and invoice number, either by summarizing this information on the form or attaching copies of bills.

Allow space for or record on a separate sheet any comments about supplier performance. A note about exceptional work will direct you to that supplier again, while comments about a missed deadline, poor quality, or overcharges will remind you to steer clear of or tighten supervision of a particular supplier next time. The comment section might also encourage notes about any production headaches caused by the design or paper or about how well the piece met its objectives.

All this information should be retained indefinitely. Even when prices become outdated, other parts of the file may still have value. The more precise you can file the information, the more room you will have to keep it and the longer you can benefit from the past experience it documents.

APPENDIX A

Trade Customs

The following trade customs describe working procedures for the various suppliers involved in print production. Although practices may vary somewhat by location and supplier, those presented here are commonly accepted. (All have been slightly edited for clarity and inclusive language from the versions provided by the organizations credited.)

PRACTICES OF THE GRAPHIC-ARTS INDUSTRY*

Agents and Studios. An artist entering into an agreement with an agent or studio for exclusive representation should not accept any order from, or permit work to be shown by, any other agent or studio. Any agreement that is not intended to be exclusive should set forth in writing the exact restrictions agreed upon between the two parties. All art submitted as samples by an artist's agent or studio representative should bear the name of the artist or artists responsible for the creation. An agent, studio, or production company should not continue to show the work of an artist as samples after termination of the association. After termination of an association between an artist and agent, the agent should be entitled to a commission on accounts that the

---

*Based on the *Code of Fair Practice of the Graphic Communications Industry* issued by the Joint Ethics Committee of the Society of Illustrators, The Art Directors Club, the American Society of Magazine Photographers, the Society of Photographer and Artist Representatives, and the Graphic Artists Guild, all of New York. Originally adopted in 1954 and last revised in 1978.

agent has secured for a period of time not exceeding six months unless otherwise specified by contract.

Alterations. All changes or additions not the fault of the artist or agent should be billed to the customer as an additional and separate charge. Other than for authorized expenses, the customer should not be charged for revisions or retakes made necessary by errors on the part of the artist or agent. Alterations should not be made without consulting the artist. When alterations or retakes are necessary and time permits and when the artist's usual standard of quality has been maintained, the artist should be given the opportunity of making such changes.

Delivery. The artist should notify the buyer of an anticipated delay in delivery. Should the artist fail to keep the contract through unreasonable delay in delivery or nonconformance with agreed-to specifications, the contract should be considered breached by the artist and the customer released from responsibility. Work stopped by a customer after it has been started should be delivered immediately and billed on the basis of the time and effort expended and expenses incurred.

Gifts. The artist or agent should not participate in secret rebates, discounts, gifts, or bonuses requested by or given to customers.

Orders. Orders to an artist or agent should be in writing and should include the specific rights that are being transferred, the price, delivery date, and a summarized description of the work.

Ownership and Copyright. Art and copyright ownership are initially vested in the hands of the artist. Original art remains the property of the artist unless it is specifically purchased and paid for, as distinct from the purchase of any reproduction rights. In cases of copyright transfers, only specified rights are transferred in any transaction, all unspecified rights remaining vested in the artist. If the purchase price of art is based specifically upon limited use and later this material is used more extensively than originally planned, the artist is to receive adequate additional remuneration. Commissioned art is not to be considered "done for hire." The publisher of any reproduction of art shall publish the artist's copyright notice if the artist so requests and has not executed a written and signed transfer of copyright ownership. The right to place the artist's signature upon art is subject to agreement between the artist and the customer. No creative art should be plagiarized. Examples of an artist's work furnished to an agent or submitted to a prospective customer shall remain the property of the artist, should not be duplicated without the artist's consent, and should be returned to the artist promptly, in good condition.

Preliminary Work. If comprehensives, preliminary work, exploratory work, or additional photographs from an assignment are subsequently published as finished art, the price should be increased to the satisfaction of artist and customer. If exploratory work, comprehensives, or photographs are bought from an artist with the intention or possibility that another artist will be assigned to do the finished work, this intention or possibility should be made clear at the time of placing the order.

Speculation. An artist should not be asked to work on speculation. However, work originating with the artist may be marketed on its merit. Such work remains the property of the artist unless purchased and paid for. Art contests for commercial purposes are not approved because of their speculative and exploitative character.

Working Hours. If an artist is specifically requested to produce any art during unreasonable working hours, fair additional remuneration should be allowed.

TRADE CUSTOMS OF THE TYPOGRAPHIC INDUSTRY*

Alterations (*any addition, change, or modification of work in progress made by the customer in copy, style, or specifications originally submitted to the typographer*). Alterations will be charged for at prevailing rates at the time of the alteration. When type selection and style are left to the best judgment of the typographer, charges will be made for customer's alterations.

Chance Order (*work produced without charge in speculation of an order*). No work will be produced on the chance an order may be placed unless agreed to beforehand. (*See also* Experimental Work.)

Charges for Work (*the basis upon which work will be performed*). Charges for work may be at an hourly rate or unit price (predicated on cost and reasonable profit).

Computer Programs and Underlying Materials. (*A computer program is a coding system consisting of a series of instructions or statements in a form acceptable to a computer prepared in order to achieve a certain result.*

---

*Courtesy of the Typographers International Association. Issued jointly by that organization and the Advertising Typographers Association of America. Originally adopted in 1920 and last revised in October 1982.

Underlying materials are those production tools necessary to the development of the computer program.) Computer programs, systems analyses, underlying materials, and output developed by the typographer in order to produce work for a customer with the use of data-/word-processing machines or computers are the property of the typographer. No use shall be made of or ideas taken from such programs, analyses, materials, and output without the express permission of the typographer and only upon payment of just compensation.

Copy (*the verbal, handwritten, typewritten, printed words, or designs prepared for typesetting by the customer*). Copy or instructions that are incomplete, inaccurate, or poorly prepared will be accepted at the typographer's discretion. Additional costs that may be incurred to produce or make alterations to the work will be charged to the customer.

Corrections (*changes in composition and other work performed by the typographer because of errors by the typographer*). Corrections of errors will be made by the typographer without charge; but no financial liability is assumed for errors beyond making the corrections.

Customer's Property (*all manuscript, data on magnetic or other media, materials, or supplies delivered to the typographer for use in the production of work*). Customer's property delivered to the typographer for use in the production of work is received and stored by the typographer without any liability for loss or damage for causes beyond the typographer's control (such as fire, water, theft, strikes, or acts of God). Customer's property will be returned upon completion of the work unless stored by the typographer by written agreement.

Delivery (*conveyance of work to points or receivers designated by the customer*). Delivery is complete upon conveyance of the work to points or receivers (such as common carriers) designated by the customer or upon deposit with the U.S. Postal Service. The typographer assumes no responsibility after delivery.

Estimate (*a preliminary projection of cost that is not intended to be binding*). Estimates are based upon prevailing wages, the anticipated hours of work, and cost of materials and supplies necessary to produce work in accordance with preliminary copy, style, and specifications and are not binding upon the typographer. (*See also* Quotation.)

Experimental Work (*innovative attempts to create unique production tools, samples, or style pages*). Experimental work performed at the request of customers will be charged for on the basis of prevailing rates. (*See also* Chance Order.)

TRADE CUSTOMS 459

Handling and Shipping (*the preparation of work for transportation and the conveyance or mailing of the work*).   This service is performed as a convenience for the customer; charges will be billed for materials, labor, and actual costs of the carrier. The typographer assumes no responsibility for publishing/media costs that are incurred as a result of late delivery, damage, or erroneous contents.

Input Data (*information that has been prepared in specialized form by the typographer for the operation of data-base, word-processing, and all other computer and data-processing equipment*).   Input data produced by or developed for the typographer from customer manuscript or other media as intermediate or permanent (data-base) material used in the production of work remains the property of the typographer. Input data may be destroyed upon completion of the work unless written instructions are received prior to completion of the work.

Layouts, Style Pages, Dummies, and Mechanical Artwork (*preliminary representations of work created as production tools or prototypes for production of the work*).   Layouts, style pages, dummies, mechanical artwork, or other specially created production tools made or developed by the typographer at the order of a customer shall be charged for on the basis of prevailing rates. No use shall be made of or ideas taken from such production tools or prototypes without express permission and only upon payment of just compensation to the typographer.

Lien (*a right to hold the property of another pending the satisfaction of an outstanding obligation*).   All materials or property belonging to the customer as well as work performed may be retained by the typographer as security until all just claims against the customer have been satisfied.

Metal (*alloys composed of lead, tin, and antimony that are commonly used in the manufacture of printers' type, rules, and spacing materials*). All metal delivered to customers in the form of composition, made-up pages, or forms shall remain the property of the typographer until the customer either returns or pays for the metal. Made-up jobs must be returned intact to receive maximum credit. Linotype and Monotype metal, except when returned intact in made-up jobs, must be separated when returned to receive maximum credit. All mixed metal not in made-up form will be credited at only the scrap-metal price. Metal that contains any brass, copper, zinc, or harmful chemicals will be classed as "junk" metal and will be credited as such.

Orders (*verbal or written requests for specific goods and/or services*).   Orders accepted by the typographer can be canceled only on terms that provide for just compensation for work commenced and necessary work-related obligations entered into pursuant to the order. All orders are accepted subject to contingencies (such as fire, water, strikes, theft, vandalism, acts of God, and other causes) beyond the control of the typographer.

Outside Purchases (*any materials purchased by the typographer as necessary to produce work in accordance with a customer's order*). All outside purchases will be charged for on the basis of cost plus a markup for handling.

Overtime (*work performed by employees in excess of the prevailing regular daily and/or weekly schedule of hours, or as provided by law*). Overtime may be charged at the typographer's prevailing rates for overtime.

Preliminary Proofs (*reproductions of composition, illustrations, or other graphic representations*). Preliminary proofs (also referred to as rough, customer, galley proofs, or proof lists) submitted to a customer for approval must be marked "OK" or "OK with corrections or alterations" (with such corrections and/or alterations indicated with standard proofreader's marks) and signed or initialed by the customer.

Quotation (*a statement of price for which specified work will be performed*). Quotations are based on the number of hours of work, the prevailing scale of wages, and the cost of materials, supplies, and services necessary to do the work. Quotations are based on copy, style, and other specifications and information originally submitted, and *any* change therein, including delivery requirements, automatically voids the quotation. Quotations are only valid for 30 days or as otherwise specified and must be in writing. (*See also* Estimate.)

Standing Type and Electrotype Forms, Mechanical Art, and Film or Paper Makeup (*metal, film, or paper assembled for reproduction into final plates, proofs, negatives, or other graphic products*). All metal, film, or paper assembled for reproduction is considered dead seven days after reproduction and may be destroyed without notice unless written instructions are provided prior to the date of destruction. All such materials will be subject to storage or rental charges.

Terms (*conditions of sale*). Work may be billed either on the basis of work completed or upon completion of the order, at the discretion of the typographer.

Type (*letters, figures, symbols, and ornaments*). The value of type used for direct printing and for molding (except fewer than five wax molds) will be charged to the customer, in addition to composition charges.

Working Negatives and Positives (*original or intermediate images on film or paper of type composition, illustrations, or other graphic representations* *used in the production of work*). Working negatives or positives made by or developed for the typographer for the production of work remain the property of the typographer.

TRADE CUSTOMS OF THE PRINTING INDUSTRY*

Alterations. Alterations represent work performed in addition to original specifications. Such additional work shall be charged at current rates and be supported with documentation, upon request.

Color Proofing. Because of differences in equipment, paper, inks, and other conditions between color proofing and production pressroom operations, a reasonable variation in color between color proofs and the completed job shall constitute acceptable delivery. Special inks and proofing stocks will be forwarded to the customer's suppliers upon request at current rates.

Condition of Copy. Estimates for typesetting are based on receipt of original copy or manuscript clearly typed, double-spaced on $8\frac{1}{2}'' \times 11''$ uncoated stock, one side only. Copy that deviates from this standard is subject to re-estimating and pricing review by the printer at the time of copy submission unless otherwise specified in the estimate.

Customer-Furnished Materials. Paper stock, camera copy, film, color separations, and other customer-furnished materials shall be manufactured, packed, and delivered to printer specifications. Additional cost because of delays or impaired production caused by specification deficiencies shall be charged to the customer.

Customer's Property. The printer will maintain fire, extended coverage, vandalism, malicious mischief, and sprinkler leakage insurance on all property belonging to the customer while such property is in the printer's possession. The printer's liability for such property shall not exceed the amount recoverable from such insurance.

Delivery. Unless otherwise specified, the price quoted is for a single shipment, without storage, F.O.B. the local-customer's place of business or F.O.B. the printer's platform for out-of-town customers. Proposals are based on continuous and uninterrupted delivery of a complete order unless specifications distinctly state otherwise. Charges related to delivery from the customer to the printer or from the customer's supplier to the printer are not included in any quotations unless specified. Special priority pick-up or deliv-

*Courtesy of the Printing Industries of America. Originally adopted in 1922 and last revised in 1974.

ery service will be provided at current rates upon the customer's request. Materials delivered from the customer or the customer's suppliers are verified with delivery tickets as to cartons, packages, or items shown only. The accuracy of quantities indicated on such tickets cannot be verified, and the printer cannot accept liability for a shortage based on a supplier's tickets. Title for finished work shall pass to the customer upon delivery, to the carrier at the shipping point, or upon mailing of invoices for finished work, whichever occurs first.

Experimental Work. Experimental work performed at the customer's request (such as sketches, drawings, composition, plates, presswork, and materials) will be charged for at current rates and may not be used without the consent of the printer.

Indemnification. The customer shall indemnify and hold harmless the printer from any and all loss, cost, expense, and damages on account of any and all manner of claims, demands, actions, and proceedings that may be instituted against the printer on grounds alleging that the printing violates any copyright or any proprietary right of any person, that it contains any matter that is libelous or scandalous, or that it invades any person's right to privacy or other personal rights, except to the extent that the printer has contributed to the matter. The customer agrees to, at the customer's expense, promptly defend and continue the defense of any such claim, demand, action, or proceeding that may be brought against the printer, provided that the printer shall promptly notify the customer with respect thereto and provided further that the printer shall give to the customer such reasonable time as the exigencies of the situation may permit in which to undertake and continue the defense thereof.

Orders. Orders regularly entered, verbal or written, cannot be canceled except upon terms that will compensate the printer against loss.

Overruns and Underruns. Overruns or underruns not to exceed 10% on quantities ordered up to 10,000 copies and/or the percentage agreed upon over or under quantities ordered above 10,000 copies shall constitute acceptable delivery. The printer will bill for the actual quantity delivered within this tolerance. If the customer requires guaranteed "no-less-than" delivery, percentage tolerance of overage must be doubled.

Preparatory Materials. Artwork, type, plates, negatives, positives, and other items supplied by the printer shall remain the printer's exclusive property unless otherwise agreed to in writing.

Preparatory Work. Sketches, copy, dummies, and all preparatory work created or furnished by the printer shall remain the printer's exclusive property, and no use of them shall be made nor any ideas obtained therefrom be used, except upon compensation to be determined by the printer.

Press Proofs. Unless specifically provided in the printer's quotation, press proofs will be charged for at current rates. An inspection sheet of any form can be submitted for customer approval at no charge, provided the customer is available at the press during the time of makeready. Any changes, corrections, or lost press time because of customer change of mind or delay will be charged for at current rates.

Production Schedules. Production schedules will be established and adhered to by the customer and the printer, provided that neither shall incur any liability or penalty for delays because of state of war, riot, civil disorder, fire, strikes, accidents, action of government or civil authority, acts of God, or other causes beyond the control of the customer or the printer.

Proofs. Proofs shall be submitted with original copy. Corrections are to be made on a master set, returned marked "OK" or "OK with corrections," and signed by the customer. If revised proofs are desired, a request must be made when proofs are returned. The printer regrets any errors that may be undetected through production but cannot be held responsible for errors if the work is printed per the customer's OK or if changes are communicated verbally. The printer shall not be responsible for errors if the customer has not ordered or has refused to accept proofs, has failed to return proofs with an indication of changes, or has instructed the printer to proceed without submission of proofs.

Quotation. A quotation not accepted within 30 days is subject to review.

Terms. Payment shall be net cash 30 days from date of invoice unless otherwise provided in writing. Claims for defects, damages, or shortages must be made by the customer in writing within a period of 30 days after delivery. Failure to make such claim within the stated period shall constitute irrevocable acceptance and an admission that the product fully complies with terms, conditions, and specifications. The printer's liability shall be limited to the stated selling price of any defective goods and shall in no event include special or consequential damages, including profits (or profit loss). As security for payment of any sum due or to become due under terms of any agreement, the printer shall have the right, if necessary, to retain possession of and shall have a lien on all customer property in the printer's possession, including work in process and finished work. The extension of credit or the acceptance of notes, trade acceptances, or guarantee of payment shall not affect such a security interest and lien.

TRADE CUSTOMS OF THE QUICK-PRINTING INDUSTRY*

Alterations. Alterations represent work performed in addition to original specifications. Such additional work shall be charged at current rates and be supported with documentation, upon request.

Color Matching. Because of differences in equipment, paper, inks, and other conditions between color-sample charts and production pressroom operations, a reasonable variation in color between color-sample charts and the completed job shall constitute acceptable delivery.

Condition of Copy. Quick-printer estimates will normally be based on the receipt of camera-ready copy. Copy that deviates from this basis by requiring opaquing, paste-up, stats, composition, and/or any other preparatory work will be subject to upcharges by the quick printer.

Customer-Furnished Materials. Additional cost because of delays or impaired production caused by the customer's paper stock or corrections to the customer's camera-ready copy will be charged for at current rates.

Customer's Property. The quick-printer's liability for any in-process property belonging to the customer while such property is in the quick-printer's possession shall not exceed such amounts recoverable from any fire, extended coverage, vandalism, malicious mischief, and sprinkler-leakage insurance which may be in force.

Delivery. Unless otherwise specified, the price quoted is for a single shipment, without storage, F.O.B. the quick-printer's pick-up area. Proposals are based on continuous and uninterrupted delivery of a complete order unless specifications distinctly state otherwise. Special-priority pick-up or delivery service may be provided at current rates upon the customer's request. Materials delivered from the customer or the customer's suppliers are verified with delivery tickets as to cartons, packages, or items shown only. The accuracy of quantities indicated on such tickets cannot be verified, and the quick printer cannot accept liability for a shortage based on a supplier's tickets. Title for finished work shall pass to the customer upon pick-up or delivery.

Experimental Work. Experimental work performed at the customer's request (such as sketches, drawings, composition, plates, presswork, and materials) will be charged for at current rates, be paid for prior to any actual printing, and thereby become the customer's property.

*Courtesy of the National Association of Quick Printers.

Orders. Orders regularly entered, verbal or written, cannot be canceled except upon terms that will compensate the quick printer against loss. Certain classes of orders (e.g., volume-produced business-card orders) are not subject to cancellation once the order has been placed. The customer is responsible for inquiring about any specific item upon placement of an order, if the customer so desires.

Overruns and Underruns. The quick printer will produce and bill for the actual quantity ordered and will guarantee that quantity to be delivered for all straight runs on the quick-printer's stock papers. All runs subject to extensive finishing will be within a 5% tolerance. Final count being within this 5% or under tolerance shall constitute acceptable delivery. When stock for the job is supplied by the customer, delivery of the full quantity ordered will be dependent upon the customer's supplying sufficient stock of suitable quality for the print order, makeready, and press adjustment, and any problems directly attributable to the conditions of the paper.

Preparatory Materials. Artwork, composition, permanent plates, negatives, positives, and other items supplied by the quick printer shall be charged to the customer. The quick printer will store these materials at the request of the customer and the quick-printer's acceptance, without liability.

Preparatory Work. Sketches, copy, dummies, and all preparatory work created or furnished by the quick printer shall remain the quick-printer's exclusive property, and no use of them shall be made nor any ideas obtained therefrom be used, except upon adequate compensation to be mutually determined by the customer and the quick printer. Actual accomplishment of a finished job derived from these ideas and/or preparatory work shall be considered adequate compensation.

Press Proofs. An inspection sheet (called a press proof) can be submitted for customer approval. However, additional charges must be made for such a proof because of the additional time and materials that, by necessity, must be consumed to produce a press proof.

Production Schedules. Production schedules delayed because of state of war, riot, civil disorder, fire, strikes, accidents, action of government or similar authority, acts of God, or other causes beyond the control of the quick printer shall relieve the quick printer of any liability or penalty.

Proofs. Typesetting proofs shall be provided when ordered by the customer. Corrections are to be made on a master set, returned marked "OK" or "OK with corrections," and signed by the customer. If revised proofs are desired, a request must be made when proofs are returned. The quick printer regrets any errors that may be undetected through production but cannot be held responsible for errors if the work is printed per the customer's OK or if changes are communicated verbally. The quick printer shall not be responsible for errors if the customer has not ordered or has refused to accept proofs, has failed to return proofs with an indication of changes, or has instructed the quick printer to proceed without submission of proofs.

Quotation. A quotation not accepted within 30 days is subject to review and modification.

Remakes. Claims by a customer that a job has been done incorrectly will be considered only upon return of the goods to the quick printer.

Terms. Payment shall be net cash on delivery unless otherwise provided in writing in the form of an approved credit application. Claims for defects, damages, or shortages must be made by the customer within a period of five days after delivery. Failure to make such claim within the stated period shall constitute irrevocable acceptance and an admission that the product fully complies with terms, conditions, and specifications. The quick-printer's liability shall be limited to the stated selling price of any defective goods and shall in no event include special or consequential damages, including profits (or profit loss). As security for payment of any sum due or to become due under terms of any agreement, the quick printer shall have the right, if necessary, to retain possession of, and shall have a lien on, all customer property in the quick-printer's possession, including work in process and finished work. The extension of credit or the acceptance of notes, trade acceptances, or guarantee of payment shall not affect such a security interest and lien.

TRADE CUSTOMS OF THE BINDING INDUSTRY*

Acceptance. Unless otherwise stated in the estimate, a quotation is subject to acceptance within 10 days, and work is to start within 30 days thereafter.

Cases and Skids. All cases, skids, boxes, and so on, furnished by the customer to the binder in connection with work become the property of the binder unless agreement has been made otherwise with an appropriate charge for their return.

Counts. The binder makes no hand count on receipt of sheets or other material unless separate and distinct agreement is made, carrying extra

*Courtesy of the Trade Binders Section, Printing Industry of America.

charges. The basis of a count shall be the folded and gathered record made as soon after receipt of sheets as convenient.

Customer's Property. The binder shall charge the customer at current rates for handling and storing the customer's property held more than 30 days. All the customer's property, whether in storage or in production, is at the customer's risk. The binder is not liable for any loss or damage thereto caused by fire, water leakage, theft, negligence, insects, rodents, or any cause beyond the binder's control.

Cutting. Jobs requiring any register shall be furnished with stock squared prior to printing, together with a cutting layout or workable dummy. Failure to do so relieves the binder of responsibility for errors.

Delivery of Goods to Binder. All jobs shall be furnished to the binder jogged, securely wrapped or skidded, and dry or otherwise protected from damage. Guide and gripper sides shall be marked.

Material. Following acceptance of the binder's quotation but before commencement of actual production by the customer, the customer shall request from the binder a dummy showing the correct imposition of forms.

Overtime. All quotations are based on work being performed on a straight-time basis. Any deliveries requiring overtime because of the customer's delay in furnishing material or short delivery required shall be billed at overtime rates.

Quantities. A quotation covers only the specified quantity stated to be bound or completed as an initial order. However, should the customer in the initial order call for a part or lot less than the entire job, the binder on such parts or lots may add any increase in the cost of labor or material to the quoted price. Overruns or underruns not to exceed 10% of the bound or printed sheets furnished shall constitute an acceptable delivery, and any excess or deficiency shall be charged or credited to the customer proportionately. Loads and the customer's order are to be plainly marked if any excess is not to be processed.

Terms. Terms are net cash within 30 days. All jobs are figured on work being produced as a unit. For preliminary deliveries and set-ups requiring additional expense or if ordered to cease operations or if delayed beyond the binder's control by the customer, the binder reserves the right to bill the customer for all additional cost incurred.

APPENDIX B

Model Releases

GENERAL ADULT OR PROPERTY RELEASE

I hereby give _____, its agents, and/or assignees
(company or photographer)
permission to use the photographs taken of me (my property) in any manner it deems proper. I relinquish all rights, title, and interest I may have in the finished pictures, negatives, and copies. I waive the right of prior approval and hereby release _____ from any and all claims for
(company or photographer)
damages of any and all kinds based on the use of said material. I am of legal age and freely sign this release, which I have read and understand.

_____
(Signature)

_____
(Witness)

_____
(Date)

SPECIFIC ADULT OR PROPERTY RELEASE

I hereby give _____, its agents, and/or assign-
(company or photographer)
ees permission to use the photographs taken of me (my property) for
_____.
I relinquish all rights, title, and interest I may have in the finished pictures, negatives, and copies for this purpose. I waive the right of prior approval and hereby release _____ from any and all claims for
(company or photographer)
damages of any and all kinds based on this use of said material. I am of legal age and freely sign this release, which I have read and understand.

_____

(Signature)

_____

(Witness)

_____

(Date)

GENERAL MINOR RELEASE

I hereby give _____, its agents, and/or assignees
                      (company or photographer)
permission to use the photographs taken of the minor named below in any manner it deems proper. I relinquish all rights, title, and interest I may have in the finished pictures, negatives, and copies. I waive the right of prior approval and hereby release _____ from any and all
                                  (company or photographer)
claims of damages of any and all kinds based on the use of said material. I am of legal age, have a right to contract for the minor, and freely sign this release, which I have read and understand.

_____

(Signature of Parent or Legal Guardian)

_____

(Name of Minor)

_____

(Witness)

_____

(Date)

Selected Bibliography

The following books and booklets are among those consulted during preparation of this handbook. They are suggested as additional sources of information.

DESIGN REFERENCES

Roy Paul Nelson. *Publication Design,* 3rd ed. Dubuque, Iowa: Wm. C. Brown, 1983.
  This and other books by the same author are written from a designer's perspective and provide useful design details and examples for the print buyer.

Jan V. White. *Editing by Design,* 2nd ed. New York: Bowker, 1982.

Jan V. White. *Mastering Graphics.* New York: Bowker, 1983.
  Both books present overview information on publication design from a designer's perspective. Details are sparse, and do's and don't's numerous; however, extensive use of illustrations makes the basics that are presented useful.

DICTIONARIES

International Typographic Composition Association. *Glossary of Typographic Terms.* Washington, DC: ITCA, 1974.
  This booklet is updated from time to time.

Patricia Barnes Mintz. *Dictionary of Graphic Arts Terms.* New York: Van Nostrand Reinhold, 1981.
  This book has the appeal of being written specifically for the print buyer.

George A. Stevenson. *Graphic Arts Encyclopedia,* 2nd ed. New York: McGraw-Hill, 1979.
  This book leans toward the technical in its detailed definitions.

GENERAL REFERENCES

EDMUND C. ARNOLD. *Ink on Paper 2*. New York: Harper & Row, 1972.
   A detailed, semi-technical book covering production from typesetting through binding.

MARK BEACH. *Editing Your Newsletter*. Portland, Oregon: Coast to Coast, 1980.
   Written for the amateur, this book covers newsletter production from preparing copy through distribution.

CLIFFORD BURKE. *Printing It*. Berkeley: Wingbow Press, 1972.
   This very general book is written for amateurs needing to produce a printed piece for a cause or candidate.

JAMES CRAIG. *Production for the Graphic Designer*. New York: Watson-Guptill, 1974.
   Well illustrated, this book is written by a designer but in rather basic language presents rudiments of the production process.

EDMUND J. GROSS. *101 Ways To Save Money on All Your Printing*. North Hollywood, California: Halls of Ivy, 1971.
   Although not especially well organized or illustrated, this book does provide a summary of many ways to cut production costs.

GERRE L. JONES. *How To Prepare Professional Design Brochures*. New York: McGraw-Hill, 1976.
   An overview of print production geared to preparation of engineering and similar brochures.

HENRY C. LATIMER. *Production Planning and Repro Mechanicals for Offset Printing*. New York: McGraw-Hill, 1980.
   This book is a compilation of information in earlier books by the same author. Although poorly organized, it does offer the persistent reader potentially useful production details.

DEAN PHILLIP LEM. *Graphics Master*, 3rd ed. Los Angeles: Dean Lem Associates, 1984.
   This notebook provides technical information and illustrations on color tints, line screens, typefaces, and other production details.

NORMAN SANDERS. *Graphic Designer's Production Handbook*. New York: Hastings House, 1982.
   Prepared for the designer, this book is heavily illustrated and laid out for easy comprehension of production techniques, especially preparation of mechanical artwork.

LARAE H. WALES. *A Practical Guide to Newsletter Editing and Design,* 2nd ed. Ames: Iowa State University Press, 1978.
   Written for the amateur, this book presents basic information on preparation of newsletters.

LEGALITIES

CONSUMER AND CORPORATE AFFAIRS CANADA. *Questions and Answers About Copyright*. Ottawa: Minister of Supply and Services Canada, 1982.
   This summary of Canadian copyright law is presented in both English and French.

TAD CRAWFORD. *Visual Artist's Guide to the New Copyright Law*. New York: Graphic Artists Guild, 1978.
   The graphic designer, photographer, and other artist will find this booklet a useful reference for legal information.

SELECTED BIBLIOGRAPHY

MARION DAVIDSON AND MARTHA BLUE. *Making It Legal.* New York: McGraw-Hill, 1979.
This book is an overview of legal concerns of the writer, artist, and craftmaker.

STYLE GUIDES

INTERNATIONAL ASSOCIATION OF BUSINESS COMMUNICATORS. *Without Bias: A Guidebook for Nondiscriminatory Communication,* 2nd ed. New York: Wiley, 1982.
Written for the practicing communicator, this handbook offers guidelines on bias-free written, visual, and oral communication related to race, sex, age, and ability.

WILLIAM STRUNK, JR., AND E. B. WHITE. *The Elements of Style,* 3rd ed. New York: Macmillan, 1979.
A succinct little book, it is a frequently consulted guide to editorial style.

TEXTBOOKS

J. MICHAEL ADAMS AND DAVID D. FAUX. *Printing Technology: A Medium of Visual Communication,* 2nd ed. North Scituate, Massachusetts: Duxbury Press, 1982.
This textbook covers the production process in semi-technical language.

JOHN E. COGOLI. *Photo-Offset Fundamentals,* 4th ed. Bloomington: McKnight, 1980.
Although a textbook, it offers very useful information for the buyer.

ERVIN A. DENNIS AND JOHN J. JENKINS. *Comprehensive Graphic Arts.* Indianapolis: Howard W. Sams, 1974.
This textbook is both comprehensive and basic in presentation of information.

TRADE PUBLICATIONS

AMERICAN SOCIETY OF MAGAZINE PHOTOGRAPHERS. *Professional Business Practices in Photography.* New York: ASMP, 1981.
Written for the practicing photographer, this book is updated periodically as a comprehensive guide to how photographers work and prefer to be treated by clients.

EASTMAN KODAK. *Exploring the World of Graphic Communications.* Rochester: Kodak, 1976.
This eight-part series of booklets includes *Basic Printing Methods, Graphic Design, Creative Photography, Copy Preparation, Photoreproduction, Film Assembly and Offset Platemaking, Offset Presswork,* and *Finishing.* Some information tends to be more technical than a print buyer might need; but it's presented in easy-to-understand language amply illustrated.

EASTMAN KODAK. *Halftone Methods for the Graphic Arts,* 3rd ed. Rochester: Kodak, 1976.
A semi-technical booklet, this trade publication provides details on use of photography in print production.

EASTMAN KODAK. *Photography and Layout for Reproduction.* Rochester: Kodak, 1978.
This booklet focuses on composition and other qualities necessary for satisfactory reproduction of photographs on the press.

INTERNATIONAL PAPER. *Pocket Pal,* 13th ed. New York: International Paper, 1983.
    A basic handbook covering most areas of production, it is written for the advanced amateur and updated frequently.

S. D. WARREN. *Ins & Outs.* Boston: S. D. Warren, 1981.
    Although overdesigned, this booklet offers the amateur useful advice on the main aspects of producing a company publication.

Glossary

Most of the terms in this glossary are explained in detail in the text of this handbook. Some, however, are terms that you might occasionally hear a supplier use and need to understand.

ABSORPTION.   A method of drying ink that relies on the vehicle's penetrating into the paper, leaving the pigment on top; also called penetration.

ACCENTS.   Marks denoting differences in pronunciation in languages other than English; either typeset with effected letters or drawn in; *see also* Acute Accent, Grave Accent, or Circumflex Accent.

ACCORDION FOLD.   A parallel fold in which panels are folded in opposite directions similar to the folds in an accordion's bellows; also called fan, Z-, or concertina fold.

ACETATE.   A clear, heavy plastic sheet used on mechanical artwork to make an overlay, for example, of art or type to be printed in a second color.

ACUTE ACCENT.   An accent mark slanting upward to the right.

ADHESIVE BINDING.   Another term for perfect binding.

ADVERTISING PRINTER.   A print shop specializing in printing of direct-mail and other sales pieces for advertisers.

ADVERTISING TYPOGRAPHER.   A typographer specializing in setting a wide array of display faces for advertisers.

AGAINST THE GRAIN.   Running across the grain of paper instead of along it.

AGATE.   A unit of type measure equivalent to $5\frac{1}{2}$-point type; 14 agates equal 1 in. or approximately 2.5 cm.

AGATE COPY.   Copy set in very small type, regardless of its actual size.

AIRBRUSH.   A tool using compressed air to spray a fine coating of ink for retouching illustrations or photographs to eliminate or alter elements or for painting original art.

ALIGNING FIGURES. Another term for lining figures.

ALIGNMENT. The straightness of type or art in relation to other elements on mechanical artwork or a printed page.

ALL CAPS. A specification that the letters or words indicated should all be set in capital letters.

ALPHANUMERIC. A set of type characters that includes letters of the alphabet, numerals, and common pronunciation, mathematical, and abbreviation symbols; a contraction of alphabetic numeric.

ALTERNATIVE CHARACTERS. Type characters that are distortions or ligatures of regular characters; offered for special design purposes on the same font as or on a separate font from regular characters.

AMPERSAND (&). Symbol for "and."

ANNOUNCEMENT ENVELOPE. A square-flapped envelope used for formal announcements and invitations.

ANTIQUE FINISH. A method of finishing paper to give it the bulky, rough look of hand-made paper.

ANTIQUE TYPE. Another term for Old Style type.

APPENDIX. Information in the back of a lengthy publication or report that provides more detail on points made in the text.

ARABIC NUMERALS. The numeral characters 1, 2, 3, and so on, as opposed to Roman numerals I, II, III, and so on.

ART. All illustrations, photographs, and other elements in a printed piece, not including body type, headlines, subheads, cutlines, or legends.

ART DIRECTION. Guidance to a photographer or illustrator about how work is to be done in relation to design and objectives of a printed piece.

ARTIST'S KNIFE. An X-Acto or other brand of knife with a fine point and sharp edge used in the execution of art or mechanical artwork.

ARTISTS' AGENT (ARTISTS' REPRESENTATIVE). A person who markets the work of artists for a salary or commission.

ARTIST'S REPRESENTATIVE. Another term for artist's agent.

ARTWORK. Another term for mechanical artwork.

ASCENDER. That portion of a type character extending above the mean line (such as the upper portion of an "h," "b," or "d").

ASTERISK. A reference mark; also called star.

ASYMMETRICAL BALANCE. Another term for informal balance.

AUTHOR. The writer of copy.

AUTHOR'S ALTERATION (AA). Proofreader's mark indicating that a change in typeset copy is the fault of the author, not an error made by the typographer.

AUTOSCREEN. A film with a built-in 133-line (54-cm) screen for making halftones.

BACK-OF-THE-BOOK. Editorial material of lesser importance appearing near the end of a publication; also called endmatter, backmatter, or reference matter.

BACK-TO-BACK. Printing on both sides of a piece of paper; also called backing up.

BACKBONE. Another term for spine.

BACKGROUND. That part of a photograph or illustration that appears behind the center of interest.

BACKING UP. Another term for back-to-back.

BACKMATTER. Another term for back-of-the-book.

BACKSLANT. Type that leans to the left, the opposite direction from italic.

BAD BREAK. Incorrect hyphenation of a word at the end of a typeset line; also division of a sentence that creates a widow or confuses understanding.

BALANCE. A principle of design calling for a visual evenness on each side of a center point; may be formal, informal, or subjective; also called opposition.

BANKERS FLAP ENVELOPE. A heavy-duty envelope with a pointed flap used for bulky correspondence.

BANNER HEADLINE. A large headline running across the top of a page.

BAR CHART. A chart comparing numerical values by using vertical or horizontal bars.

BAR MARK. The horizontal line separating numerator from denominator in a fraction; also called fraction mark.

BARONIAL ENVELOPE. A pointed-flapped envelope used for formal announcements and invitations.

BARREL FOLD. Paper folded two or more times in the same direction; also called a wraparound or letter fold.

BASELINE. The horizontal line on which type rests; capital letters and punctuation marks stand on this line, as do all lowercase letters except those with descenders.

BASELINE-TO-BASELINE (B/B). A linespacing measurement that is the distance between the baseline of one line of type and the baseline of the line below it.

BASIS SIZE (BASIC SIZE). The size in which a particular grade of paper is normally and most conveniently produced.

BASIS WEIGHT (BASIC WEIGHT). The weight of a single sheet of paper based on the weight of one ream (500 sheets) of its basis size; also called substance.

BENDAY PROCESS (BEN DAY PROCESS). A series of line or dot patterns used to shade drawings or alter photographs, type, rules, or blocks of color on the negative; named for Benjamin Day; also called mechanical tinting.

BEVEL. To cut at an angle so that the edge is less than 90° to the surface.

BIBLE PAPER. A thin, sturdy, lightweight bond paper; also called India paper.

BIBLIOGRAPHY. That section of a major printed piece that lists books and/or articles on the same topic or by the same author; called selected when not exhaustive.

BID. A firm price; the amount that will be billed if specifications don't change.

BIMETAL PLATE. An offset-printing plate made of more than one metal and used when long life is required for a lengthy press run.

BINDERY. An independent business or a department in a print shop devoted to finishing and binding printed pieces.

BINDING. The process of putting printed sheets in order and securing them in place.

BINDING EDGE. The edge of a printed sheet that is to be bound, leaving all other edges free.

BINDING POST. Another term for screw-and-post binding.

BISYMMETRICAL BALANCE. Another term for formal balance.

BLACK TEXT TYPE. Another term for Old English type.

BLACKBOX PRINTING. Another term for collotype printing.

BLACKLETTER TYPE. Another term for Old English type.

BLACKPRINT. A contact print of a halftone negative on special paper used as a proof before film assembly; also called silverprint.

BLANK-PAGE DUMMY. Another term for printing dummy.

BLANKET. The cylinder on an offset press that imprints the image onto paper.

BLEACH-OUT HALFTONE. Another term for dropout halftone.

BLEED. An image that runs off one or more edges of a printed page; can be art, solid color, or a screen or tint.

BLIND DEBOSSING. Impressing an uninked image on the front of a sheet of paper so that the design appears recessed on the front; *see also* Debossing.

BLIND EMBOSSING. Impressing an uninked image on the back of a sheet of paper so that the design is raised on the front; *see also* Embossing.

BLOCK TYPE. Another term for Sans Serif Gothic type.

BLOCKING. Mounting a plate on wood, metal, or plastic to make it type-high; also another term for stamping.

BLOW UP. To enlarge a photograph, other art, or type to the desired size.

BLUELINE (BLUEPRINT). A proof made directly from a press negative onto photosensitive paper to show how the printed piece will look; similar to a brownline or Dylux.

BOARD. Another term for mechanical artwork.

BOARD PAPER. Very heavy paper for book covers and other uses requiring exceptional stiffness and durability.

BODY COPY. The editorial portion of printed material, excluding headlines and cutlines; also called text.

BODY HEIGHT. Another term for x-height.

BODY TYPE. Type usually no larger than 12 points commonly used for body copy.

BOLDFACE (bf) A typesetting specification that the letters or words indicated should be in a heavy weight of type.

BOND PAPER. A classification of papers characterized by their light weight and suitability as writing papers; also a specific paper with a hard finish and high rag content used for business correspondence and documents.

BOOK. A bound publication of at least 48 pages; sometimes used to refer to a magazine or other substantial printed piece; also another term for portfolio.

BOOK BINDERY. A bindery specializing in binding of books and other substantial documents; also called edition bindery.

BOOK PAPER. A paper classification encompassing weights and grades commonly used for the text of printed pieces; heavier than bond paper but lighter than cover paper; called text when textured.

BOOK/MAGAZINE TYPOGRAPHER. A shop specializing in setting utilitarian type for publishers.

BOOKLET. A printed piece with fewer than 48 pages bound together; also called pamphlet.

BOOKLET ENVELOPE. An open-sided envelope used for booklets and direct-mail pieces for which its unseamed back is appropriate as a printing surface.

BORDER. The rule (plain or fancy) that defines a full or partial box.

BOUNCE. Type purposely set so that all baselines do not align.

BOURGES. A brand name for a series of transparent sheets coated with removable gray or colored ink used to modify or add color to a photograph or other art.

BOWL. That portion of a type character that is partially or fully enclosed (such as the inside of "o" or "y"); also called counter.

BOX. A design technique that highlights certain information by placing rules on all four sides to make a box or by covering the area with a rectangular screen.

BRACKETS. Type characters used to enclose parenthetical material; commonly one step up from parentheses; also called square brackets.

BRISTOL PAPER. A classification of papers characterized by their exceptional weight, making them heavier than all others; also a heavy paper manufactured as index bristol or printing bristol.

BROADSIDE FORMAT. A page designed to be turned and read longways; the opposite of portrait format; also called landscape format.

BROCHURE. A small printed piece with few pages or panels, usually folded into a compact unit and not bound.

BRONZING. Creating a metallic effect by dusting wet ink with a metallic powder.

BROWNLINE (BROWNPRINT). A proof of press negatives made with brown paper; similar to a blueline; also called Vandyke.

BRUNING. A quality proof of cold type produced by a Bruning photocopier.

BRUSH-AND-INK DRAWING. A type of line drawing using a brush and waterproof ink.

BUCKLE FOLDER. An automatic folder that uses a stop to buckle the paper and send it between folding rollers.

BUG. A small symbol appearing on a printed piece usually designating that the job was printed by union labor.

BULK. Heft or thickness of a sheet of paper.

BULKING DUMMY. Another term for printing dummy.

BULLET. A dot (usually boldface) used to call special attention to the words that follow; frequently used in lieu of numbers to highlight items in a list; called a centered dot when it is used to separate paragraphs.

BULLETIN HEADLINE. A headline that doesn't make a sentence; also called label headline.

BURNING-IN (BURNING). Exposing an area of a photographic negative to extra light so that area is more intense in the resulting photographic print; also called printing-in.

BURNISH. To smooth and secure elements on mechanical artwork so that they are firmly in place.

BUSINESS CABINET. A complete set of printed business supplies: letterhead, envelopes, business cards, and mailing labels.

BUSINESS PRINTER. A print shop specializing in the printing of forms, contracts, and other needs of business and finance.

BYLINE (BY-LINE). A credit given to the author of written material appearing at the beginning or end of an article.

CABBAGES. Another term for printer's marks.

CALENDER. To press paper between rollers during manufacture to increase smoothness and gloss and reduce bulk.

CALIPER. The measurement of paper bulk or thickness expressed in thousandths of an inch using a caliper or fraction of a millimeter using a micrometer.

CALLIGRAPHY. Fancy hand-lettering.

CAMERA-READY. Material ready to be shot for conversion into press negatives or directly into a printing plate.

CAMERA-READY ART. Designs and illustrations ready for paste-up, either purchased for this purpose or taken from non-copyrighted materials.

CAP HEIGHT. The height of a capital letter from baseline to top.

CAPITAL (cap). Another term for uppercase.

CAPS AND LOWERCASE (clc). A specification that each word is to start with a capital letter.

CAPTION. Explanatory type appearing above a photograph or illustration; now commonly synonymous with cutline.

CARBON TISSUE. A photosensitive gelatin stripping base used to make a gravure plate.

CARBONLESS PAPER. Specially treated paper sensitive to pressure; no carbon is required between sheets for image transfer; also called NCR ("no carbon required").

CARDBOARD. A type of bristol paper.

CARET. A proofreader's mark indicating where material is to be inserted.

CASE BINDING. A binding method resulting in a hard-cover book not attached directly to the cover at the spine; also called edition binding.

CASE FRACTIONS. Another term for whole fractions.

CAST COATED PAPER. A paper finished using heat and a highly polished dryer to create a bright, smooth, mirrorlike surface.

CASTING OFF. Another term for copyfitting.

CATALOG ENVELOPE. A strong envelope with exposed flaps for mailing oversized documents; also called open-ended envelope.

CATALOG/DIRECTORY TYPOGRAPHER. A shop specializing in utilitarian type for substantial documents.

CATHODE-RAY TUBE (CRT). A display and projection tube used in digital conversion of type characters into a finished galley during electronic composition.

CEDILLA. A diacritical mark resembling a "5" attached to the bottom of a letter; also called French C.

CENTER OF INTEREST. That part of a photograph or illustration to which the viewer is attracted.

CENTER QUADDING. Another term for centered.

CENTER SPREAD. The two facing pages in the center of a publication.

CENTERED. Text and/or art stacked so the midpoints align to create a symmetrical effect; also called center quadding or quad centered.

CENTERED DOT. A bullet separating paragraphs or, in mathematical equations, a substitute for a multiplication sign.

CHAIN DOT. Another term for elliptical dot.

CHARACTER. Any element that is typeset (such as letters, numerals, symbols, or punctuation marks); also called copy unit.

CHARACTER CASTING. The casting of hot-metal type by individual characters as opposed to by the line (linecasting).

CHARACTER COUNT. How many average type characters of a particular size, style, face, and weight will fit into a given horizontal space; indicated as characters per pica.

CHARACTER UNIT. A measure of a type character used to determine how much horizontal space it will occupy in relation to other characters in a headline; also called unit count.

CHARCOAL DRAWING. A type of line or continuous-tone drawing made with a charcoal pencil or block.

CHART. A visual representation of values and quantities.

CHASE. The form in which hot type is arranged to make up a page.

CHEAP-AND-DIRTY PRINTER. A slang description of a printer who produces low-quality work.

CHECKER. Another term for galley proof.

CHEMICAL PRINTING. Another term for offset printing.

CHEQUE PAPER. Another term for safety paper.

CHINESE WHITE (CHINA WHITE). An opaque paint sometimes used to outline photographs for ease in eliminating backgrounds when press negatives are made.

CHIPBOARD. A type of bristol paper.

CHROMA. The purity of a hue.

CHROMALIN. *See* Cromalin.

CHROME COATED PAPER. Another term for cast coated paper.

CIRCULAR. Advertising material in the form of handbills or newspaper inserts.

CIRCUMFLEX ACCENT. An accent mark consisting of an inverted V shape over a letter; also called hat.

CLASP ENVELOPE. A strong envelope with a metal clasp that fits through a hole on the flap for mailing of large documents; also called Columbian clasp.

CLEAN PROOF. A proof free of errors.

CLOSE UP (cu). A proofreader's mark designating that letters or words are to be brought together.

COATED PAPER. Paper that has been chemically coated to produce a specific finish (such as high-gloss enamel); it may be coated on one side (C1S) or coated on two sides (C2S).

COCKED TYPE. Another term for pied type.

COCKLE FINISH. A bond paper with a wrinkled texture.

COIN ENVELOPE. A small envelope used for paper currency and coins.

COLD TYPE. Type set by photocomposition as opposed to hot type set using hot metal.

COLLAGE. Art made by pasting several elements into a whole.

COLLATE. To assemble pages in order prior to binding them.

COLLOTYPE PRINTING. A printing process using a photosensitive gelatin plate and no halftone screen or a very fine one; also called photogelatin or blackbox printing.

COLOPHON. A listing in a book or other extensive printed piece to inform the reader about production specifications; also called a printer's trademark.

COLOR. Ink used to make a printed image; may be any hue and includes black.

COLOR BALANCING. To change the color balance of art by emphasizing or diminishing components until the desired color perception is achieved.

COLOR CORRECTION. Any retouching process that improves color.

COLOR OPAQUE PROOF. Another term for transfer proof.

COLOR SEPARATION. Breaking down color art into its component secondary colors so that individual press negatives can be made for reconstruction of the image on the press.

COLOR TRANSPARENT PROOF. Another term for overlay proof.

COLOR-KEY. A brand name for an overlay proofing system for color separations.

COLUMBIAN CLASP. A brand name for a clasp envelope.

COLUMN. A section of a printed page extending from top to bottom and having a defined width; also a regular feature in a publication.

COLUMN CHART. Another term for pie chart or bar chart when the area divided is vertical.

COLUMN HEADING. A headline and/or art appearing at the beginning of a regular column in a publication.

COLUMN INCH. A measure of the depth of a column of type, one column wide by 1 in. deep.

COLUMN RULE. A vertical rule between columns of type.

COMB BINDING. A looseleaf method of binding in which a row of slots is punched along the side of pages to be bound and a series of flat, curled, plastic teeth on a solid backbone is inserted to hold them together.

COMBINATION PLATE (COMBINE, COMBO). One printing plate with both line and halftone images.

COMMERCIAL ENVELOPE. A pointed-flapped envelope in standard sizes used for business correspondence; also called regular envelope.

COMP (COMPREHENSIVE LAYOUT, COMPREHENSIVE DUMMY). A detailed, full-sized mock-up showing how a printed piece will look, including copy, art, color, finishing, and binding; used to see the full design before type is set and art secured and possibly as a guide for paste-up.

COMPLEMENTARY STRIPPING. Making two or more flats for different treatments to be exposed to one press plate.

COMPOSING STICK. A metal tray used to hold individual metal type characters as they are hand-assembled into lines of display or text type.

COMPOSITION. The makeup of a photograph or illustration; also another term for typesetting.

COMPRESSED TYPE. Another term for condensed type.

COMPUTER GRAPHICS. Art designed on a computer terminal; also called digitalized art.

COMPUTER TYPESETTING. Another term for electronic composition.

CONCERTINA FOLD. Another term for accordion fold.

CONDENSED TYPE. Type that appears squeezed together; letters are designed to be more vertical than horizontal; also called compressed type.

CONTACT PRINTING. Making a print from a negative by exposing it in direct contact with photographic paper.

CONTACT SHEET (CONTACT). A photographic proof of black-and-white film made from negatives by contact printing to show how images would look if enlarged; also called a proofsheet.

CONTINUED HEADLINE. An abbreviated headline over text continued from another page; also called jump headline.

CONTINUED LINE. A line of type at the end of a break in text indicating on which page it continues.

CONTINUOUS TONE. A photograph or other art having different values from black to white reproduced using a halftone screen and high-contrast film; also called tone copy.

CONTINUOUS-TONE NEGATIVE. A photographic negative used to make a photographic print, as opposed to a halftone negative used to reproduce a photograph by most printing methods.

CONTRAST. The difference between darks and lights in a photograph; "high contrast" means that one or more elements stand out from the rest; "low contrast" means little distinction; also another term for emphasis.

COPPERPLATE. Line engraving on a copper plate.

COPPERPLATE GOTHIC TYPE. A variety of Square Serif Gothic type characterized by its all-capital design.

COPY. The manuscript from which type is set; also everything to be reproduced by the camera when press negatives or plates are made.

COPY CAMERA. Another term for process camera.

COPY CASTING. Another term for copyfitting.

COPY EDITING. Refining of copy prior to typesetting so that facts, grammar, spelling, and punctuation are accurate; writing style is consistent; and copy is complete, appropriate, and understandable.

COPY UNIT. Another term for character.

COPYFITTING. Estimating before typesetting how much space a manuscript will occupy when set in a given typeface, style, weight, and size; also called casting off.

CORNER MARKS. Short solid lines in the margins of mechanical artwork indicating the outside edges of the finished piece after trimming.

CORNERING. Punching away square corners to round them.

COUNTER. Another term for bowl.

COVER. Usually the front of a printed piece; also used to designate all sides of the outside sheet of a publication: Cover I (front), Cover II (inside front), Cover III (inside back), and Cover IV (back).

COVER PAPER. A classification of paper characterized by its weight and durability; heavier than book or text papers but lighter than bristol.

COVER SHEET. Another term for tissue-paper overlay.

CREDIT LINE. The typeset name of the illustrator or photographer set adjacent to the art or at the end of its cutline; also used to recognize the source of borrowed text, props, or costumes used in preparation of art.

CROMALIN. A brand name for a transfer proofing system for color separations.

CROP. To indicate by marks in the margin or on a tissue-paper overlay which portion of a photograph or other art is to appear in the printed piece.

CROP MARKS. Short lines or arrows in the margin indicating how a photograph or other art is to be cropped.

CROPPING ANGLES. L-shaped pieces of heavy paper used to frame the image area of a piece of art to check composition before crop marks are made.

CROSS-STROKE. A horizontal stroke on a type character (such as on "t" or "f").

CURSIVE TYPE. Another term for Script type.

CURVE CHART (CURVE GRAPH). A line chart or graph using a curved line to connect points.

CUT. A common term for any illustration or photograph reproduced in a printed piece; originally referred to cutting an image into a surface in preparation for inking and printing.

CUTLINE. Explanatory type appearing below or beside a photograph or illustration; now commonly also used to refer to type appearing above (caption).

CUTOFF RULE. A rule used to separate items (such as classified ads, text from advertising, or an article from its sidebar).

CUTOUT HALFTONE. Another term for outline halftone.

CUTTING. The severing of paper to prepare it for printing, finishing, and/or binding.

CYAN. The blue-green used as one color in process printing to produce full-color art; also called process blue.

DAGGER. A reference mark.

DANDY ROLLER. A roller that pushes paper fibers into a texture or watermark during manufacture.

DASH. A straight, horizontal line between the main sentence and an intervening phrase; longer than a hyphen and designated in typed copy to be set as --.

DEADLINE. When some step in production must be finished.

DEBOSSING. Impressing an image on the front of a sheet of paper so that the design appears recessed on the front; called "blind debossing" if the design is not inked.

DECK. To stack a headline and one or more subheads down a column; once very popular with city newspapers to build up interest in the article that followed.

DECKLE EDGE. The ragged, feathered edge of a quality sheet of book paper; commonly used for invitations with the deckle edge on the flap of the invitation and its envelope.

DECORATIVE TYPE. Another term for Novelty type.

DEEP-ETCH PLATE. An offset printing plate made from positives, resulting in a slight etching of the metal so that the plate can carry more ink and have a longer life.

DEEP-ETCHED HALFTONE. Another term for highlight halftone.

DEGENERATION. Converting a halftone to line art by eliminating grays; also called two-tone posterization, line conversion, linear definition, or tone line conversion.

DELETE. To take out.

DENSITY. The percentage of area in a halftone or continuous-tone illustration occupied by dots; a measure of the opacity or blackness of an area in a negative or transparency.

DEPTH. The vertical measurement of art or a column of type.

DEPTH OF FIELD. How much of the background and foreground in a photograph is in focus.

DESCENDER. That portion of a type character that extends below the baseline (such as "g" or "p").

DESIGN GROUP (DESIGN STUDIO). Two or more designers working together to offer a range of services and styles; also frequently includes mechanical artists or other related suppliers.

DESIGNER'S DELIGHT. A slang expression for a design with complex specifications that would be difficult or impossible to produce.

DIACRITICAL MARKS. Indicators for special pronunciation of a letter.

DIAERESIS. Another term for umlaut.

DIAZO. A quality proof of cold type produced by a Diazo photocopier.

DIE-CUT. A cutout used for design impact (such as a hole cut into the cover of a booklet to reveal a portion of the photograph inside); also a perforated line or box.

DIE-CUT LINE. A solid nonrepro-blue line on mechanical artwork to indicate where die-cutting is to occur.

DIE-STAMPING. Another term for stamping.

DIFFUSION TRANSFER. Another term for photomechanical transfer.

DIFFUSION-TRANSFER PLATE. An offset plate made with special negative paper exposed in contact with a diffusion-transfer paper plate.

DIGITAL-PRESS PRINTING. Another term for ink-jet printing.

DIGITALIZED ART. Another term for computer graphics.

DILITHO. Another term for direct lithography.

DINGBATS. Another term for printer's marks.

DIPHTHONG. Two vowels joined together and set as one character; *see also* Ligature.

DIRECT COLOR SEPARATION. Exposing an image directly to a filter and halftone screen in one step.

DIRECT LITHOGRAPHY. A waterless printing process that uses a planographic plate and letterpress ink to reproduce an image directly onto paper.

DIRECT SCREENING. Using photo-masking in direct contact with film as it is exposed during color separation.

DIRECT-IMAGE PLATE (DIRECT-IMAGE MASTER). A chemically treated plate on which the image is written, drawn, or proofed that becomes ink receptive in image areas.

DIRECT-IMPRESSION TYPESETTING. Another term for impact typesetting.

DIRECT-TRANSFER PROCESS. A method of making a gravure plate that results in ink cells of variable size but the same depth.

DISPLAY AD. An advertisement that uses large type, an illustration, and/or other art to call special attention.

DISPLAY MATTER. Any element on a page used to call attention.

DISPLAY TYPE. Type usually larger than 12 points used to call attention.

DISTRIBUTION. Who will receive a printed piece; also the evenness of ink coverage throughout a printed piece or the return of hot-type matrices or characters to storage.

DITTO. A printing method in which the image is typed or drawn on special carbon-backed paper and impressed on a rotary machine onto special paper until the carbon is depleted; also called spirit duplicating or hectography.

DOCTOR BLADE. The squeegee that scrapes excessive ink from a gravure plate prior to impression.

DODGING. Using hands or a tool to block light during exposure of a photographic negative so that the affected area is eliminated or subdued in the resulting photographic print.

DOMINANCE. Another term for emphasis.

DOT GAIN. The enlargement of halftone dots on the press, especially on a long press run, resulting in a loss of detail in the image.

DOT PATTERN. Distribution of the dots making up a halftone negative or printed image.

DOUBLE BUMP. Another term for duocolor.

DOUBLE BURN. Separating comprehensive artwork at the engraving stage for the making of at least two press plates, instead of preseparating on an overlay.

DOUBLE DAGGER. A reference mark.

DOUBLE NUMERATION. Labeling tables, charts, and so on, with two numbers, the first being a chapter or section number (such as "Figure 3.1").

DOUBLE TRUCK. A two-page spread.

DOUBLE-DOT HALFTONE. A halftone-conversion technique using two film positives with the screen angled between exposures to result in a composite having richer blacks, whites, and grays.

DOUBLE-THICK PAPER. Another term for duplex paper.

DOUBLE-WIRE BAND BINDING. A looseleaf method of binding in which a row of slots is punched along the edge of pages to be bound and a band of double-wire teeth on a metal base is inserted to hold them together.

DOWN STYLE. Capitalizing no words or only the first word and proper nouns in a headline or title; the opposite of up style.

DRIER. Another term for drying agent.

DRILLING. Making holes in a stack of paper during finishing by using a sharp, hollow drill of the desired size; *see also* Punching.

DRIOGRAPHY. A waterless printing process that uses a silicone-coated, relief metal plate to reproduce an image directly onto paper.

DRIVE OUT. To add more space between words so that a word or syllable moves to the next line; the opposite of run back; also called run down or run over.

DROP INITIAL. An initial cap that extends below the baseline into the text.

DROPOUT. Another term for reverse.

DROPOUT HALFTONE. A halftone consisting entirely of black and white achieved by line photography to drop gray tones; also called a bleachout halftone or facsimile; *see also* Highlight Halftone.

DROPPED HEAD. An article or chapter headline appearing far down on the page; also called sunken head.

DRY OFFSET PRINTING. Another term for letterset printing.

DRYING AGENT. An ink ingredient that promotes fast drying to prevent smudging; also called a drier.

DRYING POWDER. A fine powder sprayed between printed sheets to prevent off-setting of ink.

DULL COATED PAPER. A paper finished to be coated but not shiny.

DULTGEN PROCESS. A method for making a gravure plate that results in ink cells of variable size and depth.

DUMMY. Another term for mechanical dummy; also the process of constructing a mechanical dummy.

DUMMY SHEET. A full-sized sheet with placement guidelines used to construct a mechanical dummy.

DUOCOLOR. The printing of a color twice so that coverage is solid and intense; also called two-impression printing or double bump.

DUOGRAPH. A fake duotone achieved by printing a solid or screened second color over a halftone with no control over shadows and highlights.

DUOTONE. A halftone printed in a dark color (usually black) and tinted with a second, lighter color using two negatives so that shadows and highlights are retained; also called a duplex halftone.

DUPE AND ASSEMBLY. The grouping of color art after correction and scaling for full-page separation into press negatives; also called pre-assembly or pre-positioning.

DUPLEX HALFTONE. Another term for duotone.

DUPLEX PAPER. Paper with two plies laminated together to produce a heavy sheet with a different color on each side (color-on-white or color-on-color); also called double-thick paper.

DUPLICATE (DUP, DUPE). An exact copy of a photograph, galley, page proof, press plate, or other element of print production.

DUPLICATING PRESS (DUPLICATOR). Generally, a press no larger than 11" × 17" (279 × 432 mm) capable of producing medium- to low-quality, one-color work.

DYLUX. One type of paper used to make a blueline proof of press negatives.

EAR. A block of type on either side of a publication's nameplate; commonly used by newspapers for the date and name of the edition.

EDITING. Another term for copy editing; also the selection of art for inclusion in a printed piece.

EDITING AND CORRECTING TERMINAL. A separate display terminal used during phototypesetting for the sole purpose of editing and correcting galleys.

EDITION. One version of a printed piece; an altered or updated edition is called "second edition" or given a specific designation (such as "North American edition" or "special edition").

EDITION BINDERY. Another term for book bindery.

EDITION BINDING. Another term for case binding.

EDITOR. The person who corrects and rewrites copy for publication or who edits art.

EDITORIAL. All written material for a printed piece as opposed to art or advertising; also an article usually written by the editor stating an opinion.

EGGSHELL FINISH. A paper finish that is only slightly textured.

EGYPTIAN TYPE. Another term for Square Serif Gothic type.

ELECTRONIC COMPOSITION (ELECTRONIC TYPESETTING). A composition method that uses a computer for input, editing, and output of finished galleys; also called computer typesetting.

ELECTRONIC PRINTING. A plateless printing process that uses a laser beam and toner to reproduce a computer-generated image on paper.

ELECTROSTATIC PLATE. An offset plate made using xerography.

ELECTROTYPE. A mold of a metal plate formed under pressure from which a duplicate plate is made by using electrolysis to coat it with metal.

ELEMENT. Anything to be printed (such as text, photographs, rules, boxes, or spot color); also the type font used on a typewriter.

ELITE TYPE. Type that is 12 characters to the inch (2.54 cm) on a typewriter.

ELLIPSIS. Three periods indicating that some words in a quotation have been omitted.

ELLIPTICAL DOT. A halftone screen in which dots are elongated rather than round to soften tonal changes; also called chain dot.

EM. The width of the capital M in a given typeface, style, size, and weight; used to indicate indentation (such as "indent 1 em").

EMBOSSED COATED PAPER. Another term for gloss coated paper.

EMBOSSING. Impressing an image on the back of a sheet of paper so that the design appears raised on the front; called "blind embossing" if the design is not inked; also called relief.

EMPHASIS. A design principle calling for treatment of the most important element on a page so that it will stand out by size, color, position, or art; also called dominance or contrast.

EMULSION. A thin layer of gelatin containing silver salts used to coat one side of film to make it receptive to an image when exposed to light.

EN. The width of the capital N in a given typeface, style, size, and weight; used to indicate indentation (such as "indent 1 en").

ENAMEL PAPER. Paper coated to give a polished surface; can be glossy (shiny) or matte (dull).

ENDMATTER. Another term for back-of-the-book.

END-OF-LINE DECISIONS. Wordspacing and hyphenation decisions made during typesetting by the operator or automatically by the typesetting unit.

ENGLISH FINISH (EF).   A paper finish that is very smooth, although not as smooth as supercalendered; also called plater or plate finish.

ENGLISH GROTESQUE TYPE.   Another term for Sans Serif Gothic type.

ENGRAVER'S PROOF.   A copy of line or halftone engravings before film assembly.

ENGRAVING.   A printing plate with areas to be inked etched into the surface; used for high-quality invitations and other specialty printing; also the process of converting mechanical artwork into press negatives.

ENLARGEMENT.   An image made larger than the original.

EPIGRAPH.   A short quotation opening a book or chapter to suggest what is to follow.

ERASABLE BOND.   A bond paper specially treated to allow easy and thorough erasure.

ERRATUM SLIP.   A correction inserted into a printed piece; multiple corrections are printed on an errata slip.

E-SCALE.   A clear vinyl scale showing capital E's in a range of sizes to calculate type size.

ESTIMATE.   A proposed cost based on certain specifications.

ETCHING SCREEN.   A screen applied over art during engraving to give the appearance of the etching process used to create fine art.

ETCHTONE.   A special effect created by an etched screen, resulting in a printed photograph with the appearance of a steel engraving.

EVAPORATION.   A method of drying ink that relies on rapid evaporation of the solvent-based vehicle; also called volatilization.

EXCLUSIVE RIGHT.   The right to use art exclusively for a stated period of time.

EXPANDED TYPE.   Another term for extended type.

EXPANSION ENVELOPE.   A three-dimensional envelope for mailing of thick documents.

EXTENDED TYPE.   Type that appears spread apart; letters are designed to be more horizontal than vertical.

FACE.   Abbreviation for typeface.

FACSIMILE.   Another term for dropout halftone.

FACSIMILE PLATE.   An offset plate made by laser scanning and exposure.

FADE-OUT.   Another term for ghosting.

FAKE COLOR.   A process in which an engraver makes duplicate halftone negatives of an image and alters the dot pattern on each to replicate full-color separations so colors will be produced on the press, although they won't be as exact as process-color reproduction.

FAKE DUOTONE.   A halftone printed over a tint or block of solid color or on colored paper.

FAMILY. The entire range of styles available in one typeface, including different forms, proportions, and weights.

FAN FOLD. Another term for accordion fold.

FARM OUT. A slang expression for subcontracting a portion of a print job to another supplier.

FEATURE. An article in a publication given special treatment in art and/or length; a regular feature appears frequently or in every issue.

FEED EDGE. Another term for guide edge.

FELT SIDE. The side of a sheet of paper that was on top of the wire during manufacture.

FIGURE. A chart, table, or other illustration that appears within the text in a printed piece, as opposed to a plate that appears separately.

FILLER. A short paragraph or small piece of art used in a publication to fill up space.

FILM ASSEMBLY. Another term for stripping.

FINAL PROOF. The last proof with all changes made.

FINISH. Different looks and feels of paper created during the production process (such as laid or enamel).

FINISHED SIZE. The final size of a printed piece after folding and trimming.

FINISHING. The process of collating, folding, embossing, and executing other non-binding techniques necessary to complete a printed piece or to prepare it for binding.

FIRST RIGHTS. The right to use copy, an illustration, or a photograph for the first time.

FLAG. The title of a publication; often inaccurately called the masthead; also called logo.

FLAP. A page or panel smaller than a full page folded inside a printed piece and used, for example, to create a pocket or to carry a cover photograph over to the inside.

FLAT. Another term for press negative; also an abbreviation for goldenrod flat.

FLAT ASSEMBLY. Another term for stripping.

FLAT BOND. A bond paper made entirely of wood fibers.

FLAT-TINT HALFTONE. Another term for fake duotone.

FLATBED CYLINDER PRESS. A type of letterpress that uses a plate or actual metal type on a flat surface to impress an image onto paper attached to a cylinder.

FLEURONS. Another term for printer's marks.

FLEXOGRAPHY (FLEXO). A printing process using a flexible plate mounted on a rotary letterpress.

FLIMSY PAPER. Another term for onionskin paper.

GLOSSARY

FLOP. To turn over the negative of a photograph or other art to create a mirror image when printed.

FLOW CHART. A diagram depicting a sequence or relationship of one part to the whole series of events.

FLUORESCENT INK. A specialty ink that has bright, intense color.

FLUORESCING PRINT. A photograph taken under ultraviolet light that has exceptional highlights and weak midtones.

FLUSH. To line up so that all elements are even vertically with no indentation; for example, "flush left" means that all lines of type and all art line up on the left; also called quad left or quad right.

FLUSH COVER. A cover that is even with the pages it encloses.

FLY FOLD. Another term for four-panel fold.

FLYER. A single sheet printed on only one side and used to announce something; smaller and lighter than a poster.

FOCUS. How well defined shapes are in a photograph.

FOLD MARKS. Short broken lines outside the image area of mechanical artwork indicating where the finished piece is to be folded.

FOLDER. A single sheet printed on one or both sides and folded at least once; not as extensive or refined as a brochure.

FOLDING DUMMY. Another term for imposition dummy.

FOLDOUT. An oversized sheet bound into a publication and folded to fit within it.

FOLIO (FOLIO LINE). Page number, name, and/or date printed on each spread of a publication; also a sheet of paper folded once.

FONT. A complete set of type characters in a particular face, style, size, and weight; also a typewriter element.

FOOTNOTE SYMBOLS. Another term for reference marks.

FOREGROUND. That part of an illustration or photograph that appears in front of the center of interest.

FORMAL BALANCE. Designing a page or spread so that the left side looks exactly like the right; also called symmetrical or bisymmetrical balance.

FORMAT. Size, shape, and appearance of a printed piece.

FOUNDRY PROOF. A copy of hot type made up in page form.

FOUNDRY TYPE. Individual metal type characters used to hand-set lines of type.

FOUNTAIN. A reservoir on a printing press that holds ink or dampening solution.

FOUR-COLOR PROCESS PRINTING (FOUR-COLOR PRINTING). Use of cyan, magenta, yellow, and black inks to reproduce a full-color image; also called full-color or process printing.

FOUR-PANEL FOLD. A fold in which a sheet is simply creased down the middle creating four panels; also called fly fold.

FRACTION MARK. Another term for bar mark.

FRENCH C. Another term for cedilla.

FRENCH FOLD. A sheet printed on only one side and folded in half, and then folded in half again so that only the printing shows.

FRICTION BINDER. A looseleaf method of binding in which pages to be bound are inserted in a spring-back binder or plastic strip that holds them together.

FRONTMATTER. Information preceding the text in a large publication (such as a table of contents and a preface); also called preliminaries.

FUGITIVE COLOR. Stray spots of ink; also ink that tends to lose its color and fade when exposed to prolonged light; *see also* Hickey.

FULL-COLOR PRINTING. Another term for four-color process printing.

FULL FRAME. The entire available image area of an illustration or photograph.

GALLEY. An abbreviation for galley proof; also the tray in which hot type is assembled.

GALLEY PROOF. A copy of type made for the purpose of checking accuracy; also called a checker or reader.

GANG UP. To print more than one piece on a sheet to be cut apart later (such as several business cards on one sheet).

GANGING. Grouping photographs or other art for the making of press negatives rather than shooting each individually.

GATE FOLD. A brochure fold in which the two outside panels meet at the center creating the effect of a gate swinging open in the middle.

GATHER. To assemble signatures in order prior to binding them.

GHOST. Another term for phantom.

GHOSTING. A shadow effect created by an offset printing press when ink from a heavily inked area remains on the blanket to be carried to an adjacent area not intended to be printed; also called streaking or fadeout.

GLASSINE PROOF. A galley proof on semi-transparent paper used to check the layout of type; also called a tissue proof.

GLASSINE WINDOW. An envelope window covered with a clear piece of cellophane so that the contents are visible but protected.

GLOSS COATED PAPER. A paper finished to have a glossy surface, although not as glossy as cast coated paper; also called embossed coated paper.

GLOSSARY. An alphabetical list of words and their definitions.

GLOSSY FINISH. A shiny enamel surface on printing or photographic paper.

GLOSSY PRINT. A photograph printed on glossy paper or dried face down against a metal drum to make it glossy.

GOLDEN MEAN (GOLDEN DIVISION, GOLDEN SECTION, GOLDEN PROPORTION). The three-to-five ratio commonly used in graphic design.

GOLDENROD. Another term for masking sheet.

GOLDENROD FLAT. The press negative mounted to its masking sheet ready for platemaking; also called a flat.

GRADATION. The range of tones from black to white or their equivalents in color.

GRADES. How printing paper is classified; also simply called types of paper.

GRAIN (PAPER). The predominant direction of fibers that make up a sheet of paper.

GRAIN (PHOTOGRAPH). The visual texture of a photographic negative, transparency, or print created by silver particles in the film.

GRAMMAGE. The metric equivalent of basis weight or substance expressed as grams per square meter.

GRAPH. A visual representation of changes in one variable in relation to others.

GRAPHIC. Another term for art.

GRAPHIC ARTIST. An artist specializing in design and execution for print.

GRAPHIC-ARTS CAMERA. Another term for process camera.

GRAVE ACCENT. An accent mark slanting upward to the left.

GRAVURE PRINTING. A quality, rotary, intaglio printing process in which the image to be printed is etched into the plate which doubles as the impression cylinder; gravure is sheet-fed while rotogravure is a web process using roll paper.

GREEKING. Simulated composition used in comprehensive layouts to show actual placement and size of type; this nonsense text is usually in Latin, despite the name.

GREEN COPY. An early sheet off the press used to check ink distribution, color, and registration.

GRIPPER MARGIN (GRIPPER EDGE, GRIPPER BITE, GRIPPER SIDE). A margin of up to ½ in. (1.27 cm) on the leading edge of a printed piece left blank for the press to grip each sheet to feed it through.

GUIDE EDGE. The side of the sheet used during printing as a guide to exact placement of the image; also called feed edge.

GUM. To coat with adhesive.

GUMMED PAPER. Another term for label paper.

GUNNING FOG INDEX. A method for estimating the education level to which copy is written.

GUTTER. The blank space between two facing printed pages; also commonly used to refer to the space between any two columns of type or art.

H-HEIGHT. The height of a type character, including its ascender.

HAIRLINE. A very thin rule; also the thin portion of a type character.

HAIRLINE REGISTER. A method of closely aligning transparent colors to create the illusion of full color in process printing.

HALF-COLUMN CUT. A photograph or other art appearing one-half the width of a column of type; most commonly a photograph of the author or of a person featured in an article.

HALFTONE. A screened negative of a continuous-tone photograph or other art retaining black, white, and shades of gray.

HALFTONE NEGATIVE. A copy on film of continuous-tone original art produced by halftone photography.

HALFTONE PHOTOGRAPHY. The process of converting a continuous-tone image into a pattern of dots spaced equally apart but of different sizes so that variations in dot density establish a range from highlights to shadows when the image is reproduced.

HANG FOR TEN. A typesetting specification calling for figures in a column to be aligned vertically by ones, tens, and hundreds.

HANGING FIGURES. Numerals with ascenders and descenders; *see also* Lining Figures.

HANGING INDENTATION. Setting the first line of copy flush left and indenting the lines that follow.

HARD COPY. Typewritten or handwritten copy on a sheet of paper, as opposed to copy on a word-processing disk or tape.

HARMONY. Another term for unity.

HAT. Another term for circumflex accent.

HEAD-AND-SHOULDERS SHOT. Another term for mug shot.

HEAD-TO-FOOT. Having the top of the image on the front of a sheet corresponding with the bottom of the image on the back.

HEAD-TO-HEAD. Having tops of images corresponding on both sides of a sheet.

HEADING. Display type at the top of a column of figures; also another term for headline.

HEADLINE. Display type identifying the subject of accompanying text and usually appearing above or beside it.

HECTOGRAPHY. Another term for ditto (based on its ability to print approximately 100 suitable copies).

HICKEY. Spots on a finished printed piece caused by dust on the press negative, lint on the offset blanket, or stray ink; *see also* Fugitive Color.

HIDDEN COMB BINDING. Another term for side comb binding.

HIGH-KEY PHOTOGRAPH. A photograph in which details are recorded in mostly highlight tones.

HIGHLIGHT HALFTONE. A halftone in which the dots in a certain area have been removed for emphasis; called a dropout halftone when removal is in gray areas; also called a deep-etched halftone.

HOLDING LINE. Another term for keyline.

GLOSSARY

HOLDOVER. Type left over from one edition of a publication for possible inclusion in the next; also called overset or overmatter.

HORIZON. That part of an illustration or photograph that establishes a distinct horizontal line in the background.

HOT TYPE. Type set by a hot-metal process, as opposed to cold type set by a photographic process.

HOT-STAMPING. Another term for stamping.

HUE. The perceived color of an object.

ILLUSTRATION. Any line or continuous-tone drawing, painting, graph, chart, table, or type character used as art.

IMAGE AREA. That portion of a page available for type or art according to the format; also that portion of a mounted transparency that will appear when projected.

IMAGE ASSEMBLY. Another term for stripping.

IMAGIC PRINTING (IMITATION EMBOSSING). Other terms for thermography.

IMPACT TYPESETTING. A method of cold-type composition in which ink is transferred directly onto paper to produce a camera-ready galley, as with an office typewriter; also called strike-on or direct-impression typesetting.

IMPOSITION. The arrangement of pages of a publication for printing in one impression so that, when finished and bound, they will be in the proper sequence.

IMPOSITION DUMMY. A mock-up of a piece to be printed showing the arrangement of pages so that, when the piece is folded and/or assembled, the pages will be in the proper sequence.

IMPRESSION. The printing of a piece; also a way of expressing the number of press plates needed to complete a piece.

IMPRINTING. The printing of additional information on a finished piece (such as a local name and address on an already-printed brochure).

IN-LINE. Performing more than one production operation automatically at a time (such as folding and trimming equipment attached to a printing press so that a project can be completed in one pass).

INDENTATION. Setting the first line of a paragraph or items in a list in from the margin; also any portion of a column less than the full measure in width.

INDEX. An alphabetical listing in the back of a publication of key words and phrases and their corresponding page numbers.

INDEX BRISTOL. A moderately heavy paper with a hard finish commonly used for notebook dividers and file folders.

INDIA PAPER. Another term for bible paper.

INDIRECT COLOR SEPARATION. Exposing an image to create an intermediate separation that can be retouched before exposure to a halftone screen.

INDIRECT LETTERPRESS PRINTING (INDIRECT RELIEF PRINTING). Other terms for letterset printing.

INFERIOR TYPE. Figures or letters smaller than body type and placed below the baseline (such as the "2" in H_2O); also called subscript.

INFORMAL BALANCE. Designing elements of similar but not identical weight on each side of a centerline so that the page or spread appears balanced; also called occult balance.

INITIAL CAP. Setting the first letter in the first line of text in an oversized capital; called a rising, raised, standing, or stick-up initial when it extends above the cap height, a sunken initial when it extends below the baseline, and an inset or set-in initial when it is set into the first few lines.

INK HOLDOUT. The extent to which a given paper prevents ink absorption.

INK STARVING. Depriving a photograph or illustration of sufficient ink because the one printed directly before it on the impression cylinder used most of the ink.

INK WASH. A type of continuous-tone drawing made by using water-soluble ink to create shades of gray.

INK-JET PRINTING. A plateless printing process that uses fine jets of ink to reproduce a computer-generated image on paper; also called non-impact or digital-press printing.

IN-LINE TYPE. A type style designed with additional strokes inside a type-character's image area to give the illusion of depth.

INSERT. To add to copy or type at some place other than the end; also a special small publication or card added to a major piece after binding (such as an advertisement inserted in a magazine).

INSERT BLANK. A specification calling for the center to be left blank, with columns of type set on either side.

INSET. One piece of art placed inside another to show detail or emphasize a related image.

INSET INITIAL. An initial cut into the upper left of a paragraph so that it aligns with the x-height of text at the top and with the baseline of a lower line of text; also called a set-in initial.

INSTANT PRINTING. Another term for quick printing.

INTAGLIO PRINTING. A method of printing in which the image to be printed is etched below the surface of the plate.

INTEGRAL PROOF. Another term for transfer proof.

INTERCHARACTER SPACING. Another term for letterspacing.

INTERLEAF. Another term for slipsheet.

INTERLINE SPACING. Another term for leading.

INTERNATIONAL STANDARDS ORGANIZATION (ISO). A metrically based measurement system.

GLOSSARY

INTERNEGATIVE (INTERNEG). A negative made from a color transparency and used to convert the image into a color or black-and-white print.

INTERTYPE. A machine for linecasting hot-metal text type.

INVERTED PYRAMID. Writing copy with the most important information first so any cuts can be made from the bottom up without great loss of content; also setting a headline with the longest line on top and increasingly shorter lines centered beneath it.

ISSUE. One of a series of regular publications with a specific date (such as the November issue of a monthly newsletter).

ITALIC (ital). Type that leans to the right, the opposite direction from backslant.

JOB. One writing, design, photography, illustration, typesetting, paste-up, engraving, printing, or binding project.

JOB BINDERY. Another term for trade bindery.

JOB PRINTER. A general shop accepting a wide variety of work.

JOB TICKET. The instructions accompanying a project while it is in a supplier's shop.

JOBBER. A press used for small projects; also a shop specializing in such work.

JOBBER FOLDER. An automatic folder for high-volume folding and other finishing processes in one pass.

JOG. To straighten sheets of paper in a stack by shaking them into line.

JUMP. To continue an article on another page.

JUMP LINE. Another term for continued headline.

JUMPING THE GUTTER. Running a rule, headline, or art across the gutter of a spread.

JUSTIFY. To line up type vertically on both left and right sides.

KERN. Adjusting spacing between type characters so that the characters are closer together or appear evenly spaced; also the part of a type character that extends beyond the body (such as a swash).

KEYING STRIPPING. Making one flat but exposing it in sections to multiple press plates for multiple colors.

KEYLINE (KEY, KEY LINE). Lines sketched on a tissue-paper overlay to indicate placement, content, and approximate size of art on mechanical artwork; also called holding line; also used to refer to a precise line between elements to account for lap register.

KICKER. A short line of type above an indented main headline, sometimes underscored, used for emphasis.

KILL. To get rid of or throw away (such as outdated galleys).

KISS IMPRESSION. An impression in which plate and paper barely touch to avoid embossing of the paper; *see also* Sock Impression.

KNIFE FOLDER. An automatic folder that uses a dull blade to crease the paper and force it between folding rollers.

KNIFE-CUT STENCIL. Another term for the conventional method of making a hand-cut screen stencil.

KNOCKOUT. Another term for reverse.

KRAFT PAPER. A sturdy paper with high-pulp content.

LABEL HEADLINE. Another term for bulletin headline.

LABEL PAPER. A type of specialty paper coated with a moisture- or pressure-sensitive adhesive on one side for labels; also called gummed paper.

LACQUER. A liquid applied in the bindery to a printed sheet to add an overall high gloss; glossier than varnish.

LADDER. A block of justified type having many hyphens on the right that make it difficult to read.

LAID FINISH (LAID PAPER). A bond or book paper finished to look handmade and having faint vertical and horizontal lines throughout visible when held to the light; commonly used for letterhead.

LAMINATED PAPER. A paper with a transparent plastic coating; also two or more plies glued together (such as duplex paper).

LAMINATING. Using heat and pressure to bond film to paper; also using adhesive and pressure to bond two or more plies of paper together.

LANDSCAPE FORMAT. Another term for broadside format.

LAP REGISTER. A method of aligning two or more opaque colors that are to touch on the printed piece.

LAYOUT. The plan for a printed piece showing the arrangement of all elements; also a grouping of photographs or illustrations.

LEAD. The opening paragraph(s) of an article which catches reader interest and defines what the piece is about; pronounced "leed."

LEADER. A line of periods or hyphens used to connect elements in a list or table.

LEADING. A thin strip of metal used to space out lines of hot-metal type; also any extra points of space between lines; pronounced "ledding."

LEFTHAND ART. A photograph or other art in which the action directs the viewer's eye from left to right.

LEGEND. A brief explanation accompanying a chart, map, or other illustration, usually below.

LETTER FOLD. A form of barrel fold commonly used for letters.

LETTERGUIDE. A transparent plate for tracing letters onto mimeograph stencils.

LETTERPRESS PRINTING. A printing method in which the image to be inked is raised above the printing plate.

LETTERSET PRINTING. A type of relief printing using the raised plate of letterpress and the intermediate blanket of offset to impress the image; also called indirect relief, indirect letterpress, offset letterpress, or dry offset printing.

LETTERSHOP. A small print shop limited to letterhead, envelopes, and similar print projects.

LETTERSPACING. The distance between letters in a typeset word; also called intercharacter spacing; *see also* Optical Letterspacing.

LIBRARY BINDERY. A bindery specializing in hand binding, rebinding, and repair of books and periodicals.

LIFE. The amount of time that a printed piece or press plate is usable.

LIFT-OFF ART OR TYPE. A kind of transfer art or type that is lifted off the backing sheet.

LIFTOUT. Another term for reverse.

LIFTOUT HALFTONE. Another term for silhouette.

LIGATURE. Two or more type characters that are connected; called a diphthong when two connected letters are vowels and a triphthong when three characters are involved; also called quaint character.

LIGHT TABLE. A surface lighted from below; used for tracing or stripping so that the image beneath can be seen.

LINE ART. Art with no shades of gray; also any art converted to a press negative by line (rather than halftone) reproduction.

LINE CONVERSION. Another term for degeneration.

LINE COPY. Art and type to be reproduced as black and white only.

LINE DRAWING. An illustration drawn in black and white only.

LINE ENGRAVING. An intaglio printing process using a metal plate into which a wrong-reading image has been cut by hand or machine, paired with a right-reading die for impressing the image onto paper.

LINE GAUGE. A ruler to measure the baseline-to-baseline distance between lines of type; also called linespacing gauge.

LINE GRAPH. A graph in which precise points on a grid are connected by a line.

LINE NEGATIVE. A copy on film of original art produced by line photography.

LINE PHOTOGRAPHY. Copying on film single-color or single-tone original art using a process camera; also called simple black-and-white photography or single-color photography.

LINE SCREEN. A screen used to convert a halftone or other continuous-tone art into a dot pattern for printing; the higher the line-screen number, the finer and closer the dots.

LINE-FOR-LINE. To set as typed or written in the manuscript so that each typeset line has exactly the words shown in the copy.

LINEAR DEFINITION. Another term for degeneration.

LINECASTING. The casting of hot-metal type by the line, as opposed to individual characters (character casting).

LINEN FINISH. A bond paper with a woven or linen-like texture.

LINESPACING GAUGE. Another term for line gauge.

LINING FIGURES. Numerals the same height as capital letters with a common baseline; also called aligning figures; *see also* Hanging Figures.

LINOTYPE. A machine for linecasting hot-metal text type.

LITHOGRAPHY. A printing process in which areas containing images to be transferred are coated with a greaselike substance that attracts ink on the press.

LIVE MATTER. Elements on mechanical artwork that are to be printed.

LOGO (LOGOTYPE). Art and/or type used in a unique way to identify a business, association, publication, or service; also symbols or trademarks on a type font.

LONG GRAIN. Grain running with the long dimension of paper.

LONG INK. An ink that is very fluid and that has low viscosity.

LOOSE REGISTER. A method of aligning two or more opaque colors that are not to touch on the printed piece.

LOOSELEAF BINDING. Any binding method that secures loose pages into some kind of mechanical binder.

LOW-KEY PHOTOGRAPH. A photograph in which details are recorded in mostly shadow tones.

LOWERCASE (lc). Type set without capital letters.

LUDLOW. A machine for linecasting hot-metal display type.

M WEIGHT. The weight of 1000 sheets of paper of any size.

MACHINE FINISH (MF). A paper finish that is smooth, although not as smooth as English finish.

MACRON. A diacritical mark consisting of a straight horizontal line above a letter.

MAGAPAPER. A publication having format qualities of both a magazine and a newspaper; also less commonly called a newzine.

MAGAZINE. A bound publication of at least eight pages issued on a regular basis and containing primarily feature articles, as opposed to brief news items.

MAGENTA. The red-blue used as one color in process printing to reproduce full-color art; also called process red.

MAGNETIC INK. A specialty ink that is magnetized after printing for electronic scanning.

MAKEREADY (MAKE READY). Setting up a press to run a job; may also apply to preparing a typesetting unit for composition.

MAKEUP. The design of a printed piece; also the process of assembling type and art for conversion into a press negative.

GLOSSARY

Manifold Paper. Another term for onionskin.

Manila Paper. A smooth, sturdy paper (usually buff or gold) commonly used for large mailing envelopes.

Manuscript. The original copy from which type is set.

Map Fold. A fold similar to that used for a roadmap; used for a large brochure or insert.

Mar Coating. A lacquer coating applied to a printed sheet to give it an overall high gloss.

Margin. The white space around type or art; usually refers to the space around the outside of a printed page.

Margin Art. Photographs, illustrations, or blocks of type in the margins of a printed page.

Marker. A type of line or continuous-tone art using markers in black, shades of gray, and/or colors.

Markup. How various elements are to be typeset; also a percentage added to the cost of supplies or services secured by a supplier in the customer's behalf.

Mask Out. To block out so that whatever is masked doesn't appear on the press negative or plate.

Masking Film. Orange or red transparent plastic film used to make halftone windows or other areas to be shot as black, creating a clear hole on the line negative for insertion of a halftone negative or line screen; also called stripping film.

Masking Sheet. Orange or yellow gridded paper or plastic used to cover non-image areas of a press negative; also called a goldenrod, stripping base, or stripping vinyl.

Master. Preprinted sheets containing elements of a publication that don't change; other elements are imprinted issue by issue; also called preprint or shell.

Masthead. A block of information about a publication (such as names of editor and staff, address, date, volume and number); often mistakenly used to refer to the flag or nameplate.

Matrix (Mat). A mold used to cast hot-metal type into a mirror image (the opposite of patrix); also a mold used to make a duplicate press plate from hot metal.

Matte Finish (Matte Print). A dull untextured finish on paper or a photograph; the opposite of glossy.

Mean Line. The top of lowercase letters having no ascenders, such as "m" or "a."

Mechanical Artist. Someone who executes mechanical artwork; also called paste-up artist or paste-up person.

Mechanical Artwork (Mechancial Art, Mechanical). Precisely positioned type and other line elements on a heavy sheet, camera-ready for engraving; also called artwork or camera-ready art; *see also* Paste-Up.

MECHANICAL DUMMY. A mock-up of a printed piece after type has been set showing placement and size of all elements to guide execution of mechanical artwork; also called simply a dummy.

MECHANICAL TINTING. Another term for Benday process.

MEDIA CONVERSION. Transferring information on a computer disk to magnetic tape for electronic typesetting.

METAL ENGRAVING. Another term for line engraving.

METALLIC PAPER. A paper with a metallic luster on one side.

MEZZOTINT. A special effect created by a screen resulting in a printed photograph with obvious and irregular dots throughout.

MICROMETER. An instrument for measuring paper caliper in millimeters.

MIDTONE. Shades of gray in the middle of a black-and-white range containing much of the detail of a photograph or continuous-tone illustration.

MIMEOGRAPHY (MIMEO). A printing process using a stencil and a small rotary press for utilitarian office work; also called stencil duplicating.

MIRROR IMAGE. The reverse; may refer to a photograph that has been flopped, to the design of two facing pages with matching elements, to type matrices, or to a press negative or plate.

MISPRINT. A typographical error.

MITER. To cut rules during paste-up at 45° angles so that, when they butt up, they form perfect 90° corners of a box.

MOCK-UP. Another term for dummy.

MODEL RELEASE. A signed agreement granting permission to use someone's picture for a specific or general purpose without threat of legal action for libel or invasion of privacy.

MODEM. A computer attachment that allows transmission of a manuscript by telephone to a typesetting unit for composition.

MODERN TYPE. A type classification characterized by thinly serifed letters with distinctly thick and thin strokes.

MOIRÉ. An undesirable wavy pattern created by improper screen angles during preparation of a halftone.

MONOTYPE. A machine for character casting hot-metal text type.

MONTAGE. A grouping of illustrations or photographs into a single visual unit.

MORGUE. A publication's library or archives containing all back issues for reference.

MORTISE. A cut made in an illustration or photograph to create a blank area for insertion of type or another piece of art; when using metal type, the process is also called piercing for type.

MOTTLE. An uneven, spotty appearance of ink on a printed piece.

MOUNTAIN GRAPH. A line graph in which the gaps between lines are filled by screens or color to create a mountain-range effect; also called a surface chart.

MOVEMENT. Another term for sequence.

MS (MSS). Abbreviations for manuscript, singular and plural.

MUG SHOT. A portrait showing only the subject's head and shoulders; also called a head-and-shoulders shot.

MULTIPLE STRIPPING. Making two or more flats for different colors to be exposed to two or more press plates.

MULTIPLE-USE RIGHTS. The right to use copy, an illustration, or a photograph two or more times.

NAMEPLATE. Another term for flag.

NEGATIVE. Photographic film in which tones are reversed; blacks are white and vice versa; also a reversed photostat.

NEGATIVE ASSEMBLY. Another term for stripping.

NEWSLETTER. A publication dominated by brief news items; usually 12 or fewer pages unbound in a size smaller than a newspaper.

NEWSPAPER. A tabloid or larger-sized publication containing both news and feature items and usually unbound.

NEWSPRINT. Low-quality paper commonly used for printing newspapers.

NEWZINE. Another term for magapaper.

NON-IMPACT PRINTING. Another term for ink-jet printing.

NONPAREIL. A French measure the equivalent of $\frac{1}{2}$ pica or 6 points.

NONREPRO BLUE. A light turquoise-blue pen or pencil that cannot be reproduced photographically; commonly used for paste-up guidelines and markings.

NOVELTY FOLD. Any fold out of the ordinary done by hand or machine to achieve a special design objective.

NOVELTY PRINTER. A print shop specializing in the printing of stickers, buttons, and other novelty items.

NOVELTY TYPE. A type classification characterized by ornate or unusual designs not closely related to any other type classification; also called Special, Occasional, or Decorative.

NUMBERING MACHINE. An attachment to a press that numbers printed forms in sequence.

OCCASIONAL TYPE. Another term for Novelty type.

OCCULT BALANCE. Another term for informal balance.

OFF-SET (OFF-SETTING). Other terms for set-off.

OFFICIAL ENVELOPE. Number 10 commercial envelope used for business correspondence.

OFFSET LETTERPRESS PRINTING. Another term for letterset printing.

OFFSET PRINTING (OFFSET LITHOGRAPHY). A printing process in which the image is transferred or offset from a plate to a blanket, then printed onto paper; also known as photo-offset lithography, photolithography, or planography.

OLD COPY. All background information, manuscripts, and other materials from production of a printed piece saved for reference.

OLD ENGLISH TYPE. A type classification characterized by serifed letters resembling early hand-lettering; also called Text, Blackletter, Black Text, or Spire Gothic type.

OLD STYLE TYPE (OLD FACE TYPE). A type classification characterized by heavily serifed letters and strokes of comparable width; also called Antique type.

ONE-TIME RIGHTS. The right to use copy, an illustration, or a photograph one time but not the first time; also called reprint or subsequent rights.

ONIONSKIN PAPER. Very thin paper commonly used for carbon copies of business correspondence or lightweight letterhead; also called flimsy or manifold paper.

OPACITY. The property of paper that prevents the printed image on one side from showing through on the other; *see also* Visual Opacity and Printed Opacity.

OPAQUE. To paint out spots or other unwanted elements on mechanical artwork or press negatives so that they don't appear on the finished printed piece; also the black portion of a negative.

OPAQUE ART. Art done with oils, acrylics, or opaque watercolors.

OPEN-FACE TYPE. Another term for outline type.

OPEN MATTER. Type lines widely spaced vertically.

OPEN-ENDED ENVELOPE. Another term for catalog envelope.

OPPOSITION. Another term for balance.

OPTICAL BAR RECOGNITION (OBR). An electronic scanner that "reads" the bar code of each character in a manuscript and converts it into type.

OPTICAL CHARACTER RECOGNITION (OCR). An electronic scanner that "reads" each typewritten character in a manuscript and converts it into type; also called optical scan.

OPTICAL LETTERSPACING. The positioning of letters in typesetting so that they appear evenly spaced, regardless of individual shapes.

OPTICAL SCAN. Another term for optical character recognition.

ORGANIZATION CHART. A diagram depicting a sequence or relationship of one position to the whole organization.

ORIGINAL. The actual type galley, art, or other element; not a copy.

ORNAMENTS. Another term for printer's marks.

OUTLINE HALFTONE. A halftone in which the background is eliminated to leave only the center of interest, as opposed to a square-finish halftone; also called a cutout halftone.

GLOSSARY

OUTLINE TYPE.   A type style with extra strokes outside a character's image area to give the illusion of depth (also called silhouette) or one that is open instead of solid (also called open-face type).

OVERHANG COVER.   A cover that extends beyond the pages it encloses.

OVERLAY.   A sheet of heavy, clear plastic over mechanical artwork containing elements to be printed in another color or screened; also any sheet over art or artwork for protection.

OVERLAY PROOF.   A pre-press proof of color separations done on individual, light-sensitive, transparent sheets of acetate; Color-Key is a common brand name; also called color transparent proof.

OVERMATTER.   Another term for holdover.

OVERPRINTING.   Printing over an area that has already been printed.

OVERS.   Copies of a printed piece in excess of the number ordered.

OVERSET.   Another term for holdover.

OXIDATION (OXIDATIVE POLYMERIZATION).   A method of drying ink that relies on the oil-based vehicle's absorption of oxygen, causing the pigment to harden and gel.

OZALID.   A quality proof of cold type produced by an Ozalid photocopier.

PMT.   *See* Photomechanical Transfer.

PACKAGING PRINTER.   A print shop specializing in the printing of boxes, cans, and other needs of the packaging industry.

PADDING.   The binding of sheets into a tablet with glue on one edge.

PAGE.   One side of a sheet of paper.

PAGE MAKEUP.   The arrangement of type and art elements on a page.

PAGE PROOF.   A sample of a printed sheet for checking consisting of one or more pages; also a photocopy of mechanical-artwork pages used for a final check before conversion into press negatives.

PAGINATION.   Numbered pages in their proper sequence.

PAMPHLET.   Another term for booklet.

PANEL.   One section of a folder or brochure; corresponds with a page on a larger printed piece.

PANTONE® MATCHING SYSTEM.   A commercial system of precise formulas using base colors to get a wide range of opaque color choices.

PARALLEL.   A reference mark.

PARALLEL FOLD.   A method of folding in which all folds are parallel to one another (such as an accordion fold).

PARCHMENT PAPER.   A hard, slightly mottled bond paper commonly used for certificates.

PASTE-UP.   The process of executing mechanical artwork.

PASTE-UP ARTIST (PASTE-UP PERSON).   Other terms for mechanical artist.

PASTE-UP SHEET (PASTE-UP BOARD).   Heavy paper on which mechanical artwork is made; may be printed in nonrepro blue ink with guidelines to speed paste-up; also called a dummy sheet when used for that phase of production.

PASTED FORM. A method of binding in which pages are held together with glue along the fold.

PATRIX. A mold used to cast hot-metal type into a right-reading image (the opposite of matrix).

PEBBLE FINISH. A bond paper with a stippled finish.

PECULIARS. Another term for sorts.

PEN-AND-INK DRAWING. A type of line drawing using a drawing pen and waterproof ink.

PENALTY COPY. Copy difficult to set and for which a typographer charges extra.

PENCIL DRAWING. A type of continuous-tone drawing using a graphite or charcoal pencil.

PENETRATION. Another term for absorption.

PERFECT BINDING. Using glue and/or stitching to bind a printed piece into a booklike document with a soft cover and a distinct spine; also called adhesive binding.

PERFECTING PRESS (PERFECTOR). A press that prints both sides of a sheet during one pass.

PERFORATE (PERF). To die-cut a row of small holes or slits into a finished printed piece so that part of it can be easily torn out.

PERFORATION LINES. Dashed nonrepro-blue lines on mechanical artwork to indicate where perforations are to be made.

PHANTOM. A technique used with illustrations or photographs in which the background is screened back so that the central image stands out, yet the background is still seen; also called ghost.

PHOTO AGENCY. Another term for photo service.

PHOTO ESSAY. A group of related photographs arranged with minimal copy to express an idea or point of view; *see also* Photo story.

PHOTO SERVICE. A dealer in stock photography; also called photo agency, stock library, or stock house.

PHOTO STORY. A group of related photographs arranged with minimal copy to tell a factual story; *see also* Photo Essay.

PHOTO-DIRECT PLATE. A direct-image plate with a photosensitive emulsion for quick exposure of a typed, drawn, or typeset image; used extensively by quick printers; also called projection speed plate.

PHOTO-MASKING. Using a color mask at the time of separation to improve color, tonal range, and/or detail in a color image to be reproduced.

PHOTO-OFFSET PRINTING (PHOTO-OFFSET LITHOGRAPHY, PHOTOLITHOGRAPHY). Other terms for offset printing.

GLOSSARY

PHOTOCOMPOSING. A photocomposition method that allows manipulation of copy into a complete layout to eliminate the need for paste-up.

PHOTOCOMPOSITION (PHOTOTYPESETTING). Producing cold type by a photographic process.

PHOTOENGRAVING. A process of engraving press plates using photochemistry.

PHOTOGELATIN PRINTING. Another term for collotype printing.

PHOTOGRAPH. A continuous-tone image made from a black-and-white or color negative or from a color transparency.

PHOTOGRAPHIC ILLUSTRATION. A photograph used to show exactly how something looks.

PHOTOGRAPHIC STENCIL. A screen stencil that is presensitized for contact printing of camera-ready artwork.

PHOTOJOURNALISM. Using photographs as the primary medium to report events.

PHOTOMECHANICAL TRANSFER. A treated paper used to prescreen a halftone into a coarsely line-screened, black-and-white image ready for paste-up as line art; also called diffusion transfer or PMT.

PHOTOMECHANICS. The process of making printing plates involving exposure to light.

PHOTOPOLYMER PLATE. A type of deep-etch plate made of plastic for printing by letterpress or letterset.

PHOTOPRINT (PHOTOREPRO). Other terms for reproduction proof or repro proof.

PHOTOPROOF. Another term for proof.

PHOTOSTAT. A copy of line art and/or type made on photographic paper using a stat camera; abbreviated stat.

PI CHARACTERS. Unusual symbols not customarily on a regular type font but available on a separate font.

PIC (PIX). Abbreviations for picture, singular and plural.

PICA. A unit of measurement commonly used in typesetting and printing; 1 pica equals 12 points; 6 picas equal approximately 1 in. or 2.5 cm.

PICA POLE (PICA RULE, PICA STICK). A ruler divided into picas.

PICA TYPE. Type that is 10 characters to the inch (2.54 cm) on a typewriter.

PICK UP. To reuse type or art from a previous job.

PICKING. The pulling loose of paper fibers in heavily inked areas creating a fuzzy surface on the printed piece and lint on the offset blanket.

PICTOGRAPH. A bar chart, pie chart, or similar illustration using art as the main or background image.

PIE CHART. A chart comparing numerical values by dividing a whole into parts.

PIECE FRACTIONS. Fractions that must be constructed of separate type characters; also called split fractions; *see also* Whole Fractions.

PIED TYPE. A section of hot type in which the lines on one side are not as close as those on the other creating an effect much like a wedge of pie; also called cocked type.

PIERCING FOR TYPE. *See* Mortise.

PIGMENT. The ink component that gives it color.

PIN PERFORATION (PINHEAD PERFORATION). A die-cut row of small round holes; *see also* Slit Perforation.

PIN REGISTRATION. Aligning a series of holes on metal pins during paste-up through platemaking so that registration is assured.

PINK PROOF. A galley proof on colored paper for dummying.

PLANOGRAPHIC PLATE. A press plate with the image to be printed and the surrounding area all on one plane or level, as opposed to a raised or recessed image area.

PLANOGRAPHY (PLANOGRAPHIC PRINTING). Other terms for offset printing.

PLATE (ART). A chart, table, or other illustration that appears separate from the text in a printed piece, as opposed to a figure that appears within it.

PLATE (PRINTING). A metal, plastic, rubber, or paper sheet on which the image to be printed is exposed, ready for the press.

PLATELESS ENGRAVING. Another term for thermography.

PLATEN PRESS. A type of letterpress that uses a plate or actual metal type on a flat surface to impress an image onto flat paper.

PLATER FINISH (PLATE FINISH). Other terms for English finish.

PLY. One of the sheets pasted together to make a heavier sheet of paper.

POCKET. A receptical folded and often glued to hold documents within a presentation folder or on an inside cover.

POINT. A unit of measurement in typesetting and printing equivalent to $\frac{1}{72}$ of an inch (0.0351 cm).

POLICY ENVELOPE. A narrow envelope used for insurance policies and other legal documents.

PORTFOLIO. A collection of samples to show off the work of a writer, designer, illustrator, photographer, or mechanical artist; also called a book.

PORTRAIT FORMAT. A design in which the page is longer than it is wide; the opposite of landscape or broadside format.

POSITIVE. An image appearing as intended in the finished printed piece, white as white and black as black; the opposite of negative.

POSTAL-SAVER ENVELOPE. An envelope with one side spot-glued so it can be opened for postal inspection.

POSTER. A large announcement printed on one side of heavy paper suitable for posting on a bulletin board or wall.

GLOSSARY

PosterizATION. A screening process that converts shades of gray into one or two tones, resulting in a high-contrast printed image.

Pre-Assembly (Pre-Positioning). Other terms for dupe and assembly.

Precipitation. A method of drying ink that relies on exposing the printed piece to moisture, thus precipitating out the solvent and allowing penetration into the paper.

Preliminaries. Another term for frontmatter.

Prep (Preparatory). All production services offered by a print shop prior to actual printing.

Preprint. A piece printed ahead of the main publication and inserted during binding; also another term for master.

Prescreened Print. A halftone converted during paste-up to a coarse print that is treated as line art; PMT and Velox are types of prescreened prints.

Presentation Folder. A heavy folder, often with pockets, printed on one or both sides and used to hold proposals, brochures, and other information, usually for selling products or services.

Press Check. Proofing at the beginning of a press run to verify that inking, color, and registration are accurate.

Press Proof. An early sheet off the press.

Press Run. The quantity to be printed.

Pressboard. A type of bristol paper.

Pressure-Sensitive Paper. Paper treated on the back so that it will adhere to a surface when a protective strip is peeled away and pressure is applied.

Primary Letters. Those lowercase letters not having ascenders or descenders.

Print-and-Tumble. Another term for work-and-tumble.

Print-and-Turn. Another term for work-and-turn.

Printed Opacity. The opacity of a sheet after printing; influenced by how much ink it absorbs.

Printer. A supplier who takes actual type or press plates and produces finished printed copies; may also perform other services; also refers to the negative used to make a printing plate for a particular process color.

Printer's Error (pe). A proofreader's mark denoting that a mistake was caused by the typographer or printer, not by the author.

Printer's Marks. Designs used historically by printers to add art or emphasis to a printed piece (such as a fleur-de-lis at the end of book chapters); also called ornaments, dingbats, printer's flowers, fleurons, cabbages, or vegetables.

Printer's Trademark. Another term for colophon.

Printing Bristol. A heavy, stiff paper with a smooth finish softer than index bristol used for cards and announcements.

PRINTING DUMMY. A mock-up of a piece to be printed using the actual paper to show size, page sequence, fold, binding, and bulk; also called a blank-page or bulking dummy.

PRINTING-IN. Another term for burning-in.

PROCESS BLUE. Another term for cyan.

PROCESS CAMERA. The type of camera used to produce a press negative of art; also called copy camera or graphic-arts camera.

PROCESS PRINTING. Another term for four-color process printing.

PROCESS RED. Another term for magenta.

PRODUCTION. The complete process of converting copy and art into a finished printed piece.

PROGRESSIVE PROOFS (PROGRESSIVES, PROGS). A method of proofing four-color work whereby each color is laid down and checked in sequence.

PROJECTION SPEED PLATE. Another term for photo-direct plate.

PROOF. A copy of type, artwork, or printing for checking; also called photoproof.

PROOFING PRESS. A small hand-operated press used to check a plate before putting it on an automated press; also used to pull repro proofs of galleys of hot type.

PROOFREAD. To check against the original copy for accuracy.

PROOFREADER'S MARKS. A group of standardized indications showing how type is to be changed or corrected.

PROOFSHEET. Another term for contact sheet.

PROPERTY RELEASE. A signed agreement granting permission to use a picture of a private building or other structure for a specific or general purpose, without threat of legal action for libel or invasion of privacy.

PROPORTION. A principle of design calling for consideration of the size relationships of the page and its elements.

PROPORTION SCALE (PROPORTION WHEEL). A hand-held device for calculating reductions and enlargements of photographs and other art; shows final dimensions and may also show percent of change from the original.

PROPORTIONAL SPACING. Letterspacing in proportion to the width of each character, as opposed to allotting equal space for all characters.

PUBLISHING PRINTER. A print shop specializing in the printing of magazines, newspapers, and books for publishers.

PUNCHING. Making holes in a stack of paper during finishing by using a sharp punch of the desired size and shape; *see also* Drilling.

PUT TO BED. To complete production to the point at which all that remains is to finish the press run.

QUAD (QUADRAT). In hot-metal composition, blank type used to indent paragraphs or add extra space between characters.

QUAD CENTERED. Another term for centered or center quadding.

GLOSSARY

Quad Left (Quad Right). Other terms for flush left and flush right.

Quaint Character. Another term for ligature.

Quick Printing. While-you-wait utilitarian printing of simple one-color jobs in standard sizes on standard paper; also called instant printing.

Quote (Quotation). Another term for bid.

Rag Bond. A bond paper made entirely or predominantly of rag fibers.

Ragged Type. Type with an even edge on one side and an uneven edge on the other.

Rainbow-Fountain Printing. A type of split-fountain printing in which different-colored inks are intentionally allowed to fuse into one another on the press; *see also* Split-Fountain Printing.

Raised Initial. An initial cap extending above the cap height, as opposed to a sunken initial extending below the baseline; also called rising, standing, or stick-up initial.

Raised Printing. Another term for thermography.

Readability. Ease with which a printed piece can be read.

Reader. Another term for galley proof.

Ream. Five-hundred sheets of paper of any size.

Recto. Right-hand page of a publication, always with an odd number; the opposite of verso.

Reduction. An image smaller than the original.

Reference Marks. Symbols that refer to footnotes; the order of use is asterisk for the first reference on a page, dagger for the second, double dagger for the third, section for the fourth, and parallel for the fifth; also called footnote symbols.

Reference Matter. Another term for back-of-the-book.

Reflective Copy. Art in opaque form (such as photographs or drawings).

Registration. The exact alignment of two or more elements on a page.

Registration Mark. A circled cross used on mechanical artwork and its overlay to align elements to be screened, printed in a second color, or otherwise treated differently from the basic artwork.

Regular Envelope. Another term for commercial envelope.

Relief. Printing on a raised surface; also another term for embossing.

Remake. To make changes in a page or series of pages before it is printed or reprinted.

Remittance Envelope. An envelope with a deep square flap that folds down to prevent showthrough of the contents.

Render. To make a detailed drawing (rendering), as opposed to a rough sketch.

Reply Card. A printed return card used in direct-mail solicitation to make responding easy.

REPRINT RIGHTS. Another term for one-time rights.

REPRODUCTION PROOF (REPRO PROOF). A glossy, high-quality copy of type ready for paste-up; originally applied to copies of hot type.

RERUN. An additional printing to produce more copies of a printed piece.

RESHOOT (RETAKE). To do photographs over because the first shots were unsuitable.

RESIST. A stripping base used to make a gravure plate.

RETOUCHING. Altering a photograph or negative, usually to eliminate or diminish a scratch or other undesirable element.

REVERSE (REVERSE OUT). Light letters on a dark background; the opposite of surprint; also called dropout, liftout, or knockout.

REVISED PROOF. A copy of type, artwork, or press negatives as corrected.

RHYTHM. Another term for sequence.

RIGHT-ANGLE FOLD. A method of folding in which at least one fold is at right angles to the other(s).

RIGHTHAND ART. A photograph or other art in which the action directs the viewer's eye from right to left.

RING BINDER. A looseleaf method of binding in which holes are drilled along the edge of pages to be bound and pages are held together by two or more metal rings in a stationary binder or by tightly coiled, individual plastic rings.

RIPPLE FINISH. A bond paper finished to have a fine, overall, wavy texture.

RISING INITIAL. Another term for raised initial.

RIVERS. Distinct paths of white space through a block of type in which the spaces between words coincide vertically or diagonally.

ROMAN. Any type character that is upright, as opposed to the slanted style of italic.

ROMAN NUMERALS. The numeral characters I, II, III, and so on, as opposed to Arabic numerals 1, 2, 3, and so on.

ROTARY. A type of letterpress that uses a curved plate to impress an image onto paper attached to a rotating cylinder.

ROTO SECTION. A pictorial newspaper supplement printed on a rotogravure press.

ROTOFILM. A photosensitive silver stripping base used to make a gravure plate.

ROTOGRAVURE. A web gravure press using roll paper; *see* Gravure Printing.

ROTOPAPER. A quality newsprint manufactured for use on a rotogravure press.

ROUGH (ROUGH LAYOUT). A simple sketch indicating placement of type, copy, and art.

ROUGH PROOF. A galley proof that has not been proofed and corrected by the typographer.

RUB-OFF ART OR TYPE. A kind of transfer art or type that is rubbed off the backing sheet.

RUBBING. An illustration created by placing thin paper over an embossed or engraved surface and coloring the paper with pencil or other marker to bring out the image below.

RULE. A horizontal or vertical line used as a design technique to separate elements on a page.

RUN. An abbreviation for press run.

RUN BACK. To take up space between words so that the first word or syllable in a line of type moves up to the end of the previous line; the opposite of drive out or run over.

RUN DOWN. Another term for drive out.

RUN ON. To continue setting type without starting a new line or paragraph.

RUN OUT. Another term for hanging indentation.

RUN OVER. Another term for drive out.

RUN RAGGED. To set type flush on the left and ragged on the right.

RUN THROUGH. An unbroken rule running from one edge of the paper to the other across the gutter and margins.

RUN-IN. A design technique in which copy immediately follows the headline or subhead (such as an article starting as a continuation of the subhead in display type, and then going on to the next page in regular-sized body type).

RUNAROUND (RUN AROUND). Type surrounding art or butting up against it on two or more sides; also another term for wraparound.

RUNNABILITY. The ability of paper to pass efficiently through the press.

RUNNING HEAD (RUNNING FEET, RUNNING TITLE). A publication title, chapter title, or date appearing at the head (top) or foot (bottom) of each page of a magazine or book for easy reference.

RUSH. Performing a production step in less than customary time, as defined by individual suppliers.

SADDLE-STITCHING (SADDLE-WIRING). A binding process in which wire stitches hold pages together down the center fold.

SAFETY PAPER. A specially treated paper used for checks so that any alteration can be detected; also called cheque paper.

SAME SIZE (SS). A specification that a photograph or other art is to appear the same size as the original with no reduction or enlargement.

SANS SERIF. Type without serifs.

SANS SERIF GOTHIC TYPE. A type classification characterized by no serifs and strokes of the same or comparable width; also called English Grotesque or Block type.

SCALE. To reduce or enlarge to the desired size.

SCANNING. The "reading" of hard copy by an electronic-typesetting unit using bar-code or character recognition to bypass entering the manuscript on the keyboard; also the "reading" of a color image by computer to prepare color separations.

SCORE. A pair of creases in a W shape made in printing or binding to create a crisp fold, especially on heavy paper.

SCRATCHBOARD (SCRAPERBOARD). A heavy, white, clay-coated paper washed with black ink to be scratched away to create an illustration.

SCREEN. Breaking down a photograph or other continuous-tone art into a pattern of dots that can be printed, either a simple line screen or a special effect (such as wavy lines); also using a pattern of even dots to create a gray or light-colored area; also the sheet of film imprinted with a pattern used during engraving.

SCREEN PRINTING. A printing method that transfers images via a stencil; also called silkscreen printing or silkscreening.

SCREW-AND-POST BINDER. A looseleaf method of binding in which holes are drilled along the side of pages to be bound and covers with corresponding holes screwed into place; also called binding post.

SCRIPT TYPE. A type classification characterized by its resemblance to handwriting; also called Cursive type.

SECOND COLOR. Any color used in addition to the base color.

SECOND-COMING TYPE. Extremely large type reserved for headlines announcing the most critical news.

SECTION. A reference mark; also another term for signature.

SELF-COVER. A printed piece using the same paper for the cover as for the inside.

SELF-MAILER. A printed piece designed to be addressed, stamped, and mailed without an envelope.

SELF-SEALING ENVELOPE. An envelope with twin square flaps that adhere to one another.

SEMI-TRANSPARENT INK. A type of ink that is capable of altering the ink printed beneath it but not fully hiding it.

SEPARATRIX. Another term for shill mark.

SEPIA. A brown image achieved by toning a black-and-white photograph.

SEQUENCE. A principle of design calling for the arrangement of elements so that the reader comprehends them in the intended order; also called rhythm or movement.

SERIES. All sizes available in a particular typeface.

SERIF. The finishing strokes at the ends of letters; simplified typefaces without these strokes are called sans serif; once spelled ceriph or cerif.

SERIGRAPHY. The artistic application of screen printing.

SET. The width of a type character expressed in units; set may be condensed, normal, or extended; also the penetration of ink vehicle to bond pigment to paper during the drying process.

Set Close. To allow very little space between type characters.

Set Open. To allow considerable space between type characters.

Set Size. The set or width of a type character in addition to the space normally allowed on either side of it.

Set Solid. A specification to set lines with no points of leading between; also called solid matter.

Set-In. To add an image to a press plate.

Set-In Initial. Another term for inset initial.

Set-Off (Setting-Off). The undesirable transfer of ink from one printed sheet to another because of inadequate drying of the ink or limited absorption by the paper; also called off-set or off-setting.

Setwise. The width of a type character, as opposed to its point size.

Sewing. A binding process in which sections or signatures are sewn together with thread before being attached to a cover; *see also* Smyth Sewing or Side-Sewing.

Shading Sheet. A transparent, patterned sheet added to mechanical artwork in areas where a line pattern is desired; also called tint sheet.

Shadow Printing. A word-processor method of achieving bold headlines by printing a character a second time slightly out of registration with the first.

Sheet. A piece of paper with two sides, each side of which is a page.

Sheetwise. Printing one side of a sheet with one press plate, and then printing the back when dry with another plate using the same gripper edge.

Shell. Another term for master.

Shill Mark (Shilling Mark). The diagonal line separating numerator from denominator in a fraction or percent sign; also called a slash, slant, virgule, separatrix, solidus, or stop mark.

Shingling. Designing graduated inside margins to compensate for the fanning out that occurs when a thick document is saddle-stitched.

Short Fold. Any fold with one panel smaller than the other(s).

Short Grain. Grain running with the short dimension of a sheet of paper.

Short Ink. An ink that is very thick and that has high viscosity.

Showthrough. Seeing through on one side of a page what is printed on the other.

Shrink-Wrapping. A semi-automatic process that uses heat to envelope a printed piece in a thin film of plastic for shipment or storage.

Side Comb Binding. A method of comb binding in which the cover wraps around the comb so that it is not visible on the spine of the document; also called hidden comb binding.

Side Head. A headline set to the side of an article, rather than above it.

Side Notes. Short lines or blocks of type positioned in the margins and used for emphasis or to present brief related information.

SIDE-SEWING. A method of sewing in which signatures to be bound are sewn together on the outside on the binding edge; pages do not lay flat.

SIDE-STITCHING (SIDE-WIRING). A binding process in which printed pages are stapled or wire-stitched on the outside parallel with the binding edge.

SIDEBAR. A short article, list, or other block of information related to and placed near a main article.

SIGNATURE. All pages of a publication printed on both sides of a single sheet, folded and trimmed to finished size for final binding; also called section.

SILHOUETTE. A halftone from which the background has been removed so that only the central image remains; also called liftout halftone or (when referring to type) outline.

SILK FINISH (SILK PRINT). A photograph printed on paper with an overall stippled texture.

SILKSCREEN PRINTING (SILKSCREENING). Other terms for screen printing.

SILVERPRINT. Another term for blackprint.

SIMPLE BLACK-AND-WHITE PHOTOGRAPHY (SINGLE-COLOR PHOTOGRAPHY). Other terms for line photography.

SIMPLICITY. A principle of design calling for a design that encourages comprehension of content.

SINGLE-PLATE STRIPPING. Making a single flat to be exposed to a single press plate for a one-color job.

SINGLE-WIRE BINDING. Another term for spiral binding.

SIZE. Dimensions stated as width by depth.

SIZING. An additive to paper to give it extra body and strength and to make it water resistant for offset printing.

SKID PAPER. A frequently used paper that a printer buys in bulk (by the skid) to have available at all times.

SLAB SERIF TYPE. Another term for Square Serif Gothic type.

SLANT (SLASH). Other terms for shill mark.

SLIDE. A color transparency mounted in a paper, metal, plastic, or glass frame for viewing with a projector.

SLIPSHEET. A sheet of a different color or weight inserted as a printed piece comes off the press or out of the bindery to mark increments (such as every 100 copies); also a blank sheet inserted between printed ones to prevent setting-off of wet ink; also called interleaf.

SLIT PERFORATION. A die-cut row of slits; *see also* Pin Perforation.

SLITTING. The severing of paper during the binding process.

SLOTTING. Punching holes in paper in a rectangular or other shape.

SLUG LINE (SLUG). One or a few key words at the top of each page of manuscript or each galley used to identify it or on mechanical artwork

GLOSSARY

to indicate where type or art is to be placed; also a line of hot-metal type.

SMALL CAPS (sc). A specification that letters or words designated are to be set in capital letters equivalent to the x-height of the typeface, not as large as a full capital letter.

SMYTH SEWING. A method of sewing in which signatures to be bound are sewn together inside the gutter so that pages lay flat; *see also* Side-Sewing.

SOCK IMPRESSION. An impression made using pressure, resulting in a noticeable embossing of the paper; *see also* Kiss Impression.

SOFT FOCUS. An image photographed intentionally slightly out of focus to soften it.

SOLID MATTER. Another term for set solid.

SOLIDUS. Another term for shill mark.

SORTS. Odd characters on a type font that are used only occasionally; also called peculiars.

SPACE OUT. To add space between words or letters so that a line of type fills a given horizontal measure.

SPACING. The distance between words or letters in typesetting; *see also* Wordspacing, Letterspacing, Optical Letterspacing.

SPECIAL TYPE. Another term for Novelty type.

SPECIFICATIONS (SPECS, SPECING). Instructions to a supplier indicating exactly what is wanted.

SPECIMEN. A sample of type or of a page to give an idea of the finished product.

SPINE. The bound edge of a publication; also called backbone.

SPIRAL BINDING. A looseleaf method of binding in which a row of holes is punched along the side of pages to be bound and a wire or plastic spiral inserted to hold them together; also called single-wire binding.

SPIRE GOTHIC TYPE. Another term for Old English type.

SPIRIT DUPLICATING. Another term for ditto.

SPLIT FRACTIONS. Another term for piece fractions.

SPLIT RUN. Printing part of a run now and the rest later; also printing two different pieces on the same sheet, and then cutting them apart.

SPLIT-FOUNTAIN PRINTING. Printing two or more colors at a time by dividing the ink fountain with dams to keep inks separated; *see also* Rainbow-Fountain Printing.

SPOILAGE. Printed sheets or finished pieces that are damaged; also called waste.

SPOT COLOR. A second color used only here and there for accent.

SPOT VARNISH. Selective use of varnish to highlight a portion of a page or an art element.

SPOTTING. Retouching a photograph with special ink to remove or diminish unwanted spots.

SPREAD. Pages that will face each other in the finished piece.

SQUARE BRACKETS. Another term for brackets.

SQUARE INDENTATION. Indented both left and right so material is clearly set apart from the rest of the column.

SQUARE SERIF GOTHIC TYPE. A type classification characterized by boxy, heavy serifs; also called Slab Serif or Egyptian type.

SQUARE-FINISH HALFTONE. A halftone in a square or rectangular shape, as opposed to an outline halftone.

STAMPING. Impressing an image onto the front of a sheet of paper so that the design appears recessed on the front, using heat and gold leaf or other metallic foil; also called hot-stamping, die-stamping, or blocking.

STAND-BY TIME. The time a photographer must wait for conditions to be right for shooting.

STANDING INITIAL. Another term for raised initial.

STANDING TYPE (STANDING ART). Type or art that repeats from issue to issue (such as column headings or a masthead).

STAR. Another term for asterisk.

STAT. An abbreviation of photostat.

STAT CAMERA. A small self-contained camera for making photostats.

STEEL DIE. Line engraving on a steel plate.

STENCIL. A thin film with backing that is cut by hand, by typewriter, or by a photographic process to be open in image areas for use in screen printing or mimeography.

STENCIL DUPLICATING. Another term for mimeography.

STEP-AND-REPEAT STRIPPING. Using a mechanical process to make multiple exposures of a flat on a press plate so that several images can be printed at once.

STEREOTYPE. A paper or plastic mold of metal type used to cast a metal duplicate plate.

STET. Proofreader's mark indicating that copy or type is to be left as originally written or set; to leave alone or let stand.

STICK-UP INITIAL. Another term for raised initial.

STIPPLE FINISH. A bond paper finished to have a fine, overall, pebbly texture.

STOCK. The liquid mixture that becomes paper when moisture is extracted; also another term for paper of any kind.

STOCK HOUSE (STOCK LIBRARY). Other terms for photo service.

STOCK PHOTOGRAPHS (STOCK ART). Photographs or other art available for general use and ordered when a need arises, for a one-time use fee.

STOP MARK. Another term for shill mark.

STORY CONFERENCE. A meeting of editorial, design, and art personnel to work out details of a print project before production begins.

STORY ENDER. A miniature symbol or logo used at the end of a lengthy article to indicate its conclusion.

STRAIGHT MATTER. Body copy consisting of plain paragraphs with no tabular or other special typesetting required; type is all one size, style, face, and weight and of the same width.

STREAKING. Another term for ghosting.

STRESS. The distribution of weight or the slant of a type character.

STRIKE-ON TYPESETTING. Another term for impact typesetting.

STRIKE-THROUGH. Excess absorption of ink so that it appears on the other side of a printed sheet.

STRING-AND-BUTTON ENVELOPE. A strong envelope with a pair of heavy paper buttons and a string allowing it to be opened and closed numerous times for interoffice mailing.

STRIPPING (STRIPPING-IN). Combining two or more negatives (usually line and halftone) or substituting part of a negative so that a correct press plate can be made of the composite; also called image assembly, flat assembly, film assembly, or negative assembly.

STRIPPING BASE (STRIPPING VINYL). Other terms for masking sheet.

STRIPPING FILM. Another term for masking film.

STROKE. A line that forms a principal part of a type character.

STUFFER. A small printed announcement or advertisement inserted in an envelope with other material for distribution.

STYLE. Editorial practice for a publication.

STYLEBOOK. A guide to usage and treatment to which an editor refers in writing and other decisions to assure accuracy and consistency.

STYLUS. A drawing tool used for tracing letters or art onto a mimeograph stencil.

SUBHEAD. Copy in smaller type that follows and elaborates on the main headline; also such copy appearing within a lengthy article to break it into sections.

SUBJECTIVE BALANCE. Designing elements of similar weight on each side of a centerline, taking into account the use of bold white space as an element.

SUBLIMATION PRINTING. A printing method using heat to transfer a design to fabric (such as imprinting a company's logo on T-shirts).

SUBSCRIPT. Another term for inferior type.

SUBSEQUENT RIGHTS. Another term for one-time rights.

SUBSTANCE. Another term for basis weight.

SULFITE BOND. Another term for flat bond.

SUNKEN HEAD. Another term for dropped head.

SUNKEN INITIAL. An initial cap extending below the baseline, as opposed to a raised initial extending above the cap height.

SUPERCALENDERED PAPER (SUPER). Paper that has been given extra gloss and smoothness using pressure, steam, and friction during manufacture.

SUPERIOR TYPE (SUPERSCRIPT). Figures or letters smaller than body type and placed above the mean line (such as footnote numbers).

SURFACE CHART. Another term for mountain graph.

SURFACE PLATE. An offset printing plate with image and non-image areas on the same plane.

SURFACE-SIZING. A treatment during manufacture of uncoated offset paper that bonds surface fibers to the sheet for extra stiffness and strength.

SURPRINT. To print type over a halftone or block of color or gray so that the dark type stands out; the opposite of reverse.

SWASH. An extra flourish added to a type character for design emphasis.

SYMMETRICAL BALANCE. Another term for formal balance.

TABLE OF CONTENTS. A list of articles, section titles, or chapter headings with their starting pages appearing in the front (or sometimes on the back cover) of a publication.

TABLE TENT. Another term for tent card.

TABLOID. A newspaper format approximately half the size of a regular newspaper.

TABULAR. Type set in vertical columns (such as budget figures).

TACK. Stickiness of ink that is still damp.

TAGBOARD. A type of bristol paper.

TEARSHEET. A sheet sent to an advertiser or author as proof of publication instead of sending the entire piece.

TENT CARD (TENT). An advertisement or announcement folded into the shape of a simple tent; commonly seen on tables in a restaurant to advertise specials; also called table tent.

TEXT PAPER. A book paper finished to have a texture (such as laid or linen).

TEXT TYPE. Another term for body copy or for Old English type.

TEXTURED COATED PAPER. Another term for gloss coated paper.

THERMOGRAPHY. A heat process commonly used on business cards to raise ink from the surface; also called imitation embossing, plateless engraving, imagic printing, or raised printing.

-30-. A designation for the end of copy; long used by newspaper reporters, it is now less common than ###.

THREE-DIMENSIONAL TYPE. Outline type characters with heavier lines on one side to create the perception of depth.

GLOSSARY

THUMB INDEX. A type of index (such as is used on dictionaries) in which notches are die-cut in the edge of a book at key sections.

THUMBNAIL. A quick, small sketch showing ideas for art or format.

TICKET ENVELOPE. The smallest envelope; open-sided for theater and other tickets.

TIED LETTERS. Type characters linked by an additional stroke to resemble handwriting.

TIGHT. A refined rough sketch showing more detail; an intermediate step between rough and finished art.

TILDE. A Spanish diacritical mark consisting of a wavy line above a letter.

TIME EXPOSURE. Exposing a photographic negative for a period of time or at slow speed to create the illusion of movement.

TINT. Weak color used to shade and accent without overwhelming a halftone or block of type.

TINT SHEET. Another term for shading sheet.

TINTING. Retouching a black-and-white photograph using colored ink to add faint, flat color.

TIPPING-IN. Inserting a single sheet into a publication with glue.

TISSUE PROOF. Another term for glassine proof.

TISSUE-PAPER OVERLAY. Thin paper used over art or mechanical artwork to protect it; spot color and other specifications are frequently indicated on the tissue so the engraver understands what to do.

TITLE. The name of a printed piece or of a chapter in a book.

TO CUM (TK). A shortened version of "to come" used to indicate when a piece of copy, art, type, or other element is not yet in hand.

TOENAIL. A slang expression for quotation mark.

TOMBSTONING. Two headlines of similar size and weight adjacent to one another, thus making either difficult to read.

TONE. Another term for value.

TONE COPY. Another term for continuous tone.

TONE LINE CONVERSION. Another term for degeneration.

TONING. Chemically treating a black-and-white photograph in the darkroom to make it brown, blue, or another color.

TRADE BINDERY. A bindery that does a wide range of finishing and binding work for the printing trade; also called job bindery.

TRADE CUSTOMS. A set of terms and conditions guiding pricing and other procedures for printing and related trades.

TRADE TYPOGRAPHER. A shop that sets type for print projects; it may specialize in particular kinds of projects.

TRANSFER ART (TRANSFER LETTERING). Designs or type that can be rubbed off or lifted off a plastic sheet onto mechanical artwork or an illustration.

TRANSFER PROOF. A pre-press proof of color separations done by exposing each negative to the same opaque sheet; Cromalin is a common brand name; also called integral proof or color opaque proof.

TRANSITIONAL TYPE. A type classification characterized by oval, serifed letters with thick and thin strokes.

TRANSMISSIVE COPY. Art in transparency form.

TRANSPARENCY. A color film positive (such as 35-mm slides); also the plastic sheet on which information appears for use on an overhead projector.

TRANSPARENT. The clear portion of a negative.

TRANSPARENT ART. Art done with pastels and transparent watercolors.

TRANSPOSITION (TR). A proofreader's mark indicating that the order of letters or words is reversed.

TRAPPING. The slight overlap of different colors in lap register that prevents any air between them; also the ability of one ink to overprint another.

TRIM MARKS. Another term for corner marks.

TRIMMING. The severing of a strip of paper along the edge of a printed piece after other processing has occurred to make it the desired finished size.

TRIPHTHONG. Three characters joined together and set as one; *see also* Ligature.

TWO-COLOR PROCESS PRINTING (TWIN-COLOR PROCESS PRINTING). Using any two of the basic four colors of full-process printing in different screen values to create a range of colors on the press.

TWO-IMPRESSION PRINTING. Another term for duocolor.

TWO-TONE PAPER. A paper that has a different color on each side of the same single sheet.

TWO-TONE POSTERIZATION. Another term for degeneration.

TYPE STYLE. The form or proportion of type (such as roman, italic, extended, or condensed).

TYPE-HIGH. Metal type that stands 0.918 in. (2.33 cm) from its feet to its face.

TYPEBOOK (TYPE-SAMPLE BOOK, TYPE SPECIMENS). A collection of samples of all the type styles, faces, weights, and sizes available from a typographer.

TYPEFACE. A particular design of type (such as Chelmsford or Futura).

TYPESETTING. The process of converting copy into type; also called composition.

TYPO (TYPOGRAPHICAL ERROR). An error in typesetting.

TYPOGRAPHER. A supplier who does typesetting.

TYPOGRAPHY. The type used in a printed piece; also the art of designing and using type.

GLOSSARY

TYVEK. A tough, water-resistant blend of synthetic fibers used to make catalog envelopes for shipping purposes.

UMLAUT. A diacritical mark consisting of two dots placed horizontally over a letter; also called diaeresis.

UNCOATED PAPER. Paper without an enamel coating.

UNION BUG. *See* Bug.

UNIT COUNT. Another term for character count.

UNITY. A principle of design calling for the tying together of all elements into a cohesive whole; also called harmony.

UP. How many pages are printed on one side of a sheet (such as two up for two pages per side).

UP STYLE. Capitalizing all main words in a headline or title; the opposite of down style.

UPPER AND LOWERCASE (ulc). Another term for caps and lowercase.

UPPERCASE (uc). A specification that the letters indicated should be capital.

VACUUM FRAME. A unit for exposing flats to make proofs or press plates.

VALUE. The lightness or darkness of a hue; also called tone.

VANDYKE (VAN DYKE). Another term for brownline.

VARNISH. A liquid selectively printed onto a sheet to add sheen to photographs or other elements; not as glossy as lacquer.

VEGETABLES. Another term for printer's marks.

VEHICLE. The ink component that gives it mobility, gloss, and abrasion resistance.

VELLUM FINISH. A paper finish that is smooth and dull.

VELO-BIND. A looseleaf method of binding in which a row of holes is punched along the side of pages to be bound and plastic prongs on a strip inserted, sheared off, and fused to another plastic strip on the back to hold them together.

VELOX. A prescreened print of a halftone; *see* Prescreened Print.

VERSO. The left-hand page of a publication, always with an even number; the opposite of recto.

VIGNETTE. A halftone in which the background fades away and blends into the paper.

VIRGULE. Another term for shill mark.

VISUAL OPACITY. The opacity of a sheet before printing.

VOLATILIZATION. Another term for evaporation.

WALLET ENVELOPE. A square-flapped, heavy-duty envelope used for bulky correspondence.

WASHUP. The cleaning of a press between runs when different colors are used or at the end of the day.

WASTE. Another term for spoilage.

WATERLESS LITHOGRAPHY. Another term for letterset printing.

WATERMARK. A design made in paper while it is still wet and visible when held to the light; often used on quality letterhead as a subtle way to present a company name or logo.

WATERPROOFING. A treatment during manufacture of coated paper that assures that the paper will not be damaged by water.

WAXED GALLEYS. Galleys with a thin coating of wax on the back allowing the galley to be moved around easily on a dummy or mechanical artwork, and then burnished securely in place.

WEB GAIN. An unacceptable stretching of paper as it feeds through a web press caused by improper tension.

WEB PRESS. A press that uses a continuous roll of paper rather than individual sheets.

WEIGHT (PAPER). How heavy a particular paper is; *see* Basis Weight.

WEIGHT (TYPE). The perceived boldness of type from hairline (very light) to extra bold (very heavy); in hot type, 4 in.^{2} (25.8 cm^2) of solid type weigh approximately 1 lb (0.45 kg).

WHITE SPACE. Areas on a page not occupied by type, photographs, or other elements.

WHOLE FRACTIONS. Type characters for common fractions set in one piece; also called case fractions; *see also* Piece Fractions.

WIDOW. A partial line of type at the top of a column; sometimes used only for a partial line no more than one-fourth the column width; also sometimes used to refer to a partial line at the end of a paragraph.

WINDOW. Space set aside on a dummy, mechanical artwork, or press negative for a halftone or other art to be added.

WINDOW ENVELOPE. A pointed-flapped envelope having a glassine or open window on the front for the address to show through.

WIRE SIDE. The side of a sheet of paper that was against the wire during manufacture.

WITH THE GRAIN. Running the same direction as the grain of the paper.

WORDSPACING. The distance between words in a typeset line.

WORK-AND-TUMBLE (WORK-AND-FLOP, WORK-AND-ROLL). On a two-sided job to be cut apart lengthwise, printing two Side A's on the front of a sheet, then turning the sheet over end to end and printing two Side B's on the back using the same press plate but a different gripper edge; also called print-and-tumble.

WORK-AND-TURN. On a two-sided job to be cut apart crosswise, printing one Side A and one Side B on the front of a sheet, and then turning the sheet over side to side and again printing one Side A and one Side B on the back, using the same press plate and gripper edge; also called print-and-turn.

WORK-AND-TWIRL (WORK-AND-TWIST). On a one-sided job, printing two identical images, and then twirling the sheet end to end to print the second using the same press plate but a different gripper edge.

WORK-FOR-HIRE. Any writing, illustration, photography, or other artistic project done for pay and relinquished as the sole property of the purchaser.

WOVE FINISH. A paper finish resembling tightly woven cloth; finer than a linen finish.

WRAPAROUND (WRAP AROUND). Art on the front cover of a publication extending to the back cover or to an inside flap; also another term for runaround.

WRAPAROUND FOLD. Another term for barrel fold.

WRITING TO FIT. Writing copy to fill a given space when typeset.

WRONG FONT (wf). A proofreader's mark indicating that a typeface other than the one specified was used.

X-HEIGHT. That portion of a type character exclusive of ascender or descender; also called body height.

XEROGRAPHY. A dry, electrostatic printing process used by photocopiers.

Z-FOLD. Another term for accordion fold.

Colophon

Manuscript Preparation: Osborne 1 using WordStar.

Phototypesetting: Linotron 202.

Type Selection: 10/12 Times Roman medium for body type; Optima for display type.

Line Screen: 133 (54 cm).

Press: One-color Harris.

Paper: 50# Thorcote 634 Hi-White.

Ink: K&M black.

Binding: Arrestox B-grade cloth over 80-point board. Smyth sewn.

Index

Accents, 297, 477
Acetate, 322, 477
Agate, 263, 477
Agent, artists', 23, 237–238, 478
Airbrushing, 194–195, 477
All caps, 478
Alternative character, 262, 478
Ampersand, 293, 478
Appendix, 478
Approvals:
   how to get, 92
   need for, 92
   when to get, 71
Art:
   camera-ready, 217, 483
   computer, 221, 486
   continuous-tone, 20, 50, 486
   cost of changes to, 39
   cropping, 328–329
   defined, 166, 478
   dummying, 311
   files, 215–216
   ganging, 54, 342, 496
   grouping of, 199
   how to handle, 210
   influence of printing paper on, 196
      on color reproduction, 357
      on design, 100
   line:
      defined, 50, 503

      ganging of, 342
      pasting up, 320
   margins, 128, 505
   marking, 332
   opaque, 176, 508
   photocopies as, 186
   placement of, 197
   place in production sequence, 4
   planning use of, 167–171
   plate, 512
   rubbings as, 186, 517
   saving on, 48–49
   scaling, 329–332
   sizing of, 199
   stock, 216
   story ender, 186, 523
   transfer, 211–212, 217, 525
   using type as, 178
Art direction:
   defined, 478
   for illustration, 226
   for photography, 233–234
Artists' agent, 23, 237–238, 478
Artists' knife, 316, 478
Art sources, 215–222
Artwork, *see* Mechanical artwork
Ascender, 261, 478
Asterisk, 294, 297, 478
Audience:
   copy style for, 78

Audience (*cont.*)
    education level of, 78
    geographic distribution of, 77
    influence:
        on copy, 77
        on design, 97
        on planning, 13
    primary, 77
    secondary, 77
Author, 478
Author's alteration, 39, 291, 478
Autoscreen, 346, 478

Backbone, *see* Spine
Background, 187, 479
Backing up, *see* Back-to-back
Backmatter, *see* Back-of-the-book
Back-of-the-book, 478
Backslant, 479
Back-to-back, 479
Bad break, 479
Bar mark, 297, 479
Baseline, 261, 479
Baseline-to-baseline, 263, 479
Basis size, 420, 479
Basis weight, 420, 479
Benday, 173, 479
Bevel, 320, 479
Bibliography, 480
Bid:
    defined, 32, 480
    evaluating, 35
    when to ask for, 33
Bill:
    avoiding surprises, 41
    content of, 42
    verifying, 43
Billing, 7
Bindery:
    defined, 480
    types of, 439
Binding:
    cost of changes during, 41
    defined, 480
    evaluating, 440
    influence:
        on design, 108
        on planning, 15
    place in production sequence, 4
    planning for, 440
    saving on, 60
    specifications for, 23
    types of, 445–452
Binding edge, 480

Binding trade customs, 440, 466–467
Blackprint, 348, 370, 480
Blanket, 385, 386, 480
Bleed, 128, 480
Board, *see* Mechanical artwork
Body height, *see* X-height
Boldface, 266, 481
Book, 3, 481
Booklet, 3, 481
Border:
    defined, 481
    pasting up, 321
Bourges, 206, 481
Bowl, 261, 481
Box:
    defined, 481
    overusing, 147
    use of, 130
Brace, 297
Brackets:
    defined, 297, 481
    square, 522
Brochure:
    defined, 3, 482
    sample format for, 154
Bronzing, 482
Budget:
    annual, 30
    fixed, 27
    influence on planning, 14, 16
    open, 29
    revising specifications to match, 29
Budgeting:
    approaches to, 27–31
    for appropriate quality, 30
    as buyer responsibility, 6
    by comparison, 28
    place in production sequence, 4
Bulk, 102, 482
Bulk out, 419
Bullet, 482
Burning-in, 193, 482
Burnish, 320, 482
Business cabinet, 2, 482
Business stationery, 2
Byline, 482

Cabbages, *see* Printer's marks
Calendering, 416, 482
Caliper, 423, 482
Calligraphy:
    defined, 482
    use of, 130, 132, 244

INDEX

535

Camera-ready art:
  defined, 20, 483
  use of, 49, 217
Camera-ready galleys, 276–277
Cannibalize, 301
Cap height, 483
Capital, *see* Uppercase
Caps, and lowercase, 483
  initial:
    defined, 295, 500
    inset, 500
    raised, 515
    sunken, 524
  small, 295, 521
Caption, 310, 483
Carbon tissue, 379, 483
Caret, 483
Casting off, *see* Copyfitting
Cathode ray tube, 247, 483
Centered dot, 483
Center of interest, 187, 483
Center spread, 483
Character casting, 241–243, 484
Character count, 277–278, 281–284, 484
Character unit, *see* Unit count
Charcoal drawing, 173, 484
Charges, minimum, 53
Chart:
  defined, 177, 484
  types of, 177–178
Chase, 372, 484
Chinese white, 484
Chroma, 349, 484
Circular, 484
Clearances, *see* Approvals
Close up, 292, 484
Cold type:
  defined, 244, 484
  electronic, 250–253
  impact, 244–246
  manual, 244
  phototypeset, 246–250
Collage, 485
Collating, 441, 485
Colophon, 485
Color:
  additive, 349
  balancing, 196, 485
  correction of, 354, 485
  defined, 485
  fake, 493
  fugitive, 408, 496
  overusing, 146
  paste-up for, 332

  perception of, 359
  proofs of, 355–357
  reproduction of, 348–359
  separation of, 342, 350–354, 485, 489, 499
  spot, 125, 333, 521
  subtractive, 350
  use of multiple, 125
Color-Key, 355, 485
Column:
  defined, 18, 485
  depth of, 18
  width of, 270
Column heading, 485
Column inch, 485
Column rule, 485
Composing stick, 241, 486
Computer graphics, 221, 486
Computer typesetting, *see* Electronic composition
Contact, *see* Proof, photographic
Contact printing, 342, 486
Continued line, 486
Copy:
  agate, 263, 477
  approving, 92
  audience for, 77
  bias-free, 79
  body, 481
  condition of, 52
  defined, 486
  for electronic typesetting, 275
  hard, 247, 498
  for hot type, 272
  how to mark, 274, 276
    to existing or related materials, 79
  influence:
    of cost, 80
    on design, 99
    of schedule, 80
  line, 503
  news service, 81
  objectives of, 76
  old, 453, 508
  penalty, 272, 510
  for phototypesetting, 272
  place in production sequence, 4
  reflective, 357, 515
  in relation to design, 79
  saving on, 48
  straight matter, 523
  style of, 78
  tabular, 524
  transmissive, 357, 526
  typesetting specifications for, 285

Copy (cont.)
  typing, 272–277
  use rights to, 84
Copy camera, see Process camera
Copy casting, see Copyfitting
Copy editing, 88–92, 486
Copyfitting, 277–285, 304, 486
Copy length, estimating, 277–285
Copyright:
  camera-ready art, 218
  illustrations, 223
  mechanical artwork, 236
  photographs, 229–230
  restrictions of, 81
Copy sources, 80–83
Copy unit, 486
Copywriter:
  agent for, 237
  benefits, drawbacks of using, 82
  evaluating, 84
  fees and expenses, 83
  how to find, 82
  references, 87
  working with, 87
Cornering, 443, 487
Corner marks, 317, 487
Corrections:
  checking, 298
  pasting up, 335
Cost, influence of:
  on art, 169
  on copy, 80
  on design, 100
Counter, see Bowl
Cover:
  defined, 487
  duplex, 418
  flush, 495
  overhang, 509
  self, 449, 518
Creasing, see Scoring
Credit line, 226, 234, 237, 487
Creep allowance, 315
Cromalin, 356, 487
Crop, 311, 487
Crop marks, 332, 487
Cropping angles, 328, 487
Cross-stroke, 261, 487
Currency, photographing, 191
Cut:
  defined, 487
  letterpress, 377
Cutline:
  defined, 18, 487

  placement of, 310
  writing, 310–311
Cutting, 443, 487
Cyan, 349–350, 486

Dandy roller, 415, 488
Dash, 297, 488
Deadline, as supplier responsibility, 7
  defined, 488
  scheduling to meet, 64–74
Debossing, blind, 135, 444, 480
  defined, 135, 443, 488
  use of, 135
Degeneration, 208, 488
Delivery:
  influence on planning, 16
  of printing pieces, 452
Depth of field, 189, 488
Descender, 261, 488
Design:
  considerations of, 97–101
  cost of changes to, 39
  errors in, 137–153
  influence on copy, 79
  place in production sequence, 4
  principles of, 109–117
  saving on, 48–49
  use rights to, 159
Designer:
  agency, 158
  agent for, 237
  defined, 157
  evaluating, 159
  fees and expenses, 158
  freelance, 157
  how to find, 157
  references, 162
  studio or group, 157, 488
  working with, 162
Designer's delight, 16, 488
Design techniques, 125–135
Diacritical marks, 297, 488
Diagram, 178
Die-cutting:
  defined, 135, 444, 488
  marking, 324, 489
  use of, 135
Die-stamping, see Stamping
Diffusion transfer, see Photomechanical transfer
Digitalized art, see Computer graphics
Dingbats, see Printer's marks
Diphthong, 262, 489
Direct screening, 355, 489
Display ad, 489

INDEX

537

Display matter, 489
Doctor blade, 395, 489
Dodging, 193, 489
Dot gain, 490
Dot pattern, 344, 345, 490
Double bump, *see* Duocolor
Double burn, 490
Down style, 490
Drawings:
 continuous-tone, 175–176
 line, 171–175
Drilling, 443, 490
Drive out, 490
Dropout, *see* Reversing
Drying power, 388, 491
Dummy:
 comprehensive, 49, 99, 302, 485
 defined, 491
 folding, 104
 imposition, 301, 499
 printing, 108, 301, 514
Dummy sheet:
 defined, 491
 ruling, 303
Duocolor, 125, 491
Duograph, 125, 126, 149
Duotone, 125, 491
Dupe and assembly, 354, 491
Duplicator, *see* Offset printing, duplication
Dylux, 491

Ear, 491
Economical production, 45–61
Editing:
 defined, 491
 what to check, 89–91
Editing and correcting terminal, 491
Editor, 492
Editorial, 492
Editorial style, 89
Electronic composition:
 capabilities of, 253
 copy for, 275–276
 defined, 250, 492
 process of, 250–253
 proof of, 289
Ellipsis, 294, 492
Em, 264, 296, 492
Embossing:
 blind, 135, 444, 480
 defined, 135, 444, 492
 imitation, *see* Thermography
 use of, 135
Emulsion, 492

En, 264, 296, 492
End-of-line decisions, 247, 492
Endmatter, *see* Back-of-the-book
Engraver, 341, 359
Engraving:
 defined, 341, 493
 fake, *see* Thermography
 as illustration technique, 173
 line, 398, 503
  copperplate, 398, 486
  steel die, 398, 522
 process of, 341–381
 specifications for, 19
Envelopes:
 cost of odd size, 60
 C series, 427
 sizes of, 427–429
 styles of, 427–431
Epigraph, 493
Erratum slip, 493
E-scale, 263, 493
Estimate:
 budgetary, 17
 defined, 32, 493
 evaluating, 35
 when to ask for, 33
Etchtone, 210, 493

Face, *see* Typeface
Feed edge, *see* Guide edge
Filler, 305, 494
Film assembly, *see* Stripping
Filters, for photographs, 193
 in color separation, 350–353
Finder's fee, 46, 228
Finishing:
 cost of changes during, 41
 defined, 494
 influence on planning, 15
 place in production sequence, 4
 saving on, 60
 specifications, 23
 types of, 441–445
Flag, 494
Flap, 494
Flat, 361, 494
Flat assembly, *see* Stripping
Fleurons, *see* Printer's marks
Flexography:
 defined, 394, 494
 ink for, 434
 platemaking for, 379
Flop, 199, 321, 495
Flow chart, 178, 495

Flyer:
    defined, 2, 495
    sample formats for, 152, 153
Focus:
    defined, 189, 495
    soft, 231, 521
Folder, 2, 495
Folding:
    described, 441
    dummy, 104
    influence on design, 105
    machines, 441
Fold marks, 324, 495
Folds:
    short, 519
    tricky, 60, 147
    types of, 441–442
Font, 241, 261, 495
Footnote symbols, see Reference marks
Foreground, 187, 495
Format:
    broadside, 482
    defined, 12, 100, 495
    double truck, 490
    influence on design, 100
    portrait, 512
Forms:
    defined, 3
    design of, 49
Fountain:
    defined, 387, 495
    rainbow, 396, 515
    split, 387, 521
Fraction mark, see Bar mark
Fractions:
    piece, 512
    whole, 528
Frontmatter, 496

Galley:
    camera-ready, 276
    changing phototypeset, 250
    defined, 289, 496
    hot-type, 241
    proofreading, 291–298
    proofs of, 288–291, 496
Glossary, 496
Golden mean, 112, 496
Goldenrod, see Masking sheet
Grammage, 420, 497
Graph:
    defined, 177, 497
    types of, 177–179
Graphic, see Art

Graphic arts camera, see Process camera
Graphic arts trade customs, 455–457
Graphic designer, see Designer
Gravure:
    defined, 395, 497
    image assembly for, 373
    ink for, 434
    paste-up for, 337
    platemaking for, 379–380
    presses, 395–396
Greeking, 302, 497
Gripper margin, 315, 497
Guide edge, 497
Gunning Fox Index, 78, 497
Gutter, 497

Hairline, 261, 497
Halftone:
    defined, 344, 498
    density of, 488
    double-dot, 490
    dropout, 490
    duplex, see Duotone
    ganging, 348
    high-key, 498
    highlight, 498
    outline, 508
    paste-up for, 324
    phantom, 205, 510
    square-finish, 522
    tinted, see Duograph
Halftone assembly, 364–366
Halftone negatives, 344–348, 498
Head-to-foot, 366, 498
Head-to-head, 366, 498
Headline:
    banner, 479
    bulletin, 206, 306, 308, 482
    continued, 486
    counting, 307–310
    deck, 488
    defined, 18, 498
    dropped, 490
    types of, 304, 306, 308
    writing, 306–307
H-height, 497
Hickey, 408, 498
Holding line, see Keyline
Holdover, 306, 499
Hot stamping, see Stamping
Hot type:
    copy for, 272
    defined, 240, 499
    forms of, 240–244

INDEX

539

  image assembly for, 372
  proof of, 289
Hue, 349, 499

Illustration, with photograph, 206
  continuous tone, 171–175
  defined, 166, 499
  line, 20, 175–176
  pasting up, 320
  protecting, 50
  roughs, 226
  saving on, 50
  techniques, 171–186
  tights, 226, 525
  use rights, 223
  when to use, 170
Illustrator:
  agent for, 237
  evaluating, 223
  fees and expenses, 222
  how to find, 222
  references, 225
  working with, 225
Image, influence:
  on design, 97
  on planning, 13
Image area, *see* Live matter
Image assembly:
  cost of changes during, 40
  place in production sequence, 4
  process of, 341–381
  saving on, 54
Imagic printing, *see* Thermography
Imposition:
  defined, 103, 499
  dummying for, 313
  stripping for, 366
  types of, 56, 366–367
Impression, defined, 104, 393, 499
Imprinting, 386, 499
Indentation:
  defined, 499
  hanging, 296, 498
  square, 522
Index, thumb, 525
Ink:
  additives, 433
  color selection, 435
  coverage, 22
  drying methods, 433–434
  flexography, 434
  fluorescent, 432, 495
  ghosting, 408, 496
  gravure, 434

  holdout, 500
  influence:
    on color reproduction, 358
    on planning, 15
  letterpress, 434
  letterset, 434
  long, 432, 504
  magnetic, 432, 504
  offset, 434
  properties of, 432–433
  screen printing, 435
  semi-transparent, 518
  short, 432, 519
  starving, 500
Ink wash,:
  defined, 500
  drawings, 175
In-line, 386, 499
Insert, 445, 500
Inset, 199, 202, 500
Instant printing, *see* Quick printing
Intaglio printing:
  defined, 500
  gravure, 395
  line engraving, 398
  thermography, 398
Intercharacter spacing, *see* Letterspacing
Interleaf, *see* Slipsheeting
Interline spacing, *see* Leading
International Standards Organization, 423, 500
Internegative, 192, 501
Invoice number, 42

Job number, 42
Job ticket, 42, 501
Jump, 304, 501
Jump line, *see* Headline, continued
Justowriter, 245

Kern, 18, 262, 271, 501
Keyline, 321, 501
Kicker, 305, 501
Knockout, *see* Reversing

Lacquer, 128, 433, 502
Laminating, 445, 502
Leader, 294
Leading, 17, 263, 502
Legend:
  defined, 502
  placement of, 311
  writing, 310
Letterguide, 502
Letterhead:

Letterhead (*cont.*)
  design of, 49
  standard sizes of, 425
Letterpress printing:
  advantages of, 393
  defined, 390, 502
  disadvantages of, 394
  ink for, 434
  paste-up for, 336
  platemaking for, 377–379
  types of, 390–393
Letterspacing, 245, 249, 271, 503
  optical, 248, 508
Lift-off art or type, 212, 503
Liftout, *see* Reversing
Liftout halftone, *see* Silhouetting
Ligature, 262, 503
Linear definition, *see* Degeneration
Linecasting, 241, 243, 504
Line conversion, *see* Degeneration
Line gauge, 263, 503
Line length, *see* Column, width of
Line screens, *see* Screens, halftone
Linespacing, 241
Linespacing gauge, *see* Line gauge
Line weight, 202
Lithography:
  defined, 385, 504
  offset, *see* Offset printing
Live matter, 303, 343, 504
Logo:
  changing, 122
  defined, 117, 504
  designing with, 117
  developing, 117
  using, 123
Logotype, type, 262
Lowercase, 241, 504

Magapaper, 3, 504
Magazine, 3, 504
Magenta, 349–350, 504
Makeready, 55, 56, 106–107, 388, 394, 504
Makeup, 504
Map, as art, 183–185
Mar coating, *see* Lacquer
Markup, 46, 505
Masking film, 323, 505
Masking sheet, 360, 505
Masters:
  defined, 505
  use of, 56
Masthead, 505
Matrix, 241, 378, 505
Mean line, 261, 505

Mechanical artist:
  agent for, 237
  defined, 505
  evaluating, 236
  fees and expenses, 235
  how to find, 235
  how to work with, 237
  references, 236
Mechanical artwork:
  benefits of, 300
  cleaning and protecting, 336
  for color, 332
  cost of changes to, 40
  defined, 300, 505
  evaluating, 337
  for gravure, 337
  for letterpress, 336
  for offset, 315
  pasting up, 313–335, 336–337
  place in production sequence, 4
  proofing, 335
  saving on, 53
  for screen printing, 337
  supplier requirements for, 314–315
  use rights, 236
Mechanical dummy:
  defined, 237, 300, 506
  making, 301
  when to prepare, 302
Mechanical tinting, *see* Benday
Media conversion, 252, 506
Mezzotint, 210, 506
Micrometer, 423, 506
Miter, 322, 506
Mock-up, *see* Dummy
Model releases, 228, 469–471, 506
Models, 228
Modem, 252, 506
Morgue, 221, 506
Mortising, 127, 143, 199, 506
Mug shot, 507
M weight, 420, 504

Nameplate, *see* Flag
Negative assembly, *see* Stripping
Negatives:
  changes on, 368–369
  color, 348–359
  continuous-tone, 344, 486
  defined, 507
  halftone, 344–348
  line, 341–343, 503
  proofing of, 369–372
Newsletter:
  defined, 3, 507

INDEX

sample format for, 155
Newspaper, 3, 507
News service, 81
Newzine, *see* Magapaper
Nonpareil, 263, 507
Nonrepro blue, 211, 507
Not fewer than, 15, 22
Not for resale, 45
Numbering:
   double numeration, 490
   of pages, 153
   process of, 444, 507
Numbers:
   alphanumeric, 478
   Arabic, 478
   hanging figures, 498
   hang for ten, 498
   lining figures, 504
   Roman, 516

Objectives, art, 167
   influence:
      on copy, 76
      on design, 97
      on planning, 12
Off-setting, *see* Set-off
Offset printing:
   advantages of, 388
   defined, 385, 507
   disadvantages of, 389
   duplication, 389, 491
   ink for, 434
   paste-up for, 315–336
   platemaking for, 374–377
   process of, 385–390
   quick printing, *see* Quick printing
   special purpose, 389–390
   waterless, 390
   web, 386, 392, 528
Optical bar recognition, 247, 508
Optical character recognition, 251, 508
Optical scan, *see* Optical character recognition
Organizational chart, 178, 508
Ornaments, *see* Printer's marks
Overlay:
   acetate, 322, 477
   defined, 54, 509
   dummying for, 313
   proof, 509
   tissue-paper, 336, 525
Overmatter, *see* Holdover
Overs, 509
Overset, *see* Holdover
Overtime, use of, 74

Packaging, of printing pieces, 452
Padding, 41, 450, 509
Page, 509
Page makeup, 300, 372–373, 509
Pagination, 509
Paintings, photographing, 191
Pamphlet, *see* Booklet
Panel, 21, 509
Pantone Matching System, 435, 509
Paper:
   A series, 423
   brightness of, 196, 424
   B series, 423
   bulk of, 423
   buying, 57–58, 419–427
   cost of coated, 58
      of odd size, 60
      of uncoated, 58
   felt side of, 415, 494
   finishes of, 415–417
   grades of, 416, 497
   grain of, 103, 420, 497, 519
   influence:
      on color reproduction, 358
      on design, 101
      on planning, 15
   manila, 505
   manufacture of, 414–416
   opacity of, 424
   photographic, 191
   RC, 249
   rotopaper, 423, 516
   runnability of, 423, 517
   S, 249
   saving on, 57, 424–427
   size of, 422–423
   skid, 58, 425, 520
   surface of, 196
   textures of, 415–416
   two-tone, 526
   types of, 416–419
   weight of, 420–422, 528
   wire side of, 415, 528
Pasted form, 450, 510
Paste-up:
   defined, 300, 509
   how to do, 313–335, 336–337
   sheet, 217, 509
Patrix, 395, 510
Peculiars, 241, 261
Perforation:
   defined, 510
   types of, 444
Permissions, 81, 93
Photo agency, *see* Photo service

Photocomposing, 248, 511
　layout for, 288
Photoengraver, *see* Engraver
Photoengraving, 341–359, 511
Photo essay, 170, 510
Photographer:
　agent for, 237
　efficient use of, 51
　evaluating, 230
　fees and expenses, 227
　how to find, 226
　how to work with, 233
　references, 233
Photographic clichés, 187
Photographic illustration, 511
Photographic stencil, 381, 511
Photographic techniques, 186, 192–196, 205–208
Photographic treatments, 205–208
Photographs:
　background, 187
　black and white from color, 192
　center of interest, 187
　composition of, 186–189, 486
　defined, 166, 511
　finishes of, 192
　fluorescing, 204
　focus, 189
　foreground, 187
　framing, 192
　full-frame, 328, 496
　grain, 202, 497
　lighting, 190
　problem, 203
　processing and printing, 229
　proofs of, 229
　protecting, 51
　stock, 50, 218, 522
　textures of, 192, 203
Photography:
　halftone, 344–348
　line, 341–342
　poor use of, 142–146
　recognizing quality, 186–191
　saving on, 50–51
　use rights, 229–230
　when to use, 170
Photogravure 395
Photojournalism, 511
Photo-masking, 355, 510
Photomechanical transfer, 325, 511
Photomechanics, 511
Photo-offset printing, *see* Offset printing
Photoprint, *see* Proof, reproduction
Photoproof, *see* Proof

Photorepro, *see* Proof, reproduction
Photo service, 218, 510
Photostat, *see* Stat
Photo story, 170, 510
Phototypesetting:
　capabilities of, 249
　changing galleys, 250
　copy for, 272–274
　defined, 246, 511
　process of, 246–249
　proof of, 289
Pica, 18, 263, 511
Pica pole, 511
Pi characters, 262, 511
Picking, 276, 511
Piercing for type, *see* Mortising
Planning:
　as buyer responsibility, 6
　considerations of, 12–14
　cost of changes during, 38
　steps in, 10
Plateless engraving, *see* Thermography
Platemaking:
　gravure, 379–380
　offset, 374–377
　letterpress, 377–379
　screen, 381
Plates:
　defined, 512
　types of, 374–380
Ply, 418, 512
PMT, *see* Photomechanical transfer
Point, 263, 512
Poster:
　defined, 3, 512
　sample format for, 156
Posterization, 209, 513
　two-tone, *see* Degeneration
Pre-assembly, *see* Dupe and assembly
Preliminaries, *see* Front matter
Prep, 403, 513
Pre-positioning, *see* Dupe and assembly
Preprint, 513
Pre-printed sheets, *see* Masters
Prescreened print, 324–325, 513
Presentation folder:
　defined, 3, 513
　sample format for, 156
Press, color reproduction on, 359
　influence:
　　on art, 169
　　on cost, 55
　　on design, 103
　perfecting, 387, 510
　proofing, 389, 514

INDEX

543

summary of capabilities, 106
Press check, 408, 513
Press run:
  defined, 14, 513
  how to determine, 14–15
  not fewer than, 15, 22
  split, 56, 521
Press work, saving on, 55
Pricing, as supplier responsibility, 7
  methods of, 32
  politics of, 34
Print buyer, responsibilities of, 5
Printers:
  cheap-and-dirty, 34, 484
  evaluating, 404–406
  job, 391, 501
  organization of, 402–404
  types of, 400–402
  working with, 406–411
Printer's error, 513
Printer's flowers, *see* Printer's marks
Printer's marks, 262, 513
Printer's trademark, *see* Colophon
Printing:
  collotype, 398, 485
  cost of changes during, 41
  direct lithography, 390, 489
  ditto, 399, 489
  driography, 390, 490
  flexography, *see* Flexography
  four-color process, 348–359, 495
  gravure, *see* Gravure
  ink-jet, 399, 500
  intaglio, *see* Intaglio printing
  letterpress, *see* Letterpress printing
  letterset:
    defined, 376, 394, 502
    ink for, 434
  mimeography, 399–400, 506
  offset, *see* Offset printing
  over, 509
  place in production sequence, 4
  quick, *see* Quick printing
  raised, *see* Thermography
  relief, 390, 515
  screen, *see* Screen printing
  shadow, 277, 519
  specifications for, 21, 407–408
  sublimation, 523
  trade customs of, 406, 461–466
  two-color process, 350, 526
  two-impression, *see* Duocolor
  types of, 385–400
  unfair practices, 410
Printing-in, *see* Burning-in

Process blue, 349
Process camera, 341, 514
Process red, 349
Production:
  advice, 7
  considerations, 14–16
  coordination of, 6
  defined, 514
  documenting, 4, 453–454
  files, 453
  influences on design, 101–108
  responsibilities for, 4–7
  schedule for, 64–74
  specifications for, 16–24
  steps in, 4
Proof:
  blueline, 289, 369, 480
  brownline, 289, 370, 482
  Bruning, 289, 482
  checker, 290
  clean, 484
  color, 355–357
  defined, 514
  Diazo, 289, 488
  electronic composition, 289
  engraver's, 493
  foundry, 495
  galley, 288–291
  glassine, 290
  hot-metal, 289
  ordering, 290
  Ozalid, 289, 509
  page, 288, 509
  photocomposition, 289
  photographic, 51, 229, 486
  pink, 290, 512
  press, 57, 513
  progressive, 356, 514
  reproduction, 290, 516
  revised, 516
  rough, 290
  transfer, 526
Proofing:
  color, 355–357
  mechanical artwork, 335–336
  negatives, 369–372
  type, 288–291
Proofreader's marks, 292–298, 514
Proofreading, 291
Property release, 228, 469–470, 514
Proportional spacing, 245, 514
Proportion scale, 330, 514
Punching, 443, 514

Quad, 241, 514

Quaint character, *see* Ligature
Quick printing:
   defined, 389, 401, 515
   trade customs of, 407, 464–466
Quotation, *see* Bid

Ream, 420, 515
Recto, 153, 515
Reference marks, 297, 515
Registration, as print specification, 22
   defined, 22, 515
   pin, 323, 366, 512
   types of color, 333–335
Registration marks, 323, 515
Reply card, 515
Reprinting, permission for, 81
Reproduction photography, *see* Photography, line
Resist, 379, 516
Retakes, 51, 234
Retouching, 194–196, 516, 522
Reversing, 133–135, 516
Rivers, 271, 516
Rotofilm, 379, 516
Rotogravure, 395, 516
Roto section, 395, 516
Rub-off art or type, 211, 517
Rules, cutoff, 487
   overusing, 147
   pasting up, 321
   use of, 130
Ruling pen, 322
Runaround, 271, 517
Run back, 292, 517
Run down, *see* Drive out
Run-in, 517
Running feet, 313, 517
Running head, 313, 517
Run on, 517
Run over, *see* Drive out
Run through, 517
Rush, 517

Sans serif, 517
Scanning, 251, 517
Schedule, drafting and revising, 66–68
   influence:
      on art, 169
      on copy, 80
      on design, 100
      on planning, 16
   staying on, 69–72
Scheduling, as buyer responsibility, 6
   considerations, 64–66

economical, 45
place in production sequence, 4
shortcuts, 72–74
Scoring, 103, 444, 518
Scratchboard, 172, 518
Screen printing:
   defined, 396, 518
   paste-up for, 337
   stencils for, 381
Screens:
   defined, 20, 518
   halftone, 344–348
   paste-up for, 322–324
   types of special, 208–210
   use of, 126–127
Section mark, 297
Selectric Composer, 245
Self-mailer, 518
Separatrix, *see* Shill mark
Sepia, 193, 518
Serigraphy, 396, 518
Set:
   close, 519
   defined, 518
   open, 519
   size, 242, 519
   solid, 17, 519
Set-in, 374, 519
Set-off, 388, 519
Setwise, 519
Sewing:
   defined, 519
   side, 448, 520
   Smyth, 450, 521
Shading sheet, 173, 519
Sheet, 519
Shells, *see* Masters
Shill mark, 297, 519
Shingling, 108, 519
Showthrough, 417–418, 424, 519
Shrink-wrapping, 452, 519
Sidebar, 520
Side notes, 519
Signature, 104, 393, 520
Silhouetting, 205, 520
   paste-up for, 328
Silkscreen printing, *see* Screen printing
Silverprint, *see* Blackprint
Simple black-and-white photography, *see* Photography, line
Single-color photography, *see* Photography, line
Sizing, 418, 520
   surface, 524
Slant, *see* Shill mark

INDEX

Slipsheeting, 61, 452, 520
Slitting, 443, 520
Slotting, 520
Slug:
    defined, 241, 274
    line, 304, 520
Solidus, *see* Shill mark
Sorts, 261, 521
Specifications:
    defined, 11, 521
    place in production sequence, 4
    supplier working to, 7
Spine, 138, 445–448, 450–452, 521
Spoilage, 14–15, 521
Spread, 522
Stamping, 444, 522
Stamps, photographing 191
Stand-by time, 228, 522
Star, *see* Asterisk
Stat, 311, 320–321, 522
Stat camera, 342, 522
Statement, *see* Bill
Stencil:
    defined, 522
    mimeograph, 400
    screen printing, 381
Stet, 522
Stitching:
    saddle, 449, 517
    side, 448, 520
Stock, 415, 522
Stock house, *see* Photo service
Stock library, *see* Photo service
Stop mark, *see* Shill mark
Story conference, 170, 523
Strike-through, 523
Stripper, 360
Stripping:
    defined, 523
    halftone, 364–366
    process of, 359–369
    types of, 362–364
Stripping base, *see* Masking sheet
Stripping film, *see* Masking film
Stripping vinyl, *see* Masking sheet
Stuffer, 523
Style guide, 89, 523
Stylus, 523
Subcontracting:
    defined, 37
    influence:
        on pricing, 37
        on schedule, 74
    markup for, 46

Subhead, 18, 307, 523
Substance, *see* Basis weight
Supercalandering, 416, 514
Suppliers:
    capabilities of, 16
    contracting for directly, 46
    matched to project, 46
    responsibilities of, 6
Surprinting, 133, 524
Swash, 524

Table, as art, 181–182
Table of contents, 524
Table tent, *see* Tent card
Tabloid, 3, 524
Tack, 432, 524
Tearsheet, 524
Teletransmission, 252
Tent card, 420, 524
Thermography, 137, 398, 524
Thumbnail, 49, 99, 525
Tied letters, 262, 525
Time exposure, 193, 525
Tint, 525
Tint sheet, *see* Shading sheet
Tinting,:
    defined, 525
    paste-up for, 322
    of photographs, 106, 125, 146, 196
    of type, 126, 146
Tipping-in, 445, 525
Title, 525
To cum, 304, 525
Toenail, 525
Tombstoning, 525
Tone, *see* Value
Tone line conversion, *see* Degeneration
Toning, 193, 525
Trade customs:
    binding, 440, 466–467
    defined, 525
    graphic arts, 455–457
    influence on pricing, 37
    printing, 406, 461–463
    quick printing, 407, 464–466
    typography, 257, 457–461
Trademarks, 124
Transparency:
    cropping, 332
    defined, 526
Transposition, 526
Trimming, 443
Triphthong, 262, 526
Type:

Type (cont.)
  alternative character, 262
  anatomy of, 260–261
  appropriateness of, 269
  blocking, 373, 480
  body, 17, 246, 481
  bounce, 481
  centered, 480
  cold, 244–253
  condensed, 486
  diphthong, 262
  display, 248, 489
  dummying, 303–306
  electronic, *see* Electronic composition
  elite, 492
  extended, 493
  flush, 53
  foundry, 241–242, 495
  hand-set, 241–242
  hot, *see* Hot type
  how to select, 268–272
  inferior, 295, 500
  in-line, 500
  Intertype, 243, 501
  italic, 264, 501
  justified, 271, 273
  kern on, 262
  legibility of, 269–270
  ligature, 262
  line length of, 270–271
  Linotype, 243, 504
  logotype, 262
  Ludlow, 241, 504
  measuring, 263–264
  Monotype, 242, 506
  normal, 264
  outline, 509
  pasting up, 318–320
  peculiars, 261
  phototypeset, *see* Phototypesetting
  pica, 511
  pi character, 262
  pied, 512
  printer's marks, 262
  ragged, 53, 515
  readability of, 99, 270–272, 515
  roman, 264, 295, 516
  sample, 53, 280
  second-coming, 242, 518
  serif, 264, 518
  spacing of, 271–272
  special characters, 261
  specimen, 521
  standing, 319, 522

  superior, 295, 524
  tied letters, 262
  transfer, 211–212, 217, 525
  triphthong, 262
  unjustified, 53
  unreadable, 137–142
  up style, 527
  using as art, 180–181
  weight of, 264, 528
  wooden, 242
Typebook, 256, 263, 277, 526
Type classifications:
  Modern, 267, 506
  Novelty, 130, 268, 507
  Old English, 266, 508
  Old Style, 266, 508
  Sans Serif Gothic, 267, 517
  Script, 268, 518
  Square Serif Gothic, 267, 522
    Copperplate Gothic, 267, 486
  Transitional, 266, 526
Typeface,:
  defined, 526
  how to select, 260–272
Type family, 264, 494
Type form, 264
Type-high, 242, 261, 377, 526
Type series, 264, 518
Typesetting:
  cold type, 244–253
  cost of changes during, 39
  defined, 240, 526
  electronic, *see* Electronic composition
  hot type, 240–244
  impact, 244, 499
  line-for-line, 273, 503
  minimum charge for, 53
  photo, *see* Phototypesetting
  place in production sequence, 4
  saving on, 51–53
  specifications for, 17, 285–288
Typo, 526
Typographer:
  defined, 526
  evaluating, 254–256
  types of, 254
Typography:
  defined, 526
  trade customs, 257, 457–461
Tyvek, 527

Union bug, 372, 482
Unit count, 307
Up, 55, 527

INDEX

Uppercase, 241, 527
Upper and lowercase, *see* Caps, and lowercase
Use rights:
  copy, 84
  design, 159
  first, 223, 494
  illustration, 223
  mechanical art, 236
  multiple, 223
  one-time, 217, 223
  photography, 229
  stipulated on bill, 43
  subsequent, 217

Vacuum frame, 374, 527
Value, 349, 527
Varityper, 245
Varnish, 128, 433, 527
  spot, 521
Vegetables, *see* Printer's marks
Velox, *see* Prescreened print
Verso, 153, 527
Vignetting, 193, 527
Virgule, *see* Shill mark

Visual display terminal, 247

Washup:
  charge for, 45
  defined, 387, 527
Waste, *see* Spoilage
Waterless lithography, *see* Letterset printing
Watermark, 415, 416, 528
Waterproofing, 528
Web:
  defined, 528
  letterpress, 393
  offset, 386
Web gain, 368, 528
Widow, 271, 528
Window:
  defined, 325, 528
  making, 325–328
  slugging, 328
Wordspacing, 528
Work-for-hire, 81, 528
Wraparound, 529

Xerography, 399, 529
X-height, 261, 529